Polymer Materials in Additive Manufacturing: Modelling and Simulation

Polymer Materials in Additive Manufacturing: Modelling and Simulation

Editors

Mohammadali Shirinbayan
Nader Zirak
Khaled Benfriha
Sedigheh Farzaneh
Joseph Fitoussi

MDPI • Basel • Beijing • Wuhan • Barcelona • Belgrade • Manchester • Tokyo • Cluj • Tianjin

Editors

Mohammadali Shirinbayan
PIMM
Arts et Metiers Institute
of Technology
HESAM University
Paris
France

Nader Zirak
PIMM
Arts et Metiers Institute
of Technology
HESAM University
Paris
France

Khaled Benfriha
LCPI
Arts et Metiers Institute
of Technology
HESAM University
Paris
France

Sedigheh Farzaneh
PIMM
Arts et Metiers Institute
of Technology
HESAM University
Paris
France

Joseph Fitoussi
PIMM
Arts et Metiers Institute
of Technology
HESAM University
Paris
France

Editorial Office
MDPI
St. Alban-Anlage 66
4052 Basel, Switzerland

This is a reprint of articles from the Special Issue published online in the open access journal *Polymers* (ISSN 2073-4360) (available at: www.mdpi.com/journal/polymers/special_issues/ Additive_Manufacturing_Polymers_Polymer_Composites).

For citation purposes, cite each article independently as indicated on the article page online and as indicated below:

LastName, A.A.; LastName, B.B.; LastName, C.C. Article Title. *Journal Name* **Year**, *Volume Number*, Page Range.

ISBN 978-3-0365-6863-8 (Hbk)
ISBN 978-3-0365-6862-1 (PDF)

© 2023 by the authors. Articles in this book are Open Access and distributed under the Creative Commons Attribution (CC BY) license, which allows users to download, copy and build upon published articles, as long as the author and publisher are properly credited, which ensures maximum dissemination and a wider impact of our publications.

The book as a whole is distributed by MDPI under the terms and conditions of the Creative Commons license CC BY-NC-ND.

Contents

Mohammad Ahmadifar, Khaled Benfriha and Mohammadali Shirinbayan
Thermal, Tensile and Fatigue Behaviors of the PA6, Short Carbon Fiber-Reinforced PA6, and Continuous Glass Fiber-Reinforced PA6 Materials in Fused Filament Fabrication (FFF)
Reprinted from: *Polymers* **2023**, *15*, 507, doi:10.3390/polym15030507 . . . . **1**

Muhammad Abas, Tufail Habib, Sahar Noor, Bashir Salah and Dominik Zimon
Parametric Investigation and Optimization to Study the Effect of Process Parameters on the Dimensional Deviation of Fused Deposition Modeling of 3D Printed Parts
Reprinted from: *Polymers* **2022**, *14*, 3667, doi:10.3390/polym14173667 . . . . **21**

Fateh Enouar Mamache, Amar Mesbah, Hanbing Bian and Fahmi Zaïri
Micromechanical Modeling of the Biaxial Deformation-Induced Phase Transformation in Polyethylene Terephthalate
Reprinted from: *Polymers* **2022**, *14*, 3028, doi:10.3390/polym14153028 . . . . **43**

Laszlo Racz and Mircea Cristian Dudescu
Numerical Investigation of the Infill Rate upon Mechanical Proprieties of 3D-Printed Materials
Reprinted from: *Polymers* **2022**, *14*, 2022, doi:10.3390/polym14102022 . . . . **57**

Syed Hammad Mian, Khaja Moiduddin, Sherif Mohammed Elseufy and Hisham Alkhalefah
Adaptive Mechanism for Designing a Personalized Cranial Implant and Its 3D Printing Using PEEK
Reprinted from: *Polymers* **2022**, *14*, 1266, doi:10.3390/polym14061266 . . . . **75**

Sedigheh Farzaneh and Mohammadali Shirinbayan
Processing and Quality Control of Masks: A Review
Reprinted from: *Polymers* **2022**, *14*, 291, doi:10.3390/polym14020291 . . . . **93**

Nader Zirak, Mohammadali Shirinbayan, Michael Deligant and Abbas Tcharkhtchi
Toward Polymeric and Polymer Composites Impeller Fabrication
Reprinted from: *Polymers* **2021**, *14*, 97, doi:10.3390/polym14010097 . . . . **115**

Tomáš Tichý, Ondřej Šefl, Petr Veselý, Karel Dušek and David Bušek
Mathematical Modelling of Temperature Distribution in Selected Parts of FFF Printer during 3D Printing Process
Reprinted from: *Polymers* **2021**, *13*, 4213, doi:10.3390/polym13234213 . . . . **141**

Chil-Chyuan Kuo, Ding-Yang Li, Zhe-Chi Lin and Zhong-Fu Kang
Characterizations of Polymer Gears Fabricated by Differential Pressure Vacuum Casting and Fused Deposition Modeling
Reprinted from: *Polymers* **2021**, *13*, 4126, doi:10.3390/polym13234126 . . . . **151**

Zohreh Shakeri, Khaled Benfriha, Mohammadali Shirinbayan, Mohammad Ahmadifar and Abbas Tcharkhtchi
Mathematical Modeling and Optimization of Fused Filament Fabrication (FFF) Process Parameters for Shape Deviation Control of Polyamide 6 Using Taguchi Method
Reprinted from: *Polymers* **2021**, *13*, 3697, doi:10.3390/polym13213697 . . . . **173**

Zhongqiu Ding, Ben Wang, Hong Xiao and Yugang Duan
Hybrid Bio-Inspired Structure Based on Nacre and Woodpecker Beak for Enhanced Mechanical Performance
Reprinted from: *Polymers* **2021**, *13*, 3681, doi:10.3390/polym13213681 . . . . **191**

Article

Thermal, Tensile and Fatigue Behaviors of the PA6, Short Carbon Fiber-Reinforced PA6, and Continuous Glass Fiber-Reinforced PA6 Materials in Fused Filament Fabrication (FFF)

Mohammad Ahmadifar [1,2,*], Khaled Benfriha [1] and Mohammadali Shirinbayan [2,*]

1. Arts et Metiers Institute of Technology, CNAM, LCPI, HESAM University, F-75013 Paris, France
2. Arts et Metiers Institute of Technology, CNAM, PIMM, HESAM University, F-75013 Paris, France
* Correspondence: mohammad.ahmadifar@ensam.eu (M.A.); mohammadali.shirinbayan@ensam.eu (M.S.)

Abstract: Utilization of additive manufacturing (AM) is widespread in many industries due to its unique capabilities. These material extrusion methods have been developed extensively for manufacturing polymer and polymer composite materials. The raw material in filament form are liquefied in the liquefier section and are consequently extruded and deposited onto the bed platform. The designed parts are manufactured layer by layer. Therefore, there is a gradient of temperature due to the existence of the cyclic reheating related to each deposited layer by the newer deposited ones. Thus, the stated temperature evolution will have a significant role on the rheological behavior of the materials during this manufacturing process. Furthermore, each processing parameter can affect this cyclic temperature profile. In this study, different processing parameters concerning the manufacturing process of polymer and polymer composite samples have been evaluated according to their cyclic temperature profiles. In addition, the manufactured parts by the additive manufacturing process (the extrusion method) can behave differences compared to the manufactured parts by conventional methods. Accordingly, we attempted to experimentally investigate the rheological behavior of the manufactured parts after the manufacturing process. Thus the three-point bending fatigue and the tensile behavior of the manufactured samples were studied. Accordingly, the effect of the reinforcement existence and its direction and density on the tensile behavior of the manufactured samples were studied. Therefore, this study is helpful for manufacturers and designers to understand the behaviors of the materials during the FFF process and subsequently the behaviors of the manufactured parts as a function of the different processing parameters.

Keywords: material extrusion; rheological behavior; mechanical properties; temperature profile

1. Introduction

Additive manufacturing (AM) is growing due to the low amount of wasted material in the manufacturing process and the ability to manufacture complex shapes [1–3]. There are different techniques concerning the additive manufacturing process, while fused filament fabrication (FFF) is the most commonly utilized technique [4]. Mechanical properties and dimensional accuracy of the FFF-processed parts are affected by the utilized processing parameters during the manufacturing process [5]. The importance and the influence of the assortment of processing parameters have been studied, such as the effect of print speed [6,7], bed platform temperature [8–10], liquefier temperature [11], and layer height [12,13]. The evolution of the temperature of the deposited layers is considerably affected by the aforementioned processing parameters during the FFF process [5]. The temperature evolution caused a gradient of the temperature in the structure, which significantly affected the adhesion and the bonding of the deposited layers and consequently the strength of the manufactured parts. Several studies related to the temperature evolution during the FFF process have been conducted [5,14,15].

Christiyan et al. [16] investigated the flexural and tensile strength of the printed ABS composite materials under the different print speed values of 50, 40, and 30 mm/s and the layer height values of 0.3, 0.25, and 0.2 mm. As for the results, it was observed that the lower layer height and print speed values (0.2 mm and 30 mm/s) increased the flexural and tensile strengths. Moreover, several studies have been carried out regarding the FFF process of polymer composite materials [17–19]. Durga et al. [20] evaluated the influence of the liquefier temperature and layer height of the deposited layers on the tensile strength of the printed CF-PLA specimens. The manufactured CF-PLA specimens under the lowest selected layer height value and the highest selected liquefier temperature had the higher tensile strength. An investigation was conducted by Ding et al. [21] to determine the influence of liquefier temperature on the mechanical properties of the FFF-processed PEEK and PEI. They discovered that the flexural strength was gradually improved as the temperature increased. Berretta et al. [22] manufactured the reinforced PEEK with 1% and 5% carbon nanotubes (CNTs). They reported that the CNTs did not have a significant effect on the mechanical behaviors of the PEEK-processed specimens. They introduced the nozzle temperature as one of the most crucial parameters in the FFF process, due to its direct contact with the polymer. Yang et al. [23] studied the effect of the thermal processing condition on mechanical behaviors and crystallinity of the PEEK material. Based on the related results, crystallinity increased from 17% to 31% in response to the increase in ambient temperature from 25 to 200 °C.

Few studies have been performed to investigate the effect of the utilized fiber reinforcements on the rheological behavior of the materials during and after the manufacturing process. This study tried to investigate the rheological behavior of the materials during and after the FFF process by considering the role of the fiber reinforcements. The selected materials for this study were PA6, short carbon fiber-reinforced PA6 (CF-PA6), and continuous glass fiber-reinforced CF-PA6 composite materials.

2. Materials and Methods

2.1. Materials

The selected materials for this study were polyamide 6 (PA6) and short carbon fiber-reinforced polyamide 6 (Onyx or CF-PA6) produced by MarkForged®. The chopped carbon fibers had a mass content of 6.5% in the CF-PA6 filament based on the pyrolysis process. The characteristics of the utilized filaments as the raw materials are presented in Table 1.

Table 1. The characterizations of the utilized raw materials.

Physical and Chemical Properties	Raw Materials	PA6	CF-PA6
Density		1.1 g/cm^3	1.2 g/cm^3
Glass transition temperature (T_g)		45 °C	47 °C
Crystallization temperature (T_c)		173 °C	162 °C
Melting temperature (T_m)		205 °C	198 °C
Spool Image			

As for the investigation of the fiber-reinforcement impact and the processing parameter effects on the rheological behavior of the materials during the FFF process, a single wall layer specimen (Figure 1) was designed. This specimen let us study the effect of the selected processing parameters on the adhesion and the bonding of the deposited layers. In addition, the location of the required specimens for the subsequent characterizations are determined in Figure 1. Two different printers were utilized during our studies. To study the effects of the processing parameters, Flashforge ADVENTURER-3 (from China) was utilized. Moreover, as for studying the infill percentage, infill pattern effects, and fatigue behaviors, a Markforged-Mark Two (from the USA) printer was utilized. This is because Markforged-Mark Two printers can provide the possibility of manufacturing composite objects with continuous reinforcement.

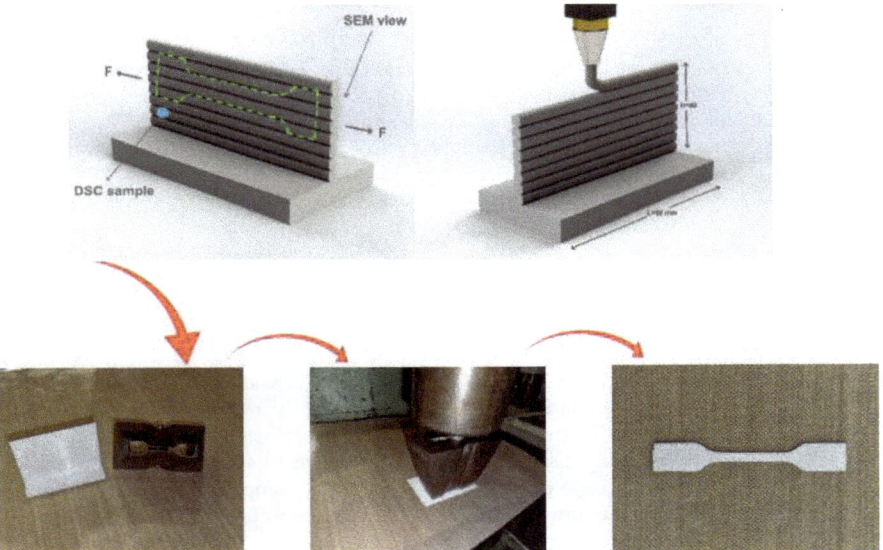

Figure 1. The printed single-wall specimens and preparation of the tensile test specimens [5].

2.2. Methods

2.2.1. In Situ Monitoring of Temperature Evolution

An Optris PI450 infrared camera was applied in the conducted study of the processing parameter impacts on the thermal and mechanical properties of the polymer and polymer-based composites using the FFF process. The stated infrared camera was positioned at a specific predetermined distance from the extrusion location of the printer. This attempt aimed to monitor and observe a consistent plain field of view (FOV) across all consecutive layers. The temperature rise during the performed fatigue test was monitored and measured by the stated infrared camera as well. Regarding some technical specifications of the infrared camera used in this experiment, frequency, accuracy values, wavelength range, optical resolution, and the rate of the frames were 32 Hz, 2%, 8–14 μm, 382 × 288 pixels, and 80 Hz, respectively.

2.2.2. Microstructural Observations

Using a scanning electronic microscope (HITACHI 4800 SEM, manufactured in Japan), observations and image analyses were performed to qualitatively analyze the composite microstructure, especially in relation to damage assessment. In this study, an optical microscope (OLYMPUS BH2, manufactured in Japan) was used to assess the quality of the manufactured samples at the various selected processing parameters with magnifications from 100 to 500 mm.

2.2.3. Differential Scanning Calorimetric (DSC)

By means of differential scanning calorimetry (Q1000), both raw filament materials and printed specimens could be assessed for their respective glass transition (T_g), crystallization temperatures (T_c), and their heat capacities based on the selected processing parameters. DSC characterization was performed on the raw filament materials (CF-PA6 and PA6) over three temperature ramps: range of 20 to 220 °C under a rate of 10 °C/min. This attempt caused an elimination of the thermal history of the filaments concerning their production process. Furthermore, the printed and fabricated specimens were analyzed by DSC in two ramps (heating and cooling).

2.2.4. Thermo-Mechanical Behavior Analysis (DMTA)

Under the multi-frequency condition, DMTA flexural tests were performed on the printed samples using the DMA Q800 instrument from the TA Company, in order to determine the major transition temperatures and viscoelastic characteristics. This characterization was utilized to study the viscoelastic behavior of the material during the fatigue test and the subsequent temperature rise. DMTA characterization was performed in the temperature range of 30 to 80 °C, frequencies of 1, 2, 5, 10, and 30 Hz, and temperature rate of 2 °C/min.

2.2.5. Quasi-Static Tensile Test

As was stated, two different conditions for studying the tensile behavior were conducted. The first condition was considered to study the effect of the processing parameters on the manufacturing of the polymer and composite specimens (Figure 1). Therefore, based on ISO 527-2, tensile test specimens were sliced/cut from the printed single-wall layer sample (Figure 1), in order to conduct the related study of the first condition. The required tensile specimens were cut from the printed single-wall samples utilizing a tensile sample-cutting die and applying the homogenous force. In addition, the homogeneity of the printed single-wall layers was ensured by using the caliper for thickness measurement. Following the cutting of the tensile test specimens from the single-wall layers, a caliper was used to ensure that gauge length dimensions were uniform. In addition, the observation of the samples under the optical microscope (OM) to control the quality of the specimens after the manufacturing and cutting process was taken into account. The second condition was applied to study the impact of fill percentage of the polymer and the different determined densities and directions of the utilized continuous reinforcements. Moreover, in order to investigate the rheological behavior of the additive manufactured specimens after the production process, the related tensile test specimens were printed based on the standard ISO 527-1. The quasi-static tensile experiment was conducted by means of the INSTRON 5966 machine with a displacement rate of 5 mm/min and a loading cell of 10 kN. A minimum of three specimens were prepared for each condition in order to conduct the quasi-static tensile tests.

2.2.6. Three Points Bending Fatigue Tests

As for the fatigue test of the fabricated specimens concerning the study of the rheological behavior of the FFF-processed composite materials (after the manufacturing process), a three-points bending fatigue test was conducted at various applied maximum strains (ε_{max}). The considered strain ratio was 0.1 ($R_\sigma = 0.1$); moreover, the related mean strain level was 0.55 ε_{max}. This test was conducted on the short carbon fiber-reinforced polyamide 6 (Onyx or CF-PA6) and the continuous glass fiber-reinforced polyamide 6 (Onyx + GF) composite specimens to be able to study the effect of the continuous glass fiber reinforcement and its subsequent impact on the rheological behavior of the FFF-processed specimens. During the three-point bending fatigue test, the temperature of the specimens was raised, which was measured by the infrared camera to determine the importance and influence of the utilized continuous glass fiber reinforcement in this phenomenon. The dimensions of the prepared rectangular composite specimens were $120 \times 10 \times 4$ mm^3. In order to perform a fatigue test, at least three specimens were prepared for each condition.

3. Results and Discussions

3.1. Study of the Rheological Behavior of the Polymer and Polymer Composite Materials during the FFF Process

The rheological behavior of the polymer and polymer composite materials during the FFF process are investigated in this section. In addition, the modifications in the rheological behavior of the materials as the consequence of the short carbon fiber existence are discussed. As part of this effort, four main processing parameters were considered to investigate their impacts on the tensile behavior of the manufactured specimens. In addition, the in situ temperature evolution during the manufacturing process was monitored. The selected processing parameters were print speed (13, 15, and 17 mm/s), liquefier temperature (220, 230, and 240 °C), layer height (0.1, 0.2, and 0.3 mm), and bed platform temperature (25, 50, 60, and 80 °C). The aforementioned processing parameters were studied individually by considering the rest of the processing parameters as constant. The reference values concerning each of the selected parameters were 15 mm/s as print speed, 240 °C as liquefier temperature, 0.1 mm as layer height, and 25 °C as bed temperature. The selected materials for this section were PA6 and CF-PA6 [5].

3.1.1. Effect of the Print Speed

In light of the importance of the print speed parameter in production time, it was studied to how it impacts the rheological behaviors. To study the impact of this manufacturing processing parameter, 13, 15, and 17 mm/s as the different print speed values were taken into consideration. According to Figure 2, the tensile strength and the crystallinity percentage pertaining to the manufactured specimens with the above-selected print speed values were evaluated. The tensile strength values of 57.78 ± 0.84, 68.92 ± 0.9, and 69 ± 0.05 MPa concerning the manufactured PA6 specimens and 65 ± 0.5, 55 ± 0.6, and 63 ± 1.3 MPa concerning the manufactured CF-PA6 specimens pertaining to the selected print speed values of 13, 15, and 17 mm/s were obtained, respectively.

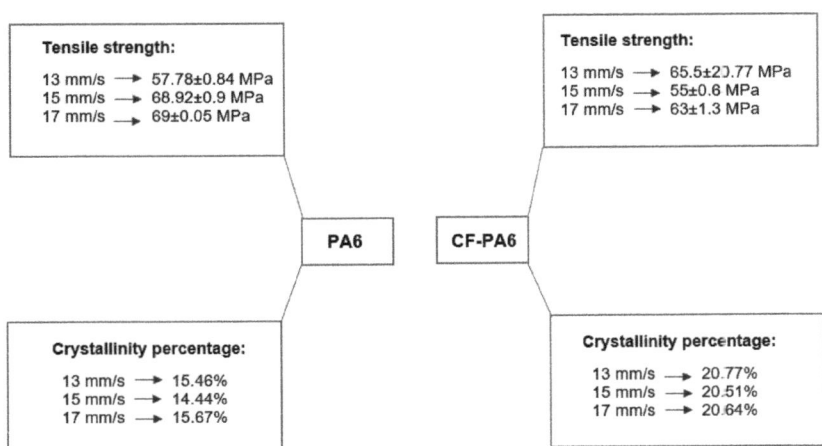

Figure 2. Obtained tensile strength and crystallinity percentage values concerning the print speed effect investigation.

Figure 3 depicts the temperature profile, obtained from the in situ monitoring of the temperature evolution concerning the PA6 and CF-PA6 specimens produced by the above-stated print speed values. As printing speed was increased, the measured temperature of the first printed layer stayed above the related crystallization temperature values in all examined PA6 and CF-PA6 specimens. The print speed affected the cooling time and polymer arrangement and the resultant crystallinity degree. All selected print speeds yielded higher crystallinity degree values for CF-PA6 specimens than PA6 specimens.

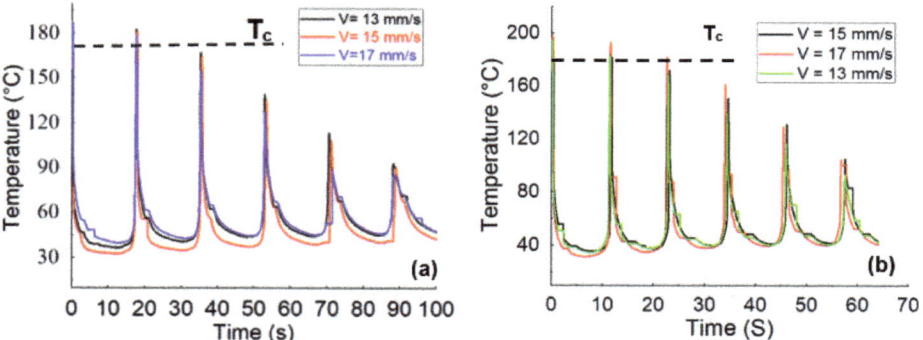

Figure 3. The obtained temperature evolution from the in situ temperature monitoring (**a**) PA6 and (**b**) CF-PA6 [5] under different print speeds.

3.1.2. Effect of the Liquefier Temperature

The impact of the liquefier temperature on the rheological behavior of the PA6 and CF-PA6 specimens were investigated. The decided liquefier temperature values for this study were 240, 230, and 220 °C. Figure 4 depicts the crystallinity percentage and tensile strength concerning the manufactured specimens with the above-selected liquefier temperature values. The tensile strength values of 55.48 ± 0.78 and 68.92 ± 0.9 MPa concerning the manufactured PA6 specimens by the liquefier temperature values of 230 and 240 °C were achieved, respectively. One can note that the lowest strength was observed in the printed specimens by the liquefier temperature values of 220 °C. This liquefier temperature value could not provide a suitable fluidity of the material during the FFF process, which caused low and inappropriate adhesion between the deposited layers. This low adhesion was revealed during the punching/slicing process of the printed single wall samples for preparation of the required tensile test specimens by a fracture that occurred in the interface of the deposited layers. This undesirable fracture did not allow us to have tensile test specimens to evaluate the tensile strength behavior of the manufactured PA6 specimens by the liquefier temperature value of 220 °C. In addition, the tensile strength values of 49 ± 1.5, 51 ± 3, and 55 ± 0.6 MPa concerning the manufactured CF-PA6 specimens pertaining to the selected liquefier temperature values of 220, 230, and 240 °C were obtained, respectively. The determined crystallinity degree of the PA6 samples printed at the liquefier temperatures of 220, 230, and 240 °C were 12.51%, 12.75%, and 14.40%, respectively, while the obtained crystallinity degree of the CF-PA6 samples printed at the liquefier temperatures of 220, 230, and 240 °C were 19.97%, 20.26%, and 20.51%, respectively.

Regarding the effect of the short fiber reinforcement and the consequent differences between the FFF-processed polymer and polymer composite material, firstly, the crystallinity percentage values were highlighted. Based on the obtained results from the DSC characterization of the FFF-processed CF-PA6 and PA6 samples, the crystallinity percentage values of the CF-PA6 specimens were higher than the PA6 specimens manufactured under the same processing parameters. The chopped carbon fibers in CF-PA6 were the superior sites and spots for the nucleation of the crystalline sections. Furthermore, another difference between the FFF-processed polymers and polymer composite materials is their tensile strength. Regarding compression between the obtained tensile strength values of CF-PA6 and PA6, it was understood that the single-wall FFF-manufactured CF-PA6 specimens exhibited lower strength. The short carbon fibers as the solid component in the molten polymer decreased the fluidity of the CF-PA6 during the FFF process compared with PA6 material. The stated decrease in the fluidity during the manufacturing process can be illustrated by the narrow width zones in the manufactured specimens (Figure 5). The stated zones are the preferable rupture sections because of their stress concentration feature in the structure.

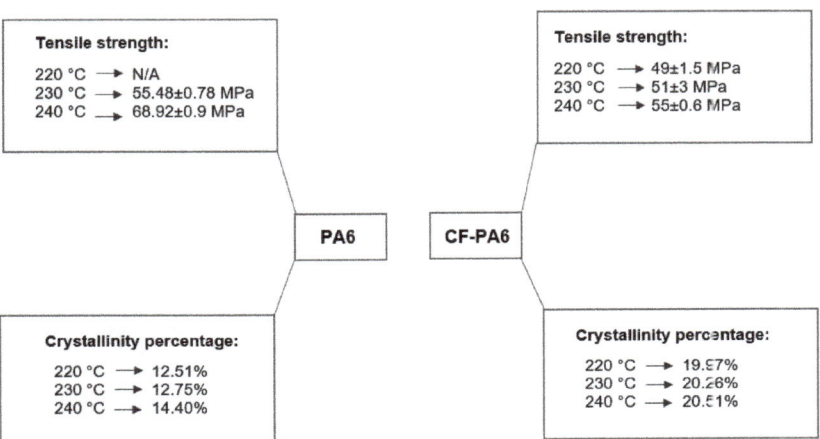

Figure 4. Obtained tensile strength and crystallinity percentage values concerning the liquefier temperature effect investigation.

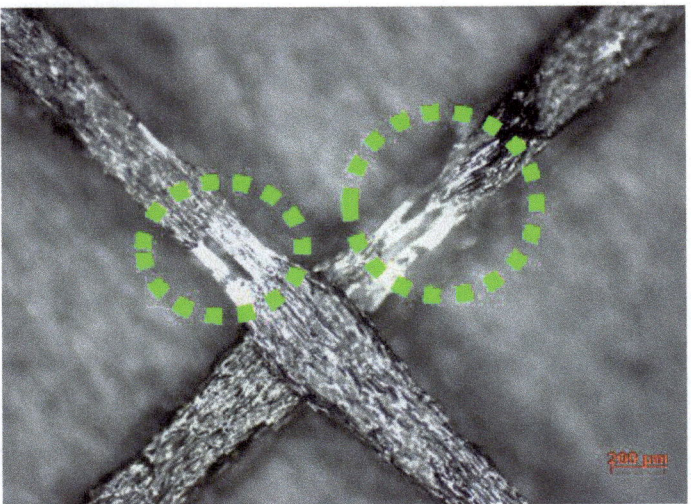

Figure 5. Narrow zone of deposited CF-PA6 filament.

3.1.3. Effect of the Layer Height

CF-PA6 and PA6 specimens were investigated for their rheological behavior influenced by the layer height. The investigation of the effect of the short fiber reinforcement on the rheological behavior of the materials was carried out by selecting the three different layer height values of 0.1, 0.2, and 0.3 mm, during and after the FFF process. Figure 6 represents the tensile strength values related to the manufactured specimens by the different selected layer height values.

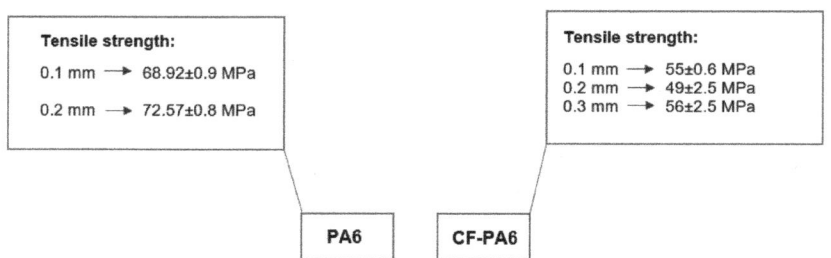

Figure 6. Obtained tensile strength values concerning the layer height effect investigation.

The tensile strength values of 55 ± 0.6, 49 ± 2.5, and 56 ± 2.5 MPa concerning the manufactured CF-PA6 specimens by the layer height values of 0.1, 0.2, and 0.3 mm were achieved, respectively. For better investigation of the effects of the layer height and the short carbon fiber reinforcements on the rheological behavior of the material during the FFF process, a thermal camera for the in situ temperature monitoring was utilized. By considering the obtained results from the tensile test and the in situ temperature monitoring during the FFF process and their subsequent correlations, the existence of the two effective phenomena was declared. The stated two phenomena were firstly the fluidity decrease in the printed materials and secondly the residual/retained temperature increase in the deposited layers during the cyclic raised temperature during the FFF process. The peaks (upper limits) of the obtained time–temperature curves during the manufacturing process revealed the first-stated phenomenon, and the lower limits of these curves revealed the second-stated phenomenon. By increasing the layer height during the manufacturing process from 0.1 to 0.2 mm, the first phenomenon was more highlighted and effective. The decreased fluidity was found according to the decrease in the measured temperature profile peaks by increasing the selected layer height from 0.1 to 0.2 mm (Figure 7a). This decrease in the fluidity of the deposited layers during the FFF process of CF-PA6 specimens resulted in a lower tensile strength of the manufactured specimens with a layer height of 0.2 mm compared to the printed specimens with a layer height of 0.1 mm. As for determining the effect of the short carbon fibers, the obtained tensile strength from printed CF-PA6-specimens and PA6 specimens in the same layer height values (0.1 and 0.2 mm) were compared. The tensile strength of PA6 specimens with layer heights of 0.1 mm and 0.2 mm were 68.92 ± 0.9 and 72.57 ± 0.8 MPa, respectively. This subsequent increase in the tensile strength with the increase in layer height demonstrated that the decreased fluidity phenomenon did not predominate in the PA6 specimens. The short carbon fibers could also be effective on the stated decreased fluidity. As further explain, the short carbon fiber reinforcements of the solid components hindered the fluidity of the polymer (matrix) as part of the liquid component during the FFF process. However, the second phenomenon, which was a residual/retained temperature increase in the deposited layers, was dominant during the FFF process of the CF-PA6 specimens, with a layer height of 0.3 mm. The residual/retained temperature increase in the deposited layers was found based on the bottom of the measured temperature profile (Figure 7b). This phenomenon during the FFF process of the CF-PA6 specimens resulted in a higher crystallinity percentage and tensile strength of the specimens with a layer height of 0.3 mm.

3.1.4. Effect of the Bed Temperature

As for the study of the influence of the last selected processing parameter, the importance of the bed platform temperature on the manufactured specimens was evaluated. The studied materials were PA6 and CF-PA6. Four different bed temperatures of 80, 60, 50, and 25 °C were taken into account. According to the monitoring of the temperature during the process, upon deposition of the first layer of the materials (PA6 and CF-PA6), the monitored temperature values ascended to the values of crystallization temperature (Figure 8).

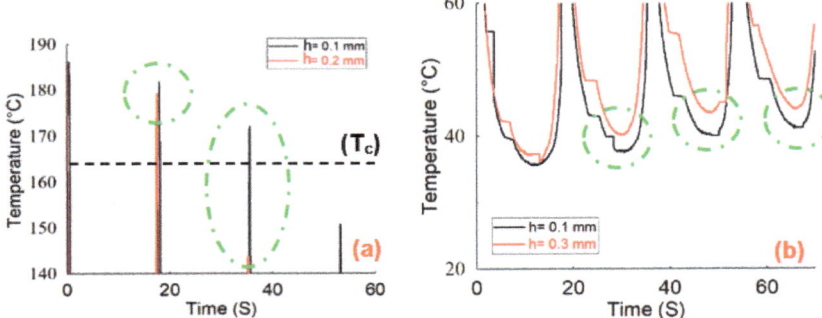

Figure 7. Monitoring of the temperature rise during the additive manufacturing process of CF-PA6 specimens with the different selected layer height values: (**a**) upper limits and (**b**) lower limits.

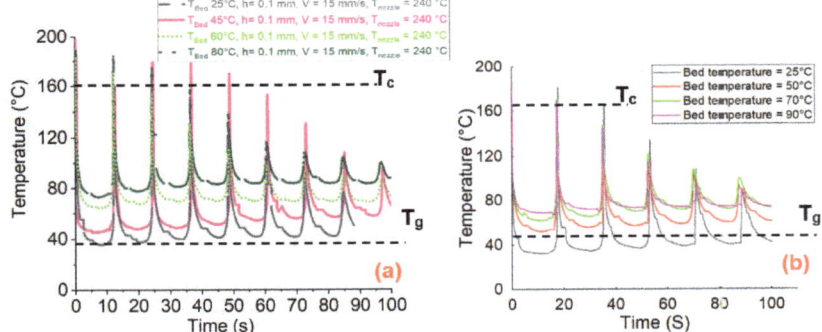

Figure 8. Monitoring of the temperature evolution during the FFF process concerning the different selected bed temperature values: (**a**) CF-PA6, (**b**) PA6.

The dimensional stability/accuracy of the printed specimens was evaluated by means of optical microscopy. Based on the performed visual comparisons, the manufactured parts under a bed temperature of 80 °C had less dimensional accuracy due to the observed inconsistency and dissonance concerning the thickness of the deposited PA6 layers (Figure 9). The stated inconsistency was observed in the FFF-processed specimens with a bed temperature of 80 °C.

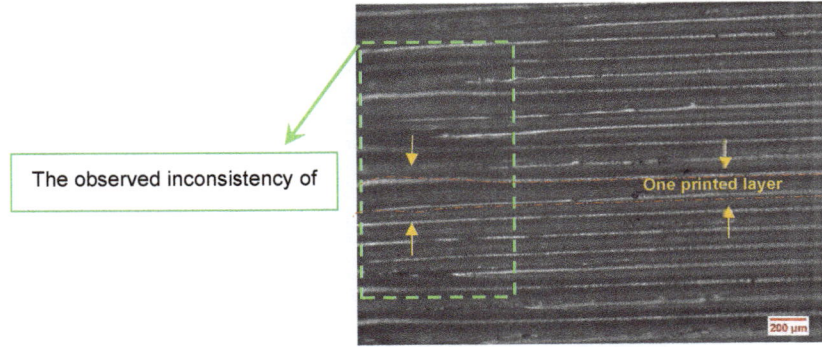

Figure 9. Optical microscopy of the printed PA6 specimen under a bed temperature value of 80 °C.

3.2. Study of the Rheological Behavior of Polymer and Polymer Composite Material after FFF Process

Investigations regarding the impact of some important FFF processing parameters of the polymer-based composites on the inter-layer adhesion (bonding) of the deposited filaments were performed. It was proven that temperature profile evaluation of the printed layers has a significant effect on the bonding of adjacent filaments. Failure stress/strain can be the indicators to determine the mechanical properties of FFF-manufactured products. In this section, using optimized processing parameters (liquefier temperature: 240 °C, print speed: 15 mm/s, layer height: 0.1 mm, platform temperature: 25 °C), the mechanical properties of the fabricated composite material under monotonic and fatigue loadings were analyzed (Figure 10). As part of the monotonic loading tests, the effect of fill pattern and fill density of the polymer matrix component as well as the density and direction of the continuous reinforcement were studied. Moreover, the effects of continuous reinforcement on the fatigue behavior of the FFF-processed specimens were investigated [24].

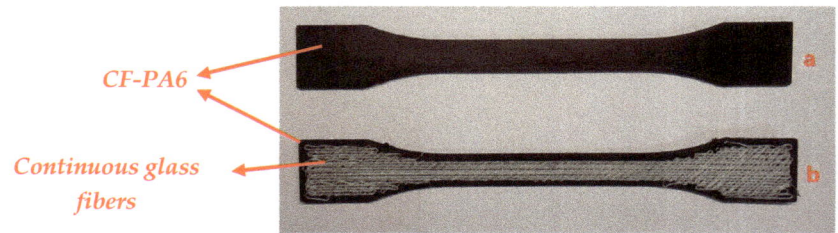

Figure 10. The manufactured CF-PA6 (**a**) and CF-PA6 reinforced with continuous glass fiber (**b**) specimens.

3.2.1. Quasi-Static Tensile Property
Influence of Fill Patterns

Using the Mark Two printer, the tensile specimens were printed into three fill patterns: triangular, rectangular, and hexagonal. Comparisons were made based on the tensile strength of the manufactured samples made of CF-PA6. Therefore, the obtained tensile strength values of the samples concerning the stated fill pattern were compared to each other as well as to the solid fill pattern (whose infill percentage was 100%) (Figure 11).

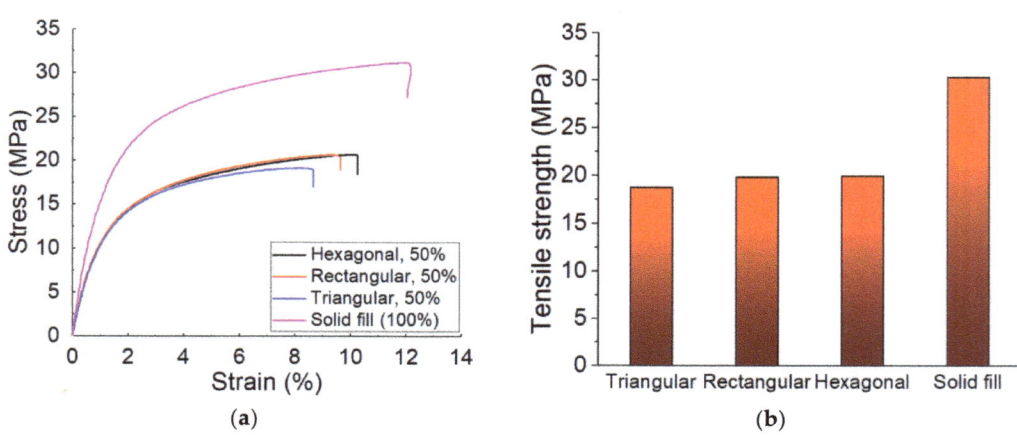

Figure 11. Obtained tensile test curves (**a**) and obtained strength (**b**) of CF-PA6 specimens [24].

The obtained tensile strength of the manufactured CF-PA6 specimens in the triangular, rectangular, hexagonal fill patterns, and the solid fill were 18.79 ± 1.19, 19.84 ± 1.66, 19.99 ± 1.32, and 30.31 ± 5.5 MPa, respectively (Figure 11). The tensile strength of the printed samples in the rectangular and hexagonal fill patterns were about 5.9% and 6.4% higher than the printed samples in the triangular fill pattern, respectively. The tensile strength of the solid fill-printed samples was about 61.3% higher than the rectangular fill pattern-printed samples. In other words, the tensile strengths of the manufactured specimens in the triangular, rectangular, and hexagonal fill patterns were close to each other. However, the highest tensile strength was related to the solid fill samples. As it is observed from Figure 12, the fracture surface of the solid filled-printed samples was more brittle and more homogenous, in comparison with the other samples.

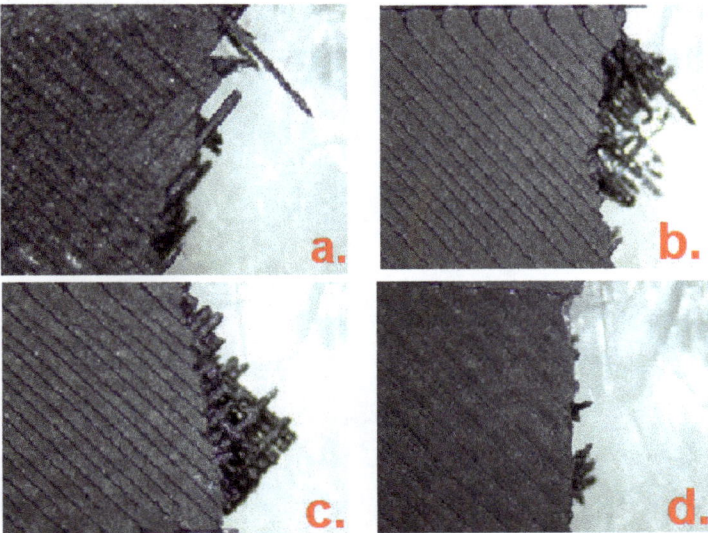

Figure 12. Macroscopic observation after quasi-static tensile test: (**a**) hexagonal, (**b**) triangular, (**c**) rectangular, and (**d**) solid fill.

Effect of the Continuous Reinforcement on Different Fill Patterns

CF-PA6 filaments with continuous glass fiber were utilized to manufacture the specimens in the stated fill patterns: triangular, rectangular, and hexagonal, as well as in the solid fill pattern (whose fill percentage was 100%). Using CF-PA6 reinforced with continuous glass fiber, the obtained tensile strengths concerning the hexagonal, rectangular, triangular fill patterns, and the solid fill specimens were 60.04 ± 4.16, 66.7 ± 2.63, 69.5 ± 3.05, and 78.35 ± 0.87 MPa, respectively. Therefore, the highest tensile strength is related to the solid fill samples. In addition, by shifting the fill pattern of the printing process of the samples, which were made of CF-PA6 reinforced with continuous glass fibers in the hexagonal infill pattern (the weakest obtained strength) and the solid fill, the tensile strength improved almost 30.5% (half of the situation in which no continuous glass fiber was used). The highest tensile strength was related to the solid fill samples (Figure 13).

Figure 14 illustrates the effect of using continuous glass fiber on the tensile strength of the FFF-processed specimens in the solid fill pattern. By utilizing the continuous glass fiber, the strength of the solid fill samples increased significantly by about 158.5%.

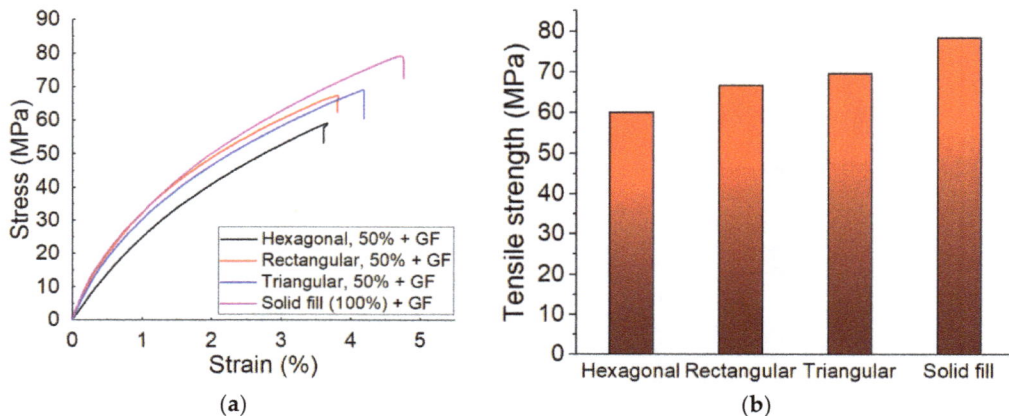

Figure 13. Obtained tensile test curves (**a**) and obtained strength (**b**) of CF-PA6 reinforced with continuous glass fiber specimens [24].

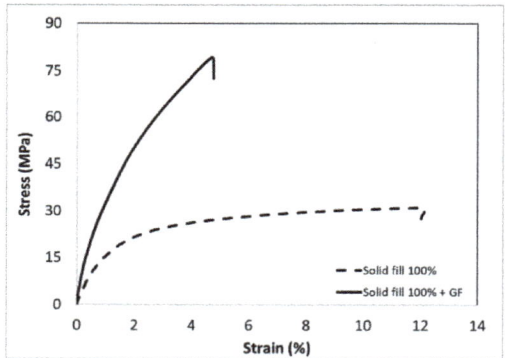

Figure 14. Quasi-static tensile curves of CF-PA6 (Onyx) and CF6-PA6 reinforced with continuous glass fiber (Onyx + GF) (solid fill pattern) [24].

Effect of the Density of Continuous Reinforcement

Regarding the effect of the density of the continuous reinforcement layers, a rectangular sample with thickness of 3.5 mm was designed as the tensile test specimens (Figure 15). The continuous fiberglass was considered for manufacturing the PA6 reinforced with continuous glass fiber composite samples. Thus, the tensile test specimens were PA6 composites reinforced with continuous glass fibers. The 0.1 mm value was selected as the layer height of the deposited PA6 layers. In addition, the layer height of the printed fiber glass was 0.1 mm. Thus, the samples were manufactured by depositing 35 layers during the layer-by-layer fabrication process. As for studying the effect of the density of the continuous glass fiber layers, different quantities of the printed continuous glass fiber layers were considered to manufacture the related samples to compare their tensile strengths. The different quantities of the continuous glass fiber layers of 2, 4, 6, 8, and 10 out of the total printed layers (35 layers) were applied in the manufactured specimens. The related tensile strengths of the specimens made up of 2, 4, 6, 8, and 10 layers were 36.38 ± 0.53, 144.26 ± 3.18, 187.55 ± 3.35, 226.34 ± 5.4, and 233.72 ± 4.32 MPa, respectively. The tensile strength of the non-reinforced samples, which were processed by 35 layers made of PA6 layers without any continuous glass fiber, was 34.63 ± 2.5 (Figure 16).

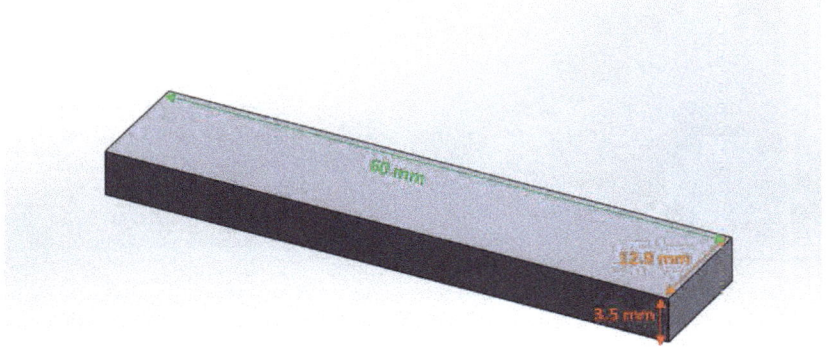

Figure 15. The scheme of the related rectangular composite.

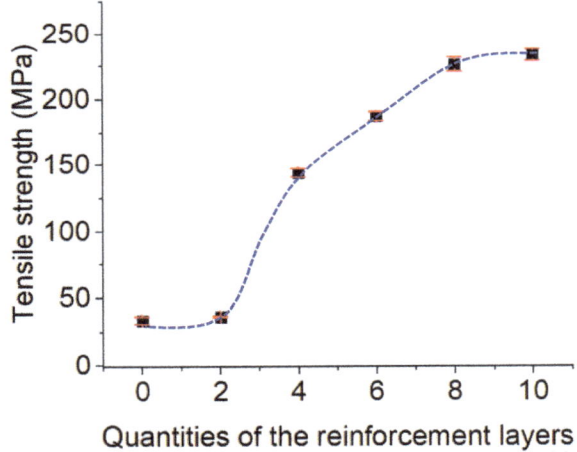

Figure 16. Effect of the density of the continuous reinforcement layers on the tensile strength.

As was found, by increasing the quantities of the continuous glass fiber layers, the tensile strengths of the manufactured samples were increased. The tensile strengths significantly increased by enhancing the quantity of the continuous glass fiber layers from two to eight. However, the tensile strength increments of the manufactured samples from eight to ten reinforced layers were less than the related increases in strength provided by enhancing the quantities of the continuous glass fiber layers from two to four. These obtained results are correlated to the weaker adhesion strength between the glass fiber reinforcement layers and the polymer matrix layers in comparison with the polymer-to-polymer layers. By enhancing quantity of the reinforcement layers from two to eight, the tensile strengths of the samples were increased because of the existence of the continuous reinforcement, which resists against the applied tensile stress. The nature of the weaker interface strength of the glass fiber reinforcement layers and the polymer matrix layers in the manufactured samples, which consists of more than eight reinforcement layers, dominated the impact of the existing reinforcement. Thus, we can consider that in our manufactured samples, the fabricated specimens with eight layers showed the optimal density in the continuous reinforcement.

Effects of the Continuous Reinforcement Direction

According to the recently stated results, concerning the performed study on the effect of the density of the continuous reinforcement, it was concluded that there is an optimal quantity layer for the continuous reinforcement layers in the manufactured reinforced polymer composites via the FFF process. In the case of our rectangular tensile test sample, eight layers of the continuous glass fiber layers prepared an optimal tensile behavior.

Therefore, the continuous glass fiber-reinforced PA6 composite samples, reinforced by eight layers, were considered for studying the tensile behavior of the manufactured samples with different continuous reinforcement directions. As for studying the effect of the continuous reinforcement direction, five direction values of 0°, 30°, 45°, 60°, and 90° were considered for continuous glass fibers. The tensile strength values of the related manufactured samples with continuous directions of 0°, 30°, 45°, 60°, and 90° were 226.34 ± 5.4, 46.8 ± 2.05, 37 ± 1.1, 36.81 ± 1.1, and 36.6 ± 0.79 MPa, respectively (Figure 17).

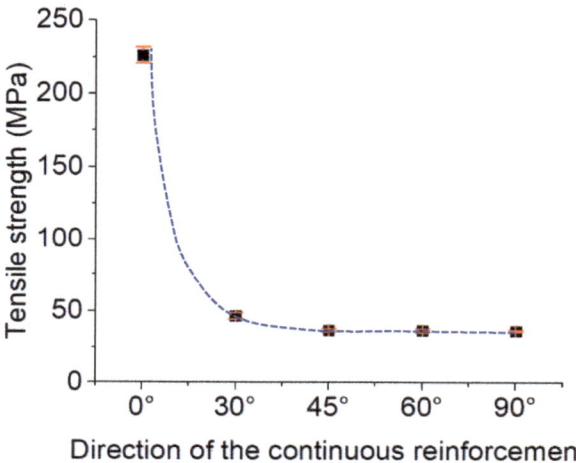

Figure 17. Effect of the direction of the continuous reinforcement on the tensile strength.

As is observed, the tensile strength of the manufactured samples with continuous reinforcement in the 0° direction had the highest strength in comparison with the other fabricated samples under the different reinforcement directions. As the direction of the continuous reinforcement increased from 0° to 90°, the tensile strength decreased. A significant drop in tensile strength was observed by the change in continuous reinforcement direction from 0° to 30°. The tensile strengths of the manufactured samples with 45°, 60°, and 90° as the reinforcement directions were close to each other. In addition, the tensile strength of the fabricated samples with a continuous reinforcement direction of 90° was close to the strength of the unreinforced samples (33.65 ± 2.8 MPa). Thus, it was found that the manufactured samples with a reinforcement direction of 0° had the highest tensile strength, while as the direction of reinforcement decreased, the tensile strength decreased as well. In the case of the samples with a reinforcement direction of 0°, the glass fibers resisted against the tensile stress well because it was in the direction of the applied strength. The fabricated samples with the reinforcement direction of 90° had the weakest strength because of the perpendicular directions of the applied stress and reinforcement. Thus, in the case of the samples with the glass fiber direction of 90°, the impact of the existence of the reinforcement on the tensile strength was not so manifest and sensible. The close tensile strength value of the specimens with a reinforcement direction of 90° to the unreinforced specimens can be evidence of this phenomenon. The macroscopic fracture observations exhibited the same fracture direction as the continuous reinforcement direction (Figure 18).

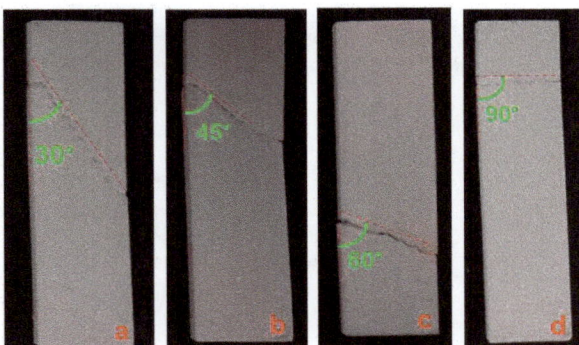

Figure 18. Macroscopic fracture view of the fabricated samples with reinforcement directions of (**a**) 30°, (**b**) 45°, (**c**) 60°, and (**d**) 90°.

3.2.2. Assessment of Fatigue Property

Influence of Continuous Reinforcement on Fatigue Behavior

Wöhler curves for CF-PA6 and CF-PA6 reinforced with continuous glass fiber samples are shown in Figure 19. Derived from three points, bending fatigue tests were conducted at a frequency of 10 Hz. CF-PA6 reinforced with continuous glass fiber specimens had a fatigue life of about 8000 cycles, while CF-PA6 specimens had a fatigue life of about 200,000 cycles in the applied strain equal to 4.5%. The obtained Wöhler curves of the CF-PA6 reinforced with continuous glass fiber specimens and the CF-PA6 specimens from the conducted fatigue test at a frequency of 10 Hz were of linear and bi-linear forms, respectively. It was found that there was a small difference between the obtained Wöhler curves of the CF-PA6 specimens and the CF-PA6 reinforced with continuous glass fiber specimens at high amplitude, while the related Wöhler curves significantly shifted at low strain amplitudes (Figure 19).

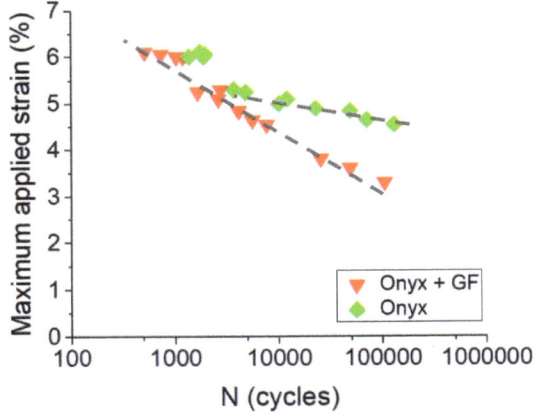

Figure 19. Obtained Wöhler curves concerning CF-PA6 reinforced with continuous glass fiber (Onyx + GF) and CF-PA6 (Onyx) specimens at 10 Hz [24].

Young's Modulus Evolution and Self-Heating Phenomenon and Relative Young's Modulus Evolution

Figure 20 illustrates the evolution of the relative stress through the conducted fatigue testing of continuous glass fiber-reinforced CF-PA6 and CF-PA6 samples at the determined strain amplitudes on the curves. For CF-PA6 and CF-PA6 reinforced with continuous glass fiber samples, mechanical fatigue (MF) mostly dominated the fatigue behavior due to damage phenomenon, while for high amplitude and low cycles, thermal fatigue (ITF) dominated the fatigue behavior of composite specimens.

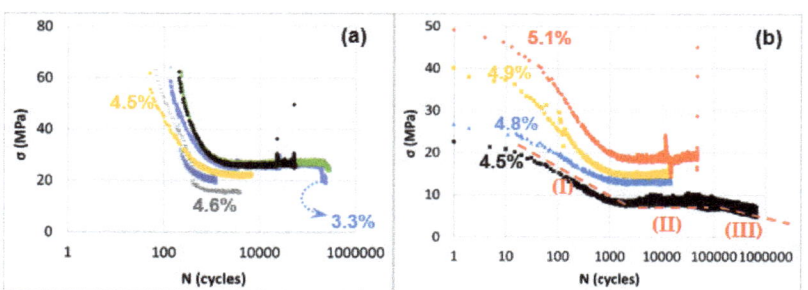

Figure 20. Relative stress (~Young's modulus) trend in the fatigue test of (**a**) CF-PA6 reinforced with continuous glass fiber (Onyx + GF) specimens and (**b**) CF-PA6 (Onyx) [24].

Based on their plots, it is evident that for high loading amplitudes, the dynamic modulus decreases rapidly in a linear regime of the logarithmic curve until the specimen fails. The dynamic modulus of applied low amplitudes deviates into three decreasing regimes–a swift one during the initial cycles (I), a gradual one (II), and a drastic decline (III) just before fracture. In addition, it was found that there is an upsurge in damage kinetics in the case of CF-PA6 reinforced with continuous glass fibers specimens compared to CF-PA6 specimens. This can be due to the existence of the weak polymer–continuous reinforcement interfaces in these specimens (Figure 21).

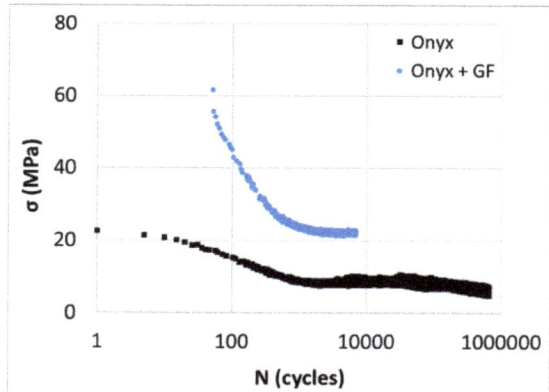

Figure 21. Relative stress (~Young's modulus) trend within the fatigue test [24].

Variation in the amplitude of the applied loading results in self-heating for CF-PA6 and continuous glass fiber-reinforced CF-PA6 during the fatigue test. This phenomenon affects the viscosity of the polymer as a function of temperature. Figure 22 depicts the temperature rise during the fatigue test concerning the utilized frequency of 10 Hz.

It was observed that the induced temperature of CF-PA6 increased up to about 60 °C, which is in the glass transition zone. Consequently, the stiffness of the matrix agent (polymer) decreased based on the obtained loss factor evolution versus the temperature curve from the performed DMTA test, as shown in Figure 23.

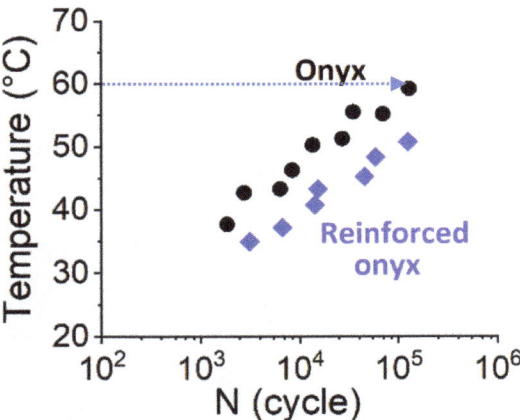

Figure 22. Evolution of the maximum induced temperature during the fatigue test [24].

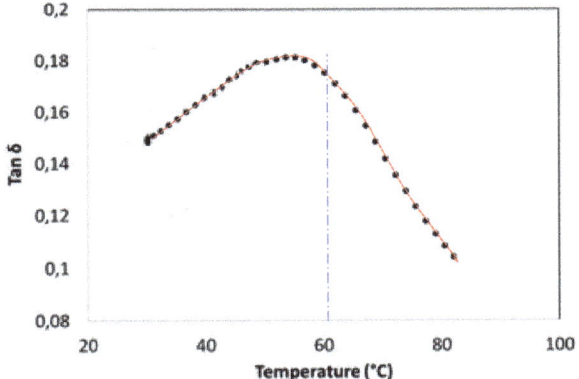

Figure 23. Obtained loss factor evolution versus temperature curve concerning CF-PA6.

Therefore, in addition to the development of damage, the fatigue failure of the CF-PA6 specimens was caused by the viscous behavior evolution of the polymer agent and the consequent brittle–ductile transition. It is vital to consider the induced temperature during the fatigue. The maximum induced temperature during the fatigue cycles is a crucial point for considering prognosticating the fatigue behavior of the printed polymer and the continuous fiber-reinforced polymer composite materials. The maximum induced temperatures during the fatigue test of the 3D-printed CF-PA6 and reinforced CF-PA6 with continuous glass fiber samples were measured, which are illustrated in Figure 22. According to this figure, the importance of the continuous fiber reinforcement in the fatigue behavior of the reinforced CF-PA6 with continuous glass fiber material is obvious. The measured induced temperature values related to the performed strains concerning the reinforced CF-PA6 with continuous glass fiber specimens were lower than in the CF-PA6 (without continuous reinforcement) specimens. In addition, the slopes of the curves concerning both composite materials, which is representative of the maximum induced temperature rate just before failure, are the same. Both mechanical fatigue (MF) and induced thermal fatigue (ITF) were observed. As shown in Figure 22, regarding the glass transition temperature zone of the utilized material, the commencement of the ITF-dominated zone can be determined.

Fatigue Fractography Analysis

Figure 24 depicts the fracture microscopic observations concerning reinforced CF-PA6 with continuous glass fiber and CF-PA6 specimens for studying the influence of the applied loading amplitude on fracture behavior. According to the conducted fractography, the fracture surface of CF-PA6 with continuous glass fiber specimens consisted of continuous glass fiber pull-out and subsequent debonding of the deposited layers (Figure 24a). The fracture surface of CF-PA6 revealed the debonding of the deposited layers and the subsequent crack propagation inside the deposited layers (Figure 24b).

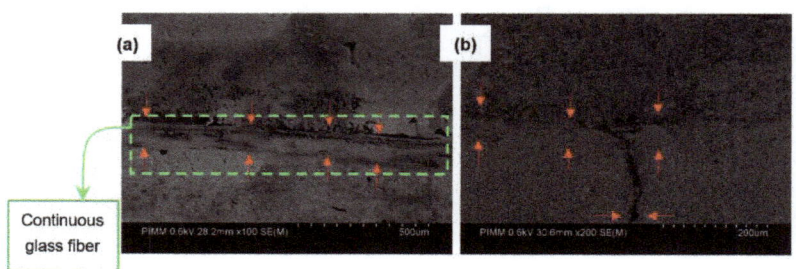

Figure 24. Fractography of the reinforced CF-PA6 with continuous glass fiber (**a**), and CF-PA6 (**b**) specimens [24].

4. Conclusions and Perspectives

The primary purpose of this research was to study the rheological characteristics of the materials during and after the FFF process. As a result of this study, we can gain a better understanding of the behaviors of polymers and polymer composite materials, as well as how reinforcements function during and after the FFF process. The obtained results will enable us to improve the FFF-processed parts. The fundamental problem of the produced parts by the FFF process is the presence of the porosities/voids inside the manufactured parts and the subsequent poor adhesion between the deposited layers. In this study, influences of the various processing parameters on the material behaviors during and after the FFF process were investigated. The physicochemical, thermal, and mechanical assessments exhibited that the crystallinity degree was influenced by the processing parameters modification. The crystallinity degree modifications could affect material diffusion during the cooling stage and the subsequent bonding of the deposited adjacent filaments. The strength of this bonding directly affects the mechanical behaviors of the manufactured specimens. Therefore, a precise and local measurement of the temperature on the scale of the diameter of the filaments during the FFF process was performed. In addition, by comparing the obtained temperature evolution curves from PA6 and CF-PA6 specimens, it was found that the existence of the chopped carbon fibers could affect the temperature evolution curves during the FFF process.

The obtained results from the three-point bending fatigue and the tensile behavior of the manufactured composite materials reveal:

- The FFF-processed CF-PA6 specimens with solid fill patterns had superior mechanical properties and stiffness under tension.
- Based on the study of the effect of the continuous glass fiber reinforcements on the fatigue behavior of the manufactured specimens, a reduction in fatigue life was observed in the manufactured CF-PA6 and continuous glass fiber-reinforced CF-PA6 specimens due to an increase in the induced temperature. By considering this induced temperature, known as self-heating, we can determine how a polymer behaves viscously, especially at the glass transition zone. It was observed that CF-PA6 reinforced with continuous glass fiber specimens exhibited a lower level of self-heating compared to CF-PA6 specimens without continuous glass fiber reinforcement dur-

ing the fatigue test. Based on the microscopic analysis of the damage mechanisms during the performed fatigue tests, the fracture surface of the continuous glass fiber-reinforced CF-PA6 specimens was characterized by continuous glass fiber pull-out and subsequent debonding of the deposited layers.

- As for the study of the density of continuous glass fiber in the FFF process of polymer-based composites, by increasing the density of the reinforcement layers, the strengths of the manufactured specimens were increased until a specific density value, and then, they became stable.
- As for studying the effect of the direction of the continuous reinforcement in the FFF-processed polymer composite materials, the five directions of 0°, 30°, 45°, 60°, and 90° were investigated. The highest strength was related to the manufactured parts, in which the reinforcements were in the applied stress direction (0°). As the direction of the utilized reinforcement deviated more from the applied stress direction, the strength of the manufactured parts decreased. The fracture of the tensile-tested samples occurred in the direction of the deposited continuous reinforcement.

In the next study, the effect of other continuous reinforcements such as carbon fiber and Kevlar fiber on the tensile and the fatigue behaviors of the printed specimens will be studied. Therefore, their effects will be compared with each other.

Author Contributions: Conceptualization, M.A., K.B. and M.S.; methodology, M.A., K.B. and M.S.; software, M.A.; validation, M.A., K.B. and M.S.; formal analysis, M.A. and M.S.; investigation, M.A., K.B. and M.S.; resources, M.A., K.B. and M.S.; data curation, M.A. and M.S.; writing—original draft preparation, M.A.; writing—review and editing, M.A. and M.S.; visualization, M.A.. K.B. and M.S; supervision, M.A., K.B. and M.S; project administration, M.A., K.B. and M.S. funding acquisition, M.A., K.B. and M.S. All authors have read and agreed to the published version of the manuscript.

Funding: This research received no external funding.

Institutional Review Board Statement: Not applicable.

Data Availability Statement: The data presented in this study are available on request from the corresponding author.

Conflicts of Interest: The authors declare no conflict of interest.

References

1. Anderegg, D.A.; Bryant, H.A.; Ruffin, D.C.; Skrip Jr, S.M.; Fallon, J.J.; Gilmer, E.L.; Bortner, M.J. In-situ monitoring of polymer flow temperature and pressure in extrusion based additive manufacturing. *Addit. Manuf.* **2019**, *26*, 76–83. [CrossRef]
2. Ahmadifar, M.; Benfriha, K.; Shirinbayan, M.; Tcharkhtchi, A. Additive manufacturing of polymer-based composites using fused filament fabrication (FFF): A review. *Appl. Compos. Mater.* **2021**, *28*, 1335–1380. [CrossRef]
3. Huang, Y.; Leu, M.C.; Mazumder, J.; Donmez, A. Additive manufacturing: Current state, future potential, gaps and needs, and recommendations. *J. Manuf. Sci. Eng.* **2015**, *137*, 014001. [CrossRef]
4. Ghazanfari, A.; Li, W.; Leu, M.C.; Hilmas, G.E. A novel freeform extrusion fabrication process for producing solid ceramic components with uniform layered radiation drying. *Addit. Manuf.* **2017**, *15*, 102–112. [CrossRef]
5. Benfriha, K.; Ahmadifar, M.; Shirinbayan, M.; Tcharkhtchi, A. Effect of process parameters on thermal and mechanical properties of polymer-based composites using fused filament fabrication. *Polym. Compos.* **2021**, *42*, 6025–6037. [CrossRef]
6. Chacón, J.M.; Caminero, M.A.; García-Plaza, E.; Núñez, P.J. Additive manufacturing of PLA structures using fused deposition modelling: Effect of process parameters on mechanical properties and their optimal selection. *Mater. Des.* **2017**, *124*, 143–157. [CrossRef]
7. Altan, M.; Eryıldız, M.; Gumus, B.; Kahraman, Y. Effects of process parameters on the quality of PLA products fabricated by fused deposition modeling (FDM): Surface roughness and tensile strength. *Mater. Test.* **2018**, *60*, 471–477. [CrossRef]
8. El Magri, A.; El Mabrouk, K.; Vaudreuil, S.; Touhami, M.E. Mechanical properties of CF-reinforced PLA parts manufactured by fused deposition modeling. *J. Thermoplast. Compos. Mater.* **2021**, *34*, 581–595. [CrossRef]
9. El Magri, A.; El Mabrouk, K.; Vaudreuil, S.; Ebn Touhami, M. Experimental investigation and optimization of printing parameters of 3D printed polyphenylene sulfide through response surface methodology. *J. Appl. Polym. Sci.* **2021**, *138*, 49760. [CrossRef]
10. Spoerk, M.; Arbeiter, F.; Cajner, H.; Sapkota, J.; Holzer, C. Parametric optimization of intra-and inter-layer strengths in parts produced by extrusion-based additive manufacturing of poly (lactic acid). *J. Appl. Polym. Sci.* **2017**, *134*, 45401. [CrossRef]
11. Yang, L.; Li, S.; Li, Y.; Yang, M.; Yuan, Q. Experimental investigations for optimizing the extrusion parameters on FDM PLA printed parts. *J. Mater. Eng. Perform.* **2019**, *28*, 169–182. [CrossRef]

12. Rajpurohit, S.R.; Dave, H.K. Effect of process parameters on tensile strength of FDM printed PLA part. *Rapid Prototyp. J.* **2018**, *24*, 1317–1324. [CrossRef]
13. Wang, L.; Gramlich, W.M.; Gardner, D.J. Improving the impact strength of Poly (lactic acid)(PLA) in fused layer modeling (FLM). *Polymer* **2017**, *114*, 242–248. [CrossRef]
14. Mackay, M.E.; Swain, Z.R.; Banbury, C.R.; Phan, D.D.; Edwards, D.A. The performance of the hot end in a plasticating 3D printer. *J. Rheol.* **2017**, *61*, 229–236. [CrossRef]
15. Peng, F.; Vogt, B.D.; Cakmak, M. Complex flow and temperature history during melt extrusion in material extrusion additive manufacturing. *Addit. Manuf.* **2018**, *22*, 197–206. [CrossRef]
16. Christiyan, K.J.; Chandrasekhar, U.; Venkateswarlu, K. A study on the influence of process parameters on the Mechanical Properties of 3D printed ABS composite. *IOP Conf. Ser. Mater. Sci. Eng.* **2016**, *114*, 012109. [CrossRef]
17. Caminero, M.Á.; Chacón, J.M.; García-Plaza, E.; Núñez, P.J.; Reverte, J.M.; Becar, J.P. Additive manufacturing of PLA-based composites using fused filament fabrication: Effect of graphene nanoplatelet reinforcement on mechanical properties, dimensional accuracy and texture. *Polymers* **2019**, *11*, 799. [CrossRef]
18. Yuan, S.; Li, S.; Zhu, J.; Tang, Y. Additive manufacturing of polymeric composites from material processing to structural design. *Compos. Part B Eng.* **2021**, *219*, 108903. [CrossRef]
19. Hmeidat, N.S.; Kemp, J.W.; Compton, B.G. High-strength epoxy nanocomposites for 3D printing. *Compos. Sci. Technol.* **2018**, *160*, 9–20. [CrossRef]
20. Rao, V.D.P.; Rajiv, P.; Geethika, V.N. Effect of fused deposition modelling (FDM) process parameters on tensile strength of carbon fibre PLA. *Mater. Today Proc.* **2019**, *18*, 2012–2018.
21. Ding, S.; Zou, B.; Wang, P.; Ding, H. Effects of nozzle temperature and building orientation on mechanical properties and microstruc-ture of PEEK and PEI printed by 3D-FDM. *Polym. Test.* **2019**, *78*, 105948. [CrossRef]
22. Berretta, S.; Davies, R.; Shyng, Y.; Wang, Y.; Ghita, O. Fused Deposition Modelling of high temperature polymers: Exploring CNT PEEK composites. *Polym. Test.* **2017**, *63*, 251–262. [CrossRef]
23. Yang, C.; Tian, X.; Li, D.; Cao, Y.; Zhao, F.; Shi, C. Influence of thermal processing conditions in 3D printing on the crystallinity and mechanical properties of PEEK material. *J. Mater. Process. Technol.* **2017**, *248*, 1–7. [CrossRef]
24. Ahmadifar, M.; Benfriha, K.; Shirinbayan, M.; Fitoussi, J.; Tcharkhtchi, A. Mechanical behavior of polymer-based composites using fused filament fabrication under monotonic and fatigue loadings. *Polym. Polym. Compos.* **2022**, *30*, 09673911221082480. [CrossRef]

Disclaimer/Publisher's Note: The statements, opinions and data contained in all publications are solely those of the individual author(s) and contributor(s) and not of MDPI and/or the editor(s). MDPI and/or the editor(s) disclaim responsibility for any injury to people or property resulting from any ideas, methods, instructions or products referred to in the content.

Article

Parametric Investigation and Optimization to Study the Effect of Process Parameters on the Dimensional Deviation of Fused Deposition Modeling of 3D Printed Parts

Muhammad Abas [1,*], Tufail Habib [1], Sahar Noor [1], Bashir Salah [2,*] and Dominik Zimon [3]

1 Department of Industrial Engineering, University of Engineering & Technology, Peshawar 25100, Pakistan
2 Industrial Engineering Department, College of Engineering, King Saud University, P.O. Box 300, Riyadh 11421, Saudi Arabia
3 Department of Management Systems and Logistics, Rzeszow University of Technology, 35-959 Rzeszow, Poland
* Correspondence: muhammadabas@uetpeshawar.edu.pk (M.A.); bsalah@ksu.edu.sa (B.S.)

Abstract: Fused deposition modeling (FDM) is the most economical additive manufacturing (AM) technology available for fabricating complex part geometries. However, the involvement of numerous control process parameters and dimensional instabilities are challenges of FDM. Therefore, this study investigated the effect of 3D printing parameters on dimensional deviations, including the length, width, height, and angle of polylactic acid (PLA) printed parts. The selected printing parameters include layer height, number of perimeters, infill density, infill angle, print speed, nozzle temperature, bed temperature, and print orientation. Three-level definitive screening design (DSD) was used to plan experimental runs. The results revealed that infill density is the most consequential parameter for length and width deviation, while layer height is significant for angle and height deviation. The regression models developed for the four responses are non-linear quadratic. The optimal results are obtained considering the integrated approach of desirability and weighted aggregated sum product assessment (WASPAS). The optimal results include a layer height of 0.1 mm, a total of six perimeters, an infill density of 20%, a fill angle of 90°, a print speed of 70 mm/s, a nozzle temperature of 220 °C, a bed temperature of 70 °C, and a print orientation of 90°. The current study provides a guideline to fabricate assistive devices, such as hand and foot orthoses, that require high dimensional accuracies.

Keywords: fused deposition modeling; polylactic acid (PLA); dimensional deviation; definitive screening design; desirability function

1. Introduction

Among the AM technologies, fused deposition modeling (FDM) is one of the most widely used additive manufacturing technologies because of its economy and ability to process a diverse range of materials, including polymers and metals [1]. However, the use of FDM printing for part fabrication is still a challenge because of the involvement of numerous process parameters and because the choice of materials affects the part quality, mechanical strength, and development time [2,3]. Depending on the application, careful consideration of process variables and material selection is necessary. According to the published reports, the process parameters can be divided into three major sets [4]. The first set of parameters includes the process-related parameters, such as infill speed, number of shells, thickness of shells, bed temperature, fill density, layer height, nozzle temperature, print speed, air gap, and raster angle. The second set of parameters includes the machine-specific parameters, such as nozzle diameter, filament width, bed adhesion type, and filament diameter. The third set of parameters is related to part geometry, such as the part's orientation and special features.

To achieve good dimensional accuracy in FDM-printed parts, the optimal process parameter settings are crucial, as they vary according to material, complexity of part geometry, material type, and chemical composition [5,6]. Therefore, finding the optimal settings and combination of parameters can be challenging and laborious. Additionally, most of the polymers used in FDM are semi-crystalline and prone to part distortion due to crystallization [7]. Therefore, the process requires trial and error experimental procedures, or application of the design of experiments (DoE), to achieve excellent quality prints with desirable mechanical properties. The most common semi-crystalline polymers are polylactic acid (PLA), polypropylene (PP), polycaprolactone (PCL), polyethylene (PE), and polybutylene terephthalate (PBL). Moreover, the dimensional specifications may vary for the same material as well as for varied materials. For instance, in PLA, positive deviation (expansion) is observed in the width and thickness direction, while negative deviation (shrinkage) is observed in the length direction [8].

PLA is considered a green material because it is made through the polymerization of lactic acid by the fermentation of renewable resources. There are four different forms of crystals, namely α, β, and γ [9]. The α crystals show two disordered modifications i.e., α' and α'' [9]. The α crystal is obtained through cold, melt, or solution crystallization at a higher temperature (i.e., above 120 °C) [10], while α' is produced at a lower temperature (i.e., below 100 °C) by mixing α and α' between 100 °C and 120 °C [11]. The α'' crystal is obtained through crystallization at a temperature (0 °C to 30 °C) under high-pressurized CO_2 [12]. The α' crystal forms the chain conformation of the PLA chain, which is more disordered than in the α form crystal [13]. Therefore, the α form provides lower elongation at break, higher Young's modulus, and better preservation against water vapor than the α' form. The α'' crystal produces poor chain packing and the lowest crystal density compared to α and α' [14]. The published studies have shown that the α form crystal is more stable compared to its other forms [9]. The β form crystal is obtained through α crystal deformation and through annealing or stretching at elevated temperatures [15]. The γ form is obtained by epitaxial growth on a hexamethyl benzene substrate [16].

The physical and mechanical properties of PLA are influenced by the degree of crystallinity. Mechanical properties can be improved by thermal annealing to increase the degree of crystallinity [17]. In FDM printing, the degree of crystallinity in the bottom layers is higher than in the top and side layers because of the bed temperature, which causes the layer to cool down slowly, thus rendering the printed part dimensionally unstable [18].

2. Literature Review

Numerous studies have been reported that investigated the effect of FDM process parameters on quality characteristics, mechanical properties, physical properties, energy consumption, and build time for diverse types of materials. For instance, Galetto et al. [4] investigated the effect of process parameters on the process efficiency and quality of PLA printed parts. Quadratic models were developed for surface roughness and dimensional accuracies. For maximizing dimensional accuracy, the design features of parts play a significant role. Kitsakis et al. [19] studied the dimensional accuracy of FDM-printed parts for medical applications. In the study, they considered different parameters, including the material type (PLA and ABS), layer height, infill rate, and the number of shells, as well as studying the dimensional accuracy. The study revealed that the best dimensional accuracy for PLA material was attained at an infill rate of 50%, with one shell, and a layer height of 0.3 mm. The study of Aslani et al. [20] showed that the extrusion temperature significantly affects the dimensional accuracy and surface roughness of PLA printed parts. The study proved that by applying grey relational analysis, high extrusion temperature (230 °C) combined with medium wall thickness values (2 mm) optimized both surface roughness and dimensional accuracy. Nathaphan and Trutassanawin [21] concluded that for good dimensional accuracy and compression strength, the layer height and print speed must be set at a low level, the nozzle temperature at a high level, while the bed temperature must be above the glass transition temperature of ABS material. Further, shrinkage occurs in the diameter of the cylinder because of the cooling and solidification of molten polymer. However, expansion was noticed in height of the cylinder due to the rounding of the number of layers to the higher integer number. Basavaraj and Vishwas [22] found that layer thickness affects the tensile strength, manufacturing time, layer thickness, shell thickness, and orientation angle. Further, the study concluded that tensile strength and dimensional accuracy decrease with an increase of the layer thickness and increase with increases of the orientation angle and shell thickness. The study of Lalegani Dezaki et al. [23] revealed that surface quality and mechanical properties are directly affected by the type of patterns. Concentric and grid patterns exhibit good surface quality and tensile strength while the zigzag pattern produces the worst surface roughness and mechanical properties. Padhi et al. [24] noted that shrinkage occurs along the width and length directions, while the thickness increases in parts printed from acrylonitrile-butadiene-styrene (ABSP 400). The shrinkage may develop inner stress upon solidification. Further, the formation of inner layer cracks and weak interlayer adhesion decrease the dimensional accuracy of final parts. Vahabli and Rahmati [25] improved the surface quality of FDM-printed parts for medical devices using artificial neural networks based on the feed-forward back propagation (FFBP) algorithm. Parts were printed from ABSplus material. The successful fabrication of medical devices such as a molar tooth, femur, skull, and stem further confirms the performance of FFBP. Deswal et al. [8] worked on FDA process parameters by applying an approach integrated with a response surface methodology, artificial neural network-genetic algorithm (ANN-GA), genetic algorithm (RSM–GA), and artificial neural network (ANN) for improving the dimensional accuracy of ABS parts. The adaptive neuro-fuzzy inference system (ANFIS) model and whale optimization algorithm (WOA) was applied by Sai et al. [26] to optimize the process parameters for printing PLA implants. Their study concluded that layer thickness followed by raster angle and infill density significantly affects the surface roughness, while layer thickness and raster angle at low level and infill density at medium level provides good surface quality. The findings of Vyavahare et al. [27] revealed that layer thickness and build orientation have a significant effect on fabrication time and surface roughness, while for dimensional accuracy, in addition to these two parameters, Camposeco-Negrete [28] optimized the process parameters to improve the dimensional accuracy, energy consumption, and the production time of FDM 3D printed acrylonitrile styrene acrylate (ASA) parts. The study showed that printing plane is the most significant parameter that helps in reducing production time and energy consumption. For dimensional accuracy, the infill pattern influences the width of the part, and layer thickness affects the length of the part significantly. Mohamed et al. [29] applied a deep neural network to analyze and optimize the dimensional accuracy of FDM PC-ABS printed parts. In the study, a total of 16 experiments were planned based on a definitive screening design (DSD). The part profile for dimensional accuracy was considered as the percentage variation in diameter and length. The quadratic model was found to be significant for both length and diameter variation. Slice thickness, print direction, interaction of print direction, and deposition angle were found to be significant for length variation. Mohanty et al. [30] applied the hybrid approach of a Taguchi- MACROS- nature-inspired heuristic optimization technique to optimize parameters affecting the dimensional precision of ABS M30 FDM-printed parts. Their results showed that part orientation significantly affected dimensional precision. All of the nature-inspired algorithms considered in the study provide comparable results for minimizing dimensional error. Garg et al. [31] studied the dimensional accuracy and surface roughness of ABS P430 FDM-printed parts under the cold vapor technique using acetone. The results revealed that chemical treatment reduces surface roughness and improves the dimensional accuracy of the final part. This may be attributed to softening of the external layer, because acetone causes rupturing of a secondary bond between the chains of ABD polymers and reaches a more stable position.

The literature review presented above shows that limited studies are available in the literature that focus on the investigation of the effect of different process parameters on dimensional accuracy or dimensional deviation (along the length, width, and height) of

FDM-printed parts. According to the best knowledge of the authors, no similar study has been published before concerning angular deviation. Parameters such as the number of shells, bed temperature, infill density, build orientation, and printing speed are studied far less compared to other parameters such as the layer height, infill angle, and extrusion temperature. Therefore, further research is needed to determine the impact of various process parameter combinations on dimensional deviation. Thus, the present study aims to cover the research gaps and offers an inclusive guide for additive manufacturing users to decide on optimal FDM process parameter settings that affect dimensional deviations. Furthermore, an integrated approach of desirability function and weighted aggregated sum product assessment (WASPAS) is proposed for simultaneous optimization of responses.

3. Materials and Methods

Test specimens were printed from commercial-grade poly lactic acid (PLA) supplied by a local manufacturer (3Dworld, Rawalpindi, Punjab, Pakistan) using an ALIFHX XC555 PRO3D printer. The diameter of the filament is 1.75 mm, having a density and a molecular weight of 1.3 g/cm^3 and 4.7–16.8 × 10^3 g/mol. The printed test specimens were prepared according to ASTM E23-12c, which is used for impact tests, as shown in Figure 1. Figure 2 shows the printing system used for the specimens.

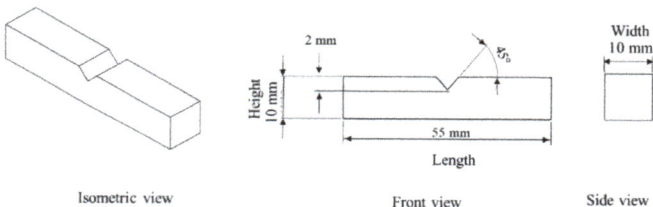

Figure 1. The geometry of the test specimen.

Figure 2. FDM 3D printer, specimen printing, and test samples.

Differential scanning calorimetry (DSC) was performed for both PLA spool material and printed PLA, as shown in Figure 3. The three key features of semi-crystalline thermoplastic PLA material represented are the heat flow at a glass transition temperature (T_g), the cold crystallization exothermic (T_c), and the melting temperature endothermic (T_m). The T_g, T_c, and T_m for spool material are 63 °C, 98 °C, and 170 °C, which agrees with the range of reported values in the literature [32]. For PLA printed material, the T_g increases slightly from 63 °C to 65 °C, and melting temperature decreases with the formation of two added peaks i.e., at 164 °C and 157 °C. This may be attributed to the formation of multiple crystalline forms, namely the α and α′ during the thermal cycling [33].

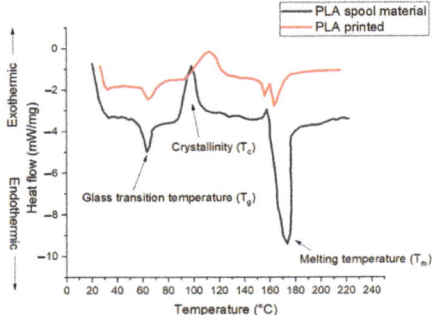

Figure 3. Comparison of differential scanning calorimetry (DSC) of PLA spool material and printed PLA.

3.1. Printing Process Parameters

The printing process parameters investigated in this study include the layer height, number of perimeters, infill density, infill angle, printing speed, nozzle temperature, bed temperature, and print orientation. Figure 4 shows the schematics of FDM printing and selected control printing parameters. For each process, parameter values at three levels were set based on the literature review and recommendation of the material manufacturer, as tabulated in Table 1. The other parameters were kept constant (given in Table 2).

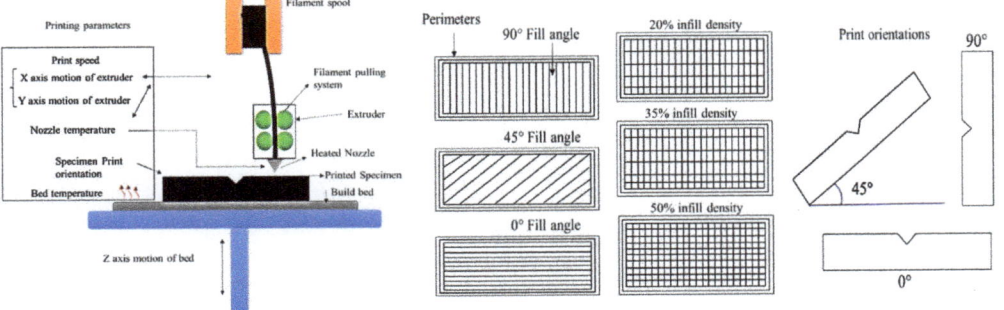

Figure 4. Schematic of FDM 3D printer with associated printing parameters and print orientations.

Table 1. Levels for printing parameters.

Printing Parameters	Symbol	Units	Levels		
			−1	0	1
Layer Height	A	mm	0.1	0.2	0.3
Number of Perimeters	B	-	2	4	6
Infill density	C	%	20	35	50
Fill angle	D	°	0	45	90
Print Speed	E	mm/s	50	60	70
Nozzle temperature	F	°C	190	205	220
Bed temperature	G	°C	70	80	90
Print orientation	H	°	0	45	90

Table 2. Printing parameters that are kept constant.

Printing Parameters	Settings
Pattern type	Rectilinear
Solid layers	3 for both top and bottom
Air gap	Negative
First layer speed	20 mm/s
Retraction speed	100/s

3.2. Experimental Design and Measurement of Responses

Due to a large number of process parameters, a systematic experimental design approach, namely the three-level definitive screening designs (DSD) is used to plan experimental runs. The purpose of using this design is to model and estimate the main effect, interaction effect, and quadratic effect in small experimental runs. A total of 17 experimental runs were designed. However, to consider the repeatability of the printing process, the experimental runs were replicated three times randomly. The final experimental design includes 51 experimental runs, tabulated in Table A1 in the Appendix A. The samples prepared according to the experimental design are shown in Figure 1.

The responses considered for the dimensional deviations include the length, width, height, and angle, as shown in Figure 1. The deviation is calculated based on the percentage variation of CAD geometry and printed geometry by using Equation (1). For this, a profile projectile (Mitutoyo PJ-A3000, Mitutoyo Corporation, Kanagawa, Japan) is used. The resolution of the instrument for linear dimensions and angle is 0.01 mm and 0.01°, respectively.

$$\Delta X\ (\%) = \left(\frac{X_c - X_e}{X_c}\right) \times 100 \tag{1}$$

where ΔX is the deviations in dimensions, X_c is CAD dimensions, and X_e is the dimension of a printed specimen.

3.3. Optimization Methodology

Responses were optimized individually as well as simultaneously. Single responses were optimized considering the desirability function. To minimize and maximize the response variable, the desirability function is used, which is expressed in Equations (2) and (3).

$$d_i(k) = \begin{bmatrix} 0, & y_i(k) \leq \min(y_i(k)), \\ \left[\frac{y_i(k) - \min(y_i(k))}{\max(y_i(k)) - \min(y_i(k))}\right]^r & \min(y_i(k)) \leq y_i \leq \max(y_i(k)) \\ 1, & y_i(k) \geq \min(y_i(k)) \end{bmatrix} \tag{2}$$

$$d_i(k) = \begin{bmatrix} 0, & y_i(k) \leq \min(y_i(k)), \\ \left[\frac{y_i(k) - \max(y_i(k))}{\min(y_i(k)) - \max(y_i(k))}\right]^r & \min(y_i(k)) \leq y_i \leq \max(y_i(k)) \\ 1, & y_i(k) \geq \max(y_i(k)) \end{bmatrix} \tag{3}$$

where $d_i(k)$ is the desirability value of each response at the *i*th experiment and *k*th response, $y_i(k)$ is the individual value of measured response k at experiment number *i*, *max* $y_i\ (k)$ and *min* $y_i(k)$ are the maximum and minimum values of data obtained for the *k*th response, and r is the weight of the desirability function.

Simultaneous optimization of responses was performed based on the proposed integrated approach of desirability function and weighted aggregated sum product assessment (WASPAS) method. It transformed the multi-response optimization problem into a single response called a relative importance score. The following procedure was adopted to optimize the process parameters:

Step 1: Compute the desirability function for responses using Equations (2) and (3). For minimization (cost criteria), apply Equation (2), while for maximization (benefit criteria) use Equation (3).

Step 2: Calculate the weighted sum of desirability functions (WSD) using Equation (4).

$$WSD_i = \sum_{j=1}^{r} d_i w_j \qquad (4)$$

where w_j stands for the weight of jth response.

Step 3: Calculate the weighted product of desirability functions (WPD) using Equation (5).

$$WPD_i = \prod_{j=1}^{r} d_i^{w_j} \qquad (5)$$

Step 4: Determine the relative importance score (RIS) of each experimental run using Equation (6) [34].

$$RIS_i = \lambda \cdot WSD_i + (1-\lambda) \cdot WPD_i \qquad (6)$$

where λ is a constant with a minimum value of 0 and a maximum value of 1, however in the reported studies, a value of 0.5 is proposed for good accuracy [35,36]. The highest RIS value is the best experimental run.

Step 5: Finally, the optimal parameter settings are obtained considering the average values of the RIS for each process parameter at each level. Higher average values of RIS represent better response performances.

4. Results and Discussion

4.1. Regression Models for Dimensional Deviation

Regression models computed for dimensional deviation in uncoded units are given in Equations (A1)–(A4) in Appendix B. The adequacy of these models is assessed based on the coefficient of determination (R^2), adjusted R^2, predicted R^2, and lack of fit. Table 3 is the summary of dimensional deviations. The results show that the developed regression models are adequate and fit well with the experimental data due to their higher R^2 values, which are near 100%, and their p-values are larger than the alpha value of 0.05. Further, the models have good prediction accuracy, as the adjusted R^2 and predicted R^2 are closer to each other (the percentage difference is less than 20%).

Table 3. Models summary of dimensional deviation.

Responses	R^2 (%)	Adjusted R^2 (%)	Predicted R^2 (%)	Lack of Fit Based on p-Value
LD (%)	95.36	93.89	91.2	0.848
WD (%)	95.71	94.64	92.96	0.899
HD (%)	98.74	98.3	97.5	0.901
AD (%)	97.18	96.19	94.51	0.545

The effect of the process parameters on individual responses was studied through the Pareto chart for standardized effect. Figure 5a–d show that all the terms that crossed the reference line at 2.02 are significant at an alpha value of 0.05. Figure 5a illustrates that for length deviation, infill density is the most influential factor, followed by bed temperature, quadratic effect of print orientation, print speed, print orientation, nozzle temperature, layer height, number of perimeters, fill angle, quadratic effect of layer height, interaction of infill density and print orientation, and quadratic effect of the number of perimeters. Accordingly, for width deviation, as shown in Figure 5b, the most influential factors are infill density followed by print orientation, bed temperature, print speed, interaction of infill density and print orientation, quadratic effect of a number of perimeters, quadratic effect of bed temperature, nozzle temperature, number of parameters, and layer height. For height deviation, as shown in Figure 5c, the most influential factors include layer height, interaction of layer height, square of print speed, bed temperature, interaction of the number of perimeters and print orientation, number of perimeters, print orientation, interaction of layer height and print speed, print speed, nozzle temperature, infill density, fill angle and interaction of layer height, and nozzle temperature. For angle deviation, as shown in Figure 5d, the most influential factors are layer height, fill angle, infill density, square of infill density, interaction of infill density and fill angle, bed temperature, number of perimeters, interaction of layer height and infill density, print speed, interaction of infill density and bed temperature, number of perimeters and fill angle, print orientation, and nozzle temperature.

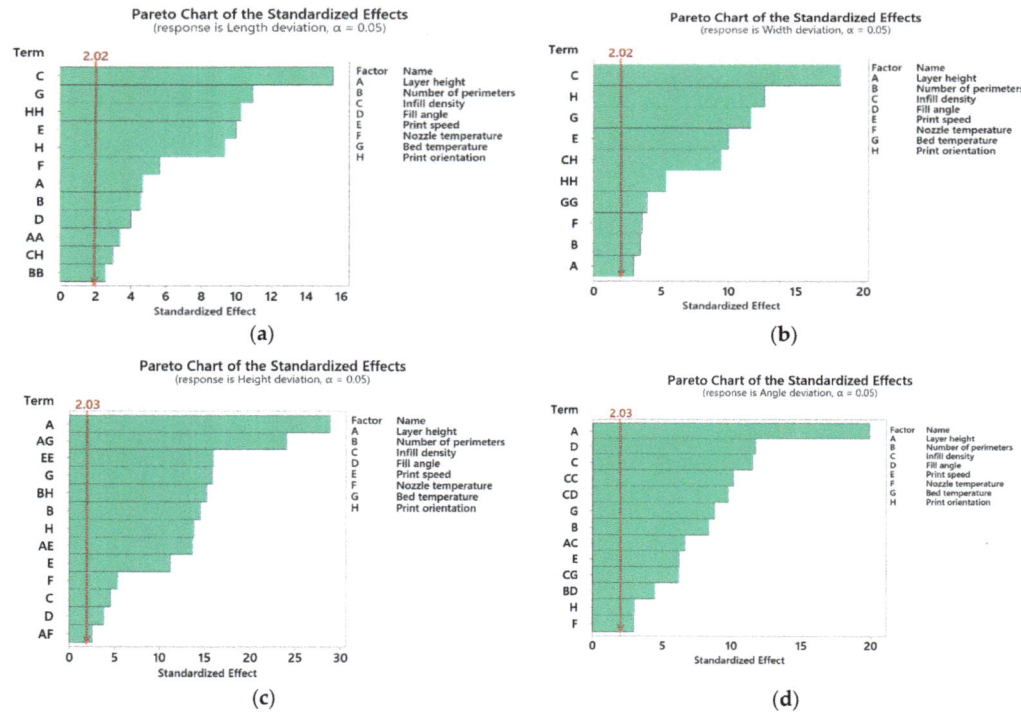

Figure 5. Pareto chart of standardized effect of dimensional deviation (**a**) length deviation, (**b**) width deviation, (**c**) height deviation, (**d**) angle deviation.

4.2. Main Effect and Interaction Plots

Main effect and interaction plots are generated for the modeled process parameters to study the effect of process parameters on dimensional deviations i.e., length deviation, width deviation, height deviation, and angular deviation.

4.2.1. Main Effect and Interaction Effect of Process Parameters on Length Deviation

The main effect plot given in Figure 6 shows that with an increase of the layer height from 0.1 mm to 0.2 mm, the mean length deviation decreases from 0.7% to 0.6%, and then it increases from 0.6% to 0.8%, with an increase of layer height from 0.2 mm to 0.3 mm. These results are in line with the findings of Deswal et al. [8], Agarwal et al. [37], and Nancharaiah et al. [38]. A high layer causes the formation of an air gap between the layers that reduces interlayer bonding and results in inner stresses that cause deformation and distortion of layers [24]. The cooling time of the material also decreases with an increase of the layer thickness, affecting the adhesion between layers. This increases dimensional deviation and also reduces the mechanical properties [24,39]. The optical inverted metallurgical microscope (Model No: M-41X, Lab Testing Technology Shanghai Co., Ltd., Shanghai,

China) images in Figure 7a,b further confirms these conclusions. Its shows the formation of cracks and pores in printed parts at a layer height of 0.3 mm.

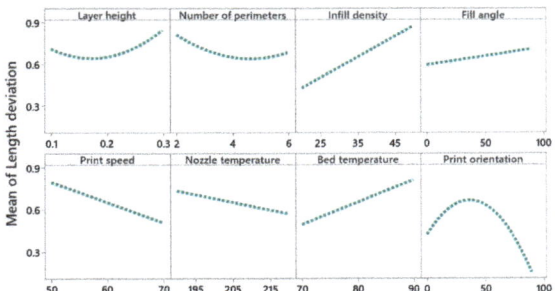

Figure 6. Main effect plot for length deviation.

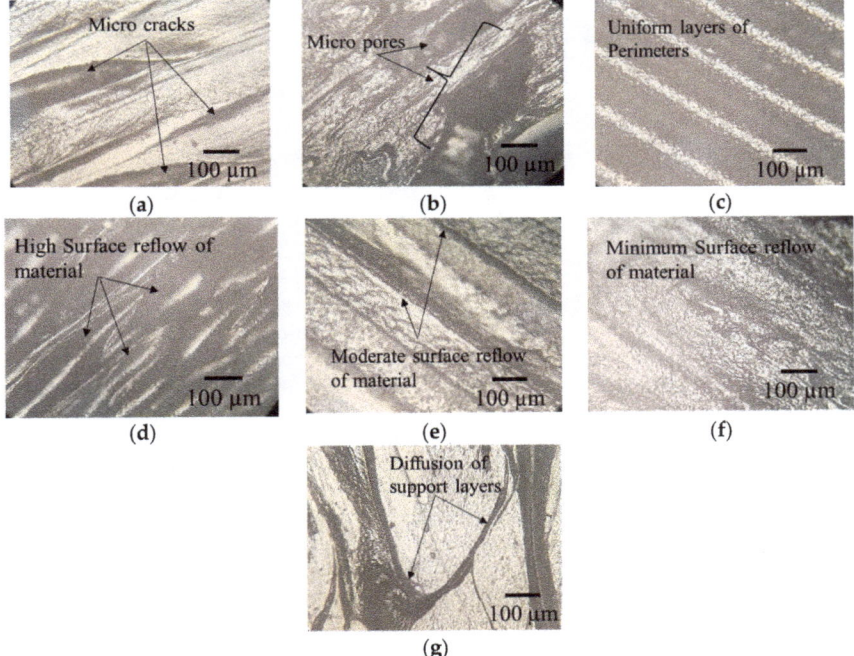

Figure 7. Optical images (**a**) formation of microcracks at 0.3 mm layer height, (**b**) formation of micropores at 0.3 mm layer height, (**c**) uniform layers of perimeters, (**d**) high surface reflow of material at 90 °C bed temperature, (**e**) moderate surface reflow of material at 80 °C bed temperature, (**f**) minimum surface reflow of material at 70 °C bed temperature, (**g**) diffusion of support layers at 45° build orientation.

An increase of the number of perimeters from two to six decreases the deviation in length from 0.8% to 0.6%. According to Mohamed et al. [40], an increase of the number of perimeters increases the dimensional accuracy of the part length, as they are built parallel to the length [29]. A larger number of contours provide a dense filling in parts and make the part structure uniform with low dimensional deviation, as shown in Figure 7c.

Increase of infill percentage from 20% to 50% increases length deviation by 0.4% to 0.9%. These findings are in line with Akande et al. [41] and Agarwal [37]. A low infill density helps in transferring heat and cools down the material from glass transition temperature to ambient temperature without creating thermal stresses [8]. The interaction of infill density and the print orientation is found to be significant, as shown in Figure 8. It shows that the length deviation is minimal at lower infill density (20%) and higher print orientation (90°). At 90° print orientation, the mean width deviation is much lower for all values of infill density compared to other orientations.

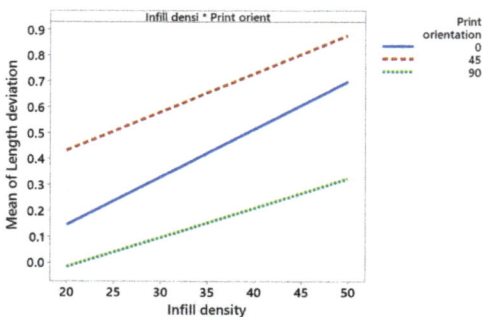

Figure 8. Interaction plot for length deviation.

An increase of the fill angle from 0° to 90° increases the length deviation by 0.6% to 0.75%. This may be due to the staircase effect that increases with an increase of raster angle [29]. At higher raster angles, voids are formed between the deposited raster and the perimeter walls causing incomplete filling and weak interlayer bonding which results in distortion and causes dimensional inaccuracies [40,42].

Increase of printing speed from 50 mm/s to 70 mm/s decreases length deviation by 0.8% to 0.5%. The same results were concluded by Agarwal et al. [37]. Low print speed allows more time for the deposition of material, and therefore, increases the dimensional deviation [27]. Generally, the polymer expands upon extrusion, however, by increasing the nozzle speed, the shear rate of polymer increases as a material is dragged by the nozzle tip and bed, thereby reducing the width of filament [43,44]. According to Brydson [45], by increasing the shear rate beyond a certain critical value, the extrusion swell decreases.

Length deviation decreases by 0.7% to 0.6% with an increase of nozzle temperature from 190 °C to 220 °C. High nozzle temperature maintains a consistent flow (good fluidity) of material that improves the fusion between layers and reduces the air gap, which helps in reducing distortion [46]. According to Afonso et al. [47], at an extrusion temperature between 210 °C to 230 °C, the PLA material becomes thermally and rheologically stable, and provides a good bonding mechanism between layers through a reduction of mesostructure voids, thereby improving the dimensional accuracy of printed parts.

An increase of the bed temperature from 70 °C to 90 °C increases the length deviation by 0.4% to 0.8%. This may be attributed to the glass transition temperature of PLA at about 60 °C. Near glass transition temperature, the mobility of macromolecules is higher, which improves the diffusion of polymer onto the glass and increases the adhesive forces [48]. According to Spoerk et al. [49], in the PLA material, adhesive forces increase with an increase of bed temperature, causing the bending of parts and damaging the bed surface upon cooling. Figure 7d shows a high layer diffusion and surface reflow of material, which results in dimensional deviations. However, it is moderate to minimum at bed temperatures of 80 °C and 70 °C, as illustrated in Figure 7e,f.

Length deviation increases by 0.4% to 0.7% with the increase of the build orientation from 0° to 45°, and then decreases by 0.7% to 0.1% from 45° to 90°. This is in line with the finding of Abdelrhman et al. [50]. This could be due to the diffusion of support material with part-built layers, which increases the surface roughness and induces dimensional inaccuracies, as shown in Figure 7g. The increase of length deviation may also be attributed to an increase of the staircase effect along the inclined surface (up to 45°), while the staircase effect is reduced by increasing the build orientation from 45° to 90° [31].

4.2.2. Main Effect and Interaction Effect of Process Parameters for Width Deviation

Figure 9 shows that the mean width deviation increases (from 1.45% to 1.6%) with the increase of the layer height. High layer height causes an uneven temperature gradient along the built axis, which causes inner residual stresses and results in distortion of the layer [51]. Increasing the number of perimeters reduces the width deviation (from 1.6% to 1.5%), while a larger number of perimeters provide dense filling in parts and make the part structure uniform [40]. Increasing the infill density increases the width deviation significantly (from 1.1% to 2%) compared to the layer height and the number of perimeters. Further, its interaction with print orientation is also found to be significant, as explained in Figure 10. It shows that at an infill density of 20% and print orientation of 90°, the minimum mean width deviation is 0.8%; at 0° it is approximately 0.85%, and at 45° it is higher i.e., approximately 1.8%. Further, it illustrates that at 90° print orientation, the mean width deviation is much lower for all values of infill density compared to 0° and 45° print orientations. An increase of print speed decreases the mean width deviation from 1.8% to 1.4%. However, increasing the print speed beyond some critical value decreases the cooling cycle of deposited materials and causes a thermal gradient that results in poor dimensional accuracies [52]. Increasing nozzle temperature reduces the mean width deviation from 1.6% to 1.4%. With an increase of extrusion temperature, polymer viscosity reduces and facilitates a good deposition process with reduced voids due to a greater flow of material through the nozzle tip [53]. At low extrusion temperatures, the layers are not completely fused, and cracks and pores are produced between each layer, which causes stress concentration near the pores and affects the mechanical and dimensional accuracy of the part [53]. An increase of bed temperature significantly increases the mean width deviation from 1% to 1.6%. According to Srinivas et al. [18], in FDM printing, the degree of crystallinity is higher in the bottom layers than in the top and side layers because of bed temperature, which causes the layer to cool down slowly, resulting in dimensional inaccuracies. Benwood et al. [54] reported that a bed temperature of 90 °C increases the crystallinity of PLA printed parts to a greater extent, thereby increasing its mechanical strength. However, the high degree of crystallinity causes poor dimensional accuracies due to shrinkage and residual stresses [55].

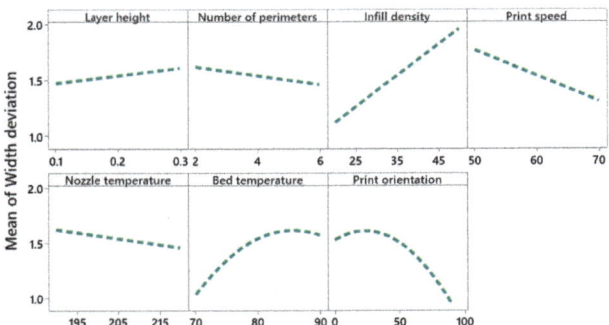

Figure 9. Main effect plot for width deviation.

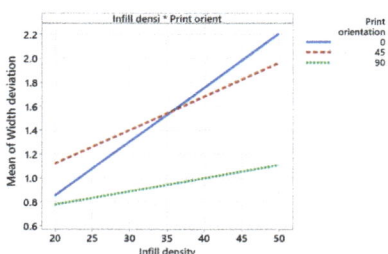

Figure 10. Interaction plot for width deviation.

4.2.3. Main Effect and Interaction Effect of Process Parameters for Height Deviation

Figure 11 shows that the mean height deviation decreases significantly (i.e., 2% to 0.2%) with the increase of layer height from 0.1 to 0.3 mm. These results are in line with studies by Deswal et al. [8], Camposeco-Negrete [56], and Peng et al. [57], which report that a high layer height reduces the deviation in thickness or the height of printed parts. The effect of layer height on mean height deviation is not sufficient information to interpret the results, due to the significant interaction between layer height and print speed, nozzle temperature, and bed temperature, as shown in the interaction plots in Figure 12. Figure 12 illustrates that at 60 mm/s print speed and 0.3 mm layer height, a minimum height deviation of 0.5% can be achieved. An interaction plot of layer height and nozzle temperature shows that at 0.3 mm layer height, a nozzle temperature of 190 °C minimizes the height deviation to 0.4%. Interaction plots of layer height and bed temperature prove that a bed temperature of 90 °C and a layer height of 0.3 mm reduces the height deviation to 0.09%. The interaction plot of the number of perimeters and print orientation depicts that a higher number of perimeters (i.e., six) and low print orientation (i.e., 0°) results in a minimum height deviation of 0.5%. The main effect plot of infill density and fill angle shows that the height deviation increases with an increase of infill density from 1% to 1.3%, and from 1% to 1.2% for fill angle.

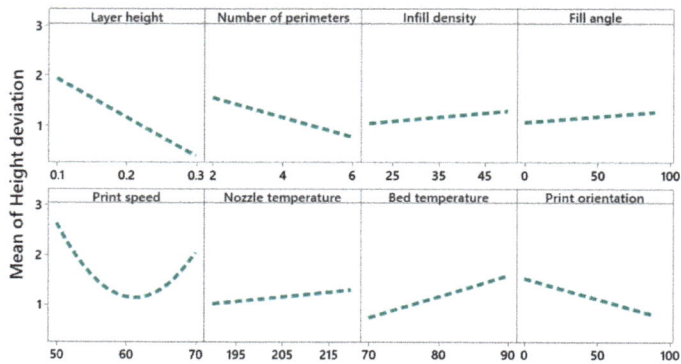

Figure 11. Main effect plot for height deviation.

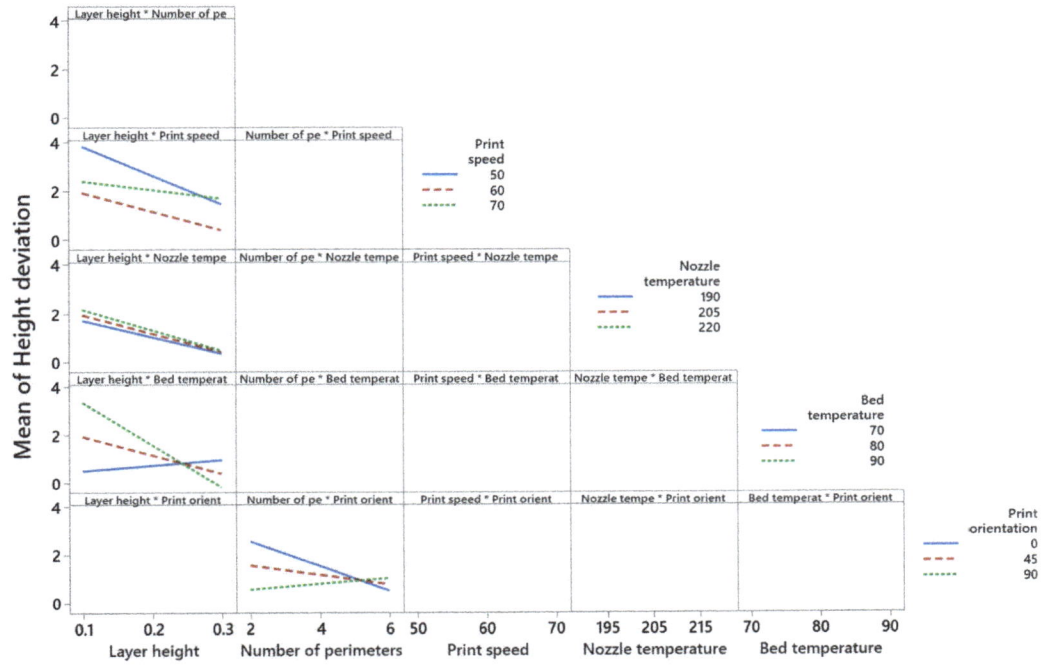

Figure 12. Interaction plot for height deviation.

4.2.4. Main Effect and Interaction Effect of Process Parameters for Angle Deviation

Figure 13 illustrates that the mean angular deviation increases with an increase of layer height. However, the interaction of layer height with infill density (as shown in Figure 14) implies that an infill density of 35% (medium level) and layer height of 0.1 mm (low level) give minimum angular deviation compared to 20% (low level) and 50% (elevated level) infill densities. The interaction plot of the number of perimeters and infill angle proves that the number of perimeters at a low level (i.e., two) and infill angle at a high level (i.e., 90°) reduces the angular deviation (i.e., from 2% to 1.5%). This may be due to the staircase effect that is more prominent in higher layer heights compared to lower layer heights, as shown in Figure 15a,b. Interaction plots of infill density with fill angle show that lower infill density (between 25% and 35%) and higher fill angle (90°) minimize the angular deviation. Figure 15c shows the thermal distortion of layers at high layer height and high infill density. An interaction plot of infill density with bed temperature proves that a bed temperature of 80 °C and infill density of 30% reduces the angular deviation to 1.7%. With an increase of print speed, the mean angular deviation decreases from 2.8% to 2%. However, with the increase of nozzle temperature, mean angular deviation increases from 2.2% to 2.4%, and a similar trend is seen for the print orientation.

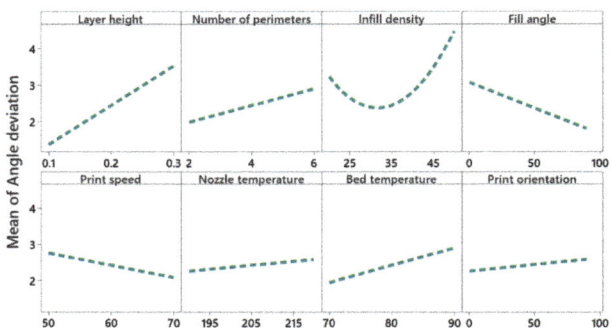

Figure 13. Main effect plot for angular deviation.

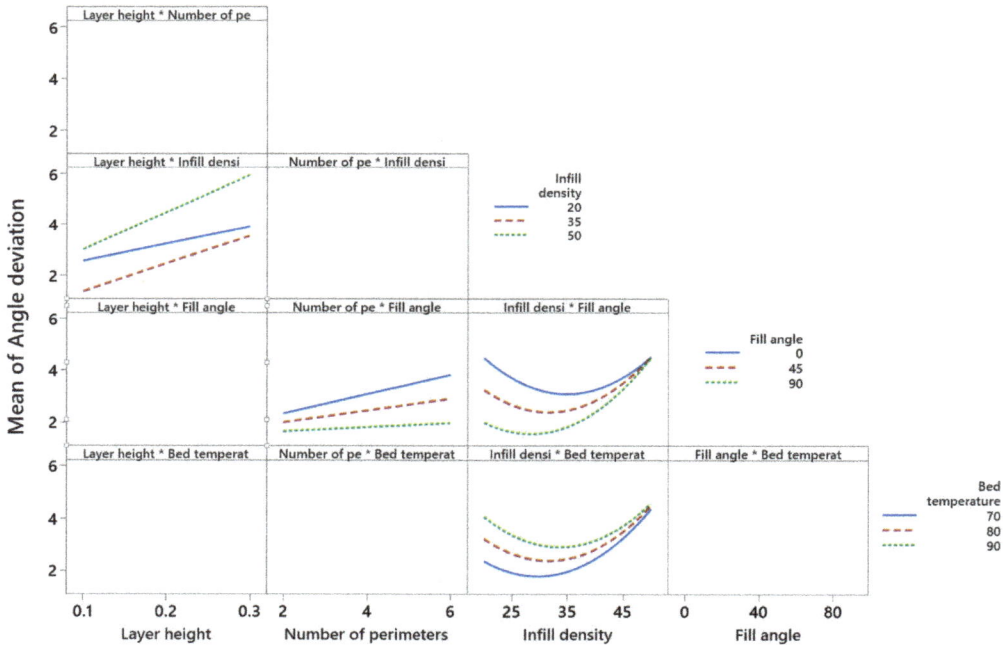

Figure 14. Interaction plot for angular deviation.

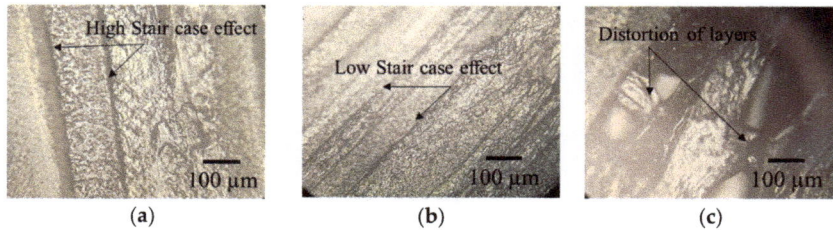

Figure 15. Optical microscope images (**a**) high staircase effect at 0.3 mm layer height, (**b**) high staircase effect at 0.1 mm layer height, (**c**) distortion of layers due to thermal stresses at 0.3 mm layer height and 50% infill density.

4.3. Optimization

For individual response optimization, the desirability function is used. Equation (2) is applied to optimize the dimensional deviations i.e., the length deviation (DL), width deviation (WD), height deviation (HD), and angle deviation (AD) are minimized. The desirability values computed after optimization are tabulated in Table A2 and given in Appendix A. The optimal setting of process parameters was identified by calculating the average desirability values of each process parameter at each level, as shown in Table 4. The highest average desirability value denotes the best levels for the process parameters. Aimed at length deviation, the highest average desirability values computed for the layer height and the number of perimeters is 0.76 and 0.75 at level 0, respectively. For infill density, fill angle, and bed temperature, the values are 0.85, 0.75, and 0.80 at level −1, respectively. For print speed, nozzle temperature, and print orientation, the values are 0.61, 0.77, and 0.82 at level 1, respectively. Thus, the optimal settings to minimize the length deviation are calculated. Similar optimal settings were obtained for width deviation. The optimal setting for height deviation is layer height, number of perimeters and print orientation at level 1, infill density, fill angle, print speed, nozzle temperature at level 0, and bed temperature at level −1. The optimal setting for angle deviation is layer height, number of perimeters, nozzle temperature, and print orientation at level 1, infill density and bed temperature at level 0, and fill angle at level 1.

Table 4. Optimal levels for individual responses.

Process Parameters	Levels			Optimal Levels of Responses			
	−1	0	1	DL	DW	DH	DA
	DL, DW, DH, DA						
Layer Height	0.72, 0.61, 0.51, 0.68	0.76, 0.64, 0.57, 0.58	0.66, 0.55, 0.84, 0.40	0	0	1	−1
Number of Perimeters	0.65, 0.55, 0.65, 0.62	0.75, 0.63, 0.47, 0.56	0.73, 0.62, 0.75, 0.49	0	0	1	−1
Infill density	0.85, 0.77, 0.65, 0.62	0.68, 0.58, 0.70, 0.52	0.57, 0.42, 0.65, 0.42	−1	−1	0	0
Fill angle	0.75, 0.62, 0.66, 0.47	0.67, 0.50, 0.67, 0.52	0.68, 0.61, 0.65, 0.66	−1	−1	0	1
Print Speed	0.61, 0.52, 0.59, 0.48	0.67, 0.49, 0.75, 0.68	0.81, 0.71, 0.69, 0.57	1	1	0	0
Nozzle temperature	0.67, 0.59, 0.67, 0.59	0.66, 0.49, 0.68, 0.50	0.77, 0.64, 0.61, 0.45	1	1	0	−1
Bed Temperature	0.80, 0.71, 0.72, 0.60	0.70, 0.56, 0.68, 0.72	0.61, 0.49, 0.60, 0.45	−1	−1	−1	0
Print orientation	0.66, 0.49, 0.62, 0.59	0.54, 0.48, 0.65, 0.49	0.82, 0.75, 0.71, 0.55	1	1	1	−1

The optimal levels vary for the individual responses; therefore, it is important to perform simultaneous optimization of responses. An integrated approach of desirability and weighted aggregated sum product assessment (WASPAS) was implemented for multi-response optimization, as discussed in Section 3.3. Table 5 shows the optimal levels obtained based on the relative importance score (RIS). The RIS values are presented in Table A2. Higher mean values of the RIS at any level represent the best set of process parameters. The most optimal setting obtained for combined responses to reduce dimensional deviations are layer height, infill density, and bed temperature at level −1, having higher RIS mean values of 1.369, 1.578, and 1.545; the number of perimeters, fill angle, print speed nozzle temperature, and print orientation at level 1, with RIS mean values of 1.392, 1.425, 1.506, 1.374, and 1.535, respectively. The encoded values of these optimal settings are layer height at 0.1 mm, number of perimeters at six, infill density at 20%, fill angle at 90°, print speed at

70 mm/s, nozzle temperature at 220 °C, bed temperature at 70 °C, and print orientation at 90°.

Table 5. Optimal levels for combined responses based on the grey relational grade.

Process Parameters	Levels			Optimal Levels
	−1	0	1	
Layer Height	1.369	1.360	1.323	−1
Number of Perimeters	1.346	1.254	1.391	1
Infill density	1.578	1.463	1.069	−1
Fill angle	1.314	1.249	1.425	1
Print Speed	1.157	1.426	1.506	1
Nozzle temperature	1.365	1.248	1.374	1
Bed Temperature	1.545	1.431	1.116	−1
Print orientation	1.243	1.157	1.535	1

Based on confirmatory experiments, the dimensional deviations obtained based on optimal settings are length deviation of 0.052%, width deviation of 0.086%, height deviation of 0.425%, and angle deviation of 0.211%.

5. Conclusions

One of the main challenges for 3D printing using fused deposition modeling (FDM) is the reduction of dimensional deviation. This is because of the numerous control parameters that must be considered in 3D printing. The present study analyzed and optimized the dimensional deviations, specifically in length deviation, width deviation, height deviation, and angle deviation. Based on experimental results and statistical analysis, the following conclusions are drawn:

The process parameters, including the layer height, number of perimeters, infill density, infill angle, print speed, nozzle temperature, bed temperature, and print orientation significantly affect the dimensional deviation. The most influential process parameters for length deviation are infill density followed by bed temperature and printing speed. For width deviation are infill density followed by print orientation and bed temperature. For height deviation are layer height followed by bed temperature and nozzle temperature. For angle deviation are layer height followed by fill angle and infill desnity.

An increase of the layer height from 0.1 to 0.3 mm causes an increase of the length deviation, width deviation, and angle deviation. An increase of the number of the perimeter (2 to 6) decreases dimensional deviation, however, it increases the angle deviation. Length, width, and angle deviation decrease with an increase of the print speed, while height deviation first decreases (from 50 to 60 mm/s) and then shows an increasing trend (from 60 to 70 mm/s). An increase of the fill angle from 0 to 90° increases the length, width, and height deviation, while it decreases the angle deviation. For print orientation from 0 to 45°, an increasing trend is observed for length deviation, while from 45 to 90°, it shows a decreasing trend.

From the obtained results, a definitive screening design was found for an efficient approach to model the non-linear quadratic models for dimensional deviation in smaller experimental runs.

The optimal settings for length deviation, width deviation, height deviation, and angle deviation vary according to the PLA material used. For length deviation, the optimal settings are layer height at 0.2 mm, number of perimeters at 4, infill angle at 0°, infill density at 25%, print speed at 70 mm/s, nozzle temperature at 220 °C, bed temperature at 70 °C, and print orientation at 90°. Similar optimal settings are obtained for width deviation. The optimal setting for the height deviation is layer height at 0.3 mm, number of perimeters at six, infill angle at 45°, infill density at 35%, print speed at 60 mm/s, nozzle temperature at 205 °C, bed temperature at 190 °C, and print orientation at 90°. For angle deviation, the optimal settings are layer height at 0.1 mm, number of perimeters at two, infill angle at 45°, infill density at 50%, print speed at 60 mm/s, nozzle temperature at 190 °C, bed temperature at 80 °C, and print orientation at 0°.

According to the proposed integrated approach of desirability and weighted aggregated sum product assessment (WASPAS), the optimal settings are layer height at 0.1 mm, number of perimeters at six, infill density at 20%, fill angle at 90°, print speed at 70 mm/s, nozzle temperature at 220 °C, bed temperature at 70 °C, and print orientation at 90°. The dimensional deviations based on these optimal settings are length deviation of 0.052%, width deviation of 0.086%, height deviation of 0.425%, and angle deviation of 0.211%.

This study provides a guideline for the practitioner to choose the right set of FDM printing process parameters.

In future work, the optimized results of the current study will be utilized for fabricating assistive devices that need control dimensions, including hand orthoses and foot orthoses.

Author Contributions: Conceptualization, M.A. and B.S.; methodology, M.A. and T.H.; software, M.A. and D.Z.; validation, B.S. and S.N.; formal analysis, M.A. and S.N.; investigation, M.A. and T.H.; resources, S.N. and D.Z.; data curation, T.H. and D.Z.; writing—original draft preparation, M.A.; writing—review and editing, B.S. and D.Z.; supervision, S.N.; project administration, S.N. and B.S.; funding acquisition, B.S. All authors have read and agreed to the published version of the manuscript.

Funding: This study received funding from King Saud University, Saudi Arabia through researchers supporting project number (RSP-2021/145). Additionally, the APCs will be funded (after acceptance) by King Saud University, Saudi Arabia through researchers supporting project number (RSP-2021/145).

Institutional Review Board Statement: Not applicable.

Informed Consent Statement: Not applicable.

Data Availability Statement: The data presented in this study are available on request from the corresponding author.

Acknowledgments: The authors extend their appreciation to King Saud University, Saudi Arabia for funding this work.

Conflicts of Interest: The authors declare no conflict of interest.

Appendix A

Calculation of integrated approach of desirability and weighted aggregated sum product assessment (WASPAS).

In Table A2, weighted sum desirability (WSD) is calculated using Equation (4). Weighted product desirability (WPD) is computed based on Equation (5). In the present study, equal weights are considered so it is set at 1. WSD and WPD are combined using Equation (6) to get a relative importance score (RIS).

Table A1. Experimental design and measured responses.

Exp. No.	A	B	C	D	E	G	H	I	Length Deviation (LD)	Width Deviation (WD)	Height Deviation (HD)	Angle Deviation (AD)
1	0.2	6	50	90	70	220	90	90	0.273	0.600	2.900	4.556
2	0.1	2	50	90	70	205	70	0	0.618	1.360	2.600	2.333
3	0.3	6	20	90	70	190	80	0	0.309	0.680	1.200	2.156
4	0.1	6	35	90	50	190	70	90	0.182	0.400	2.000	0.644
5	0.3	6	50	45	50	220	70	0	0.836	1.840	1.300	6.867
6	0.3	6	20	0	50	205	90	90	0.491	1.080	0.400	7.511
7	0.3	6	20	90	70	190	80	0	0.545	0.804	0.800	2.289
8	0.1	2	50	0	50	220	80	90	0.600	1.320	3.500	2.867
9	0.3	2	20	90	60	220	70	90	0.164	0.360	0.200	2.067
10	0.3	6	20	0	50	205	90	90	0.455	1.000	0.700	7.067
11	0.1	2	50	0	50	220	80	90	0.673	1.480	3.600	2.778
12	0.1	6	50	0	60	190	90	0	1.036	2.280	2.400	2.844
13	0.1	2	20	45	70	190	90	90	0.218	0.480	2.600	2.400
14	0.1	6	20	0	70	220	70	45	0.055	0.120	0.618	3.800
15	0.1	6	20	0	70	220	70	45	0.073	0.160	0.636	4.333
16	0.1	2	50	90	70	205	70	0	0.691	1.520	2.500	2.044
17	0.1	6	50	0	60	190	90	0	0.964	2.120	2.700	4.111
18	0.3	2	35	0	70	220	90	0	0.673	1.480	2.400	3.600
19	0.1	6	20	0	70	220	70	45	0.091	0.200	0.588	2.311
20	0.1	2	50	0	50	220	80	90	0.473	1.040	3.400	3.067
21	0.1	2	20	45	70	190	90	90	0.273	0.600	2.900	2.489
22	0.1	6	35	90	50	190	70	90	0.473	0.674	2.400	0.489
23	0.2	4	35	45	60	205	80	45	0.673	1.480	1.109	2.333
24	0.1	4	20	90	50	220	90	0	0.473	1.040	6.000	2.867
25	0.3	2	35	0	70	220	90	0	0.582	1.280	2.700	3.511
26	0.3	6	20	90	70	190	80	0	0.345	0.760	1.000	1.956
27	0.2	2	20	0	50	190	70	0	0.309	0.680	3.100	3.133
28	0.1	4	20	90	50	220	90	0	0.273	0.925	5.700	2.600
29	0.3	6	50	45	50	220	70	0	0.782	1.720	1.500	6.556
30	0.1	2	50	90	70	205	70	0	0.745	1.640	2.800	2.267
31	0.2	4	35	45	60	205	80	45	0.709	1.560	1.055	2.711
32	0.2	2	20	0	50	190	70	0	0.273	0.600	3.400	2.689
33	0.1	6	50	0	60	190	90	0	0.927	2.040	2.500	3.600
34	0.3	2	50	90	50	190	90	45	1.782	2.720	1.455	5.689
35	0.3	6	50	45	50	220	70	0	0.873	1.920	1.900	6.111
36	0.3	4	50	0	70	190	70	90	0.200	0.440	1.700	5.511
37	0.3	4	50	0	70	190	70	90	0.091	0.200	1.900	5.667
38	0.2	6	50	90	70	220	90	90	0.345	0.760	2.500	4.933
39	0.3	2	50	90	50	190	90	45	1.673	2.480	1.436	6.267
40	0.1	4	20	90	50	220	90	0	0.673	1.008	5.800	2.400
41	0.2	6	50	90	70	220	90	90	0.382	0.840	2.800	4.778
42	0.3	2	35	0	70	220	90	0	0.636	1.400	2.300	3.422
43	0.1	6	35	90	50	190	70	90	0.382	0.840	2.100	0.978
44	0.2	2	20	0	50	190	70	0	0.345	0.760	3.300	3.000
45	0.3	2	50	90	50	190	90	45	1.545	2.320	1.473	6.111
46	0.1	2	20	45	70	190	90	90	0.418	0.920	3.100	2.533
47	0.2	4	35	45	60	205	80	45	0.636	1.400	1.200	2.511
48	0.3	6	20	0	50	205	90	90	0.436	0.960	0.600	7.222
49	0.3	2	20	90	60	220	70	90	0.218	0.480	0.400	1.822
50	0.3	4	50	0	70	190	70	90	0.400	0.880	2.000	5.867
51	0.3	2	20	90	60	220	70	90	0.109	0.240	0.700	1.933

Table A2. Desirability and relative importance score values of responses.

Exp. No.	Desirability Values of Responses				Multi-Response Optimization		
	DL	WD	HD	AD	WSD	WPD	RIS
1	0.865	0.797	0.554	0.421	2.636	0.161	1.398
2	0.652	0.475	0.661	0.737	2.524	0.151	1.337
3	0.843	0.763	0.905	0.763	3.273	0.444	1.859
4	0.808	0.881	0.714	0.978	3.381	0.497	1.939
5	0.517	0.271	0.839	0.092	1.719	0.011	0.865
6	0.73	0.593	1	0	2.324	0	1.162
7	0.806	0.71	0.973	0.744	3.233	0.414	1.824
8	0.663	0.492	0.446	0.661	2.262	0.096	1.179
9	0.933	0.898	0.991	0.775	3.597	0.644	2.12
10	0.753	0.627	0.946	0.063	2.39	0.028	1.209
11	0.618	0.511	0.429	0.674	2.231	0.091	1.161
12	0.393	0.085	0.643	0.592	1.713	0.013	0.863
13	0.899	0.847	0.553	0.728	3.027	0.307	1.667
14	1	1	0.616	0.528	3.145	0.326	1.735
15	0.989	0.983	0.611	0.545	3.128	0.324	1.726
16	0.607	0.407	0.696	0.778	2.488	0.134	1.311
17	0.438	0.153	0.589	0.484	1.664	0.019	0.842
18	0.618	0.424	0.734	0.557	2.333	0.107	1.22
19	0.978	0.966	0.618	0.611	3.172	0.356	1.764
20	0.674	0.528	0.464	0.633	2.299	0.105	1.202
21	0.865	0.797	0.554	0.715	2.931	0.273	1.602
22	0.742	0.765	0.675	1	3.182	0.383	1.782
23	0.618	0.424	0.659	0.737	2.438	0.127	1.283
24	0.742	0.61	0	0.661	2.013	0	1.007
25	0.674	0.508	0.759	0.57	2.511	0.148	1.33
26	0.82	0.729	0.938	0.791	3.278	0.443	1.861
27	0.843	0.763	0.518	0.623	2.747	0.208	1.477
28	0.736	0.659	0.054	0.699	2.147	0.018	1.083
29	0.551	0.322	0.804	0.136	1.812	0.019	0.916
30	0.573	0.356	0.643	0.747	2.319	0.098	1.208
31	0.596	0.39	0.645	0.684	2.314	0.102	1.208
32	0.865	0.797	0.464	0.687	2.813	0.22	1.516
33	0.461	0.186	0.625	0.557	1.829	0.03	0.929
34	0.017	0.011	0.68	0.259	0.967	0	0.484
35	0.494	0.237	0.795	0.199	1.726	0.019	0.872
36	0.91	0.864	0.768	0.285	2.827	0.172	1.5
37	0.934	0.828	0.732	0.263	2.757	0.149	1.453
38	0.82	0.729	0.625	0.367	2.541	0.137	1.339
39	0	0	0.684	0.177	0.861	0	0.43
40	0.736	0.624	0.036	0.728	2.123	0.012	1.067
41	0.798	0.695	0.571	0.389	2.453	0.123	1.288
42	0.64	0.458	0.777	0.582	2.457	0.133	1.295
43	0.798	0.794	0.696	0.93	3.218	0.41	1.814
44	0.82	0.729	0.482	0.642	2.674	0.185	1.429
45	0.03	0.068	0.666	0.199	0.963	0	0.481
46	0.839	0.753	0.518	0.709	2.819	0.232	1.526
47	0.64	0.458	0.652	0.712	2.462	0.136	1.299
48	0.764	0.644	0.964	0.041	2.414	0.02	1.217
49	0.899	0.847	1	0.81	3.556	0.617	2.087
50	0.885	0.827	0.714	0.234	2.661	0.122	1.392
51	0.966	0.949	0.978	0.794	3.687	0.712	2.2

Appendix B

Regression equations for length deviation (LD), width deviation (WD), height deviation (HD) and angle deviation (AD)

$$\text{LD}(\%) = 1.340 - 4.450\ \text{A} - 0.229\ \text{B} + 0.015\ \text{C} + 0.0013\ \text{D} - 0.0144\ \text{E} - 0.00545 + 0.0158\ \text{G} \qquad (\text{A1})$$
$$+0.0161\text{H} + 12.82\ \text{A}^2 + 0.0245\ \text{B}^2 - 0.000181\ \text{H}^2 - 0.000079\ \text{C·H}$$

$$\text{WD}(\%) = -11.05 + 0.685\ \text{A} - 0.0400\ \text{B} + 0.04502\ \text{C} - 0.0231\ \text{E} - 0.0055\ \text{F} + 0.3598\ \text{G} \qquad (\text{A2})$$
$$+0.02016\ \text{H} - 0.002378\ \text{G}^2 - 0.00015\ \text{H}^2 - 0.00038\ \text{C·H}$$

$$\text{HD}(\%) = 33.29 + 48.40\ \text{A} - 0.5065\ \text{B} + 0.00813\ \text{C} + 0.002286\ \text{D} - 1.5427\ \text{E} + 0.02070\ \text{F} + 0.24078\ \text{G} \qquad (\text{A3})$$
$$-0.03596\ \text{H} + 0.011916\ \text{E}^2 + 0.4148\ \text{A·E} - 0.0560\ \text{A·F} - 0.9930\ \text{A·G} + 0.006939\ \text{B·H}$$

$$\text{AD}(\%) = -0.87 + 1.38\ \text{A} + 0.3716\ \text{B} - 0.3166\ \text{C} - 0.03263\ \text{D} - 0.03370\ \text{E} + 0.01072\ \text{F} + 0.1355\ \text{G} + 0.00361\ \text{H} \qquad (\text{A4})$$
$$+0.006277\ \text{C}^2 + 0.2697\ \text{A·C} - 0.003207\ \text{B·D} + 0.000895\ \text{C·D} - 0.002505\ \text{C·G}$$

References

1. Dezaki, M.L.; Ariffin, M.K.A.M.; Hatami, S. An Overview of Fused Deposition Modelling (FDM): Research, Development and Process Optimisation. *Rapid Prototyp. J.* **2021**, *27*, 562–582. [CrossRef]
2. Popescu, D.; Zapciu, A.; Amza, C.; Baciu, F.; Marinescu, R. FDM Process Parameters Influence over the Mechanical Properties of Polymer Specimens: A Review. *Polym. Test.* **2018**, *69*, 157–166. [CrossRef]
3. Liu, Z.; Wang, Y.; Wu, B.; Cui, C.; Guo, Y.; Yan, C. A Critical Review of Fused Deposition Modeling 3D Printing Technology in Manufacturing Polylactic Acid Parts. *Int. J. Adv. Manuf. Technol.* **2019**, *102*, 2877–2889. [CrossRef]
4. Galetto, M.; Verna, E.; Genta, G. Effect of Process Parameters on Parts Quality and Process Efficiency of Fused Deposition Modeling. *Comput. Ind. Eng.* **2021**, *156*, 107238. [CrossRef]
5. Azdast, T.; Hasanzadeh, R. Polylactide Scaffold Fabrication Using a Novel Combination Technique of Fused Deposition Modeling and Batch Foaming: Dimensional Accuracy and Structural Properties. *Int. J. Adv. Manuf. Technol.* **2021**, *114*, 1309–1321. [CrossRef]
6. Rajan, K.; Samykano, M.; Kadirgama, K.; Harun, W.S.W.; Rahman, M.M. Fused Deposition Modeling: Process, Materials, Parameters, Properties, and Applications. *Int. J. Adv. Manuf. Technol.* **2022**, *120*, 1531–1570. [CrossRef]
7. Samy, A.A.; Golbang, A.; Archer, E.; McIlhagger, A.T. A Comparative Study on the 3D Printing Process of Semi-Crystalline and Amorphous Polymers Using Simulation. In Proceedings of the UKACM 2021 Conference, Loughborough University, Online, 14–16 April 2021; Loughborough University: Loughborough, UK, 2021.
8. Deswal, S.; Narang, R.; Chhabra, D. Modeling and Parametric Optimization of FDM 3D Printing Process Using Hybrid Techniques for Enhancing Dimensional Preciseness. *Int. J. Interact. Des. Manuf.* **2019**, *13*, 1197–1214. [CrossRef]
9. Hsieh, Y.-T.; Nozaki, S.; Kido, M.; Kamitani, K.; Kojio, K.; Takahara, A. Crystal Polymorphism of Polylactide and Its Composites by X-Ray Diffraction Study. *Polym. J.* **2020**, *52*, 755–763. [CrossRef]
10. Wasanasuk, K.; Tashiro, K. Crystal Structure and Disorder in Poly (l-Lactic Acid) δ Form (A' Form) and the Phase Transition Mechanism to the Ordered α Form. *Polymer* **2011**, *52*, 6097–6109. [CrossRef]
11. Di Lorenzo, M.L.; Rubino, P.; Immirzi, B.; Luijkx, R.; Hélou, M.; Androsch, R. Influence of Chain Structure on Crystal Polymorphism of Poly (Lactic Acid). Part 2. Effect of Molecular Mass on the Crystal Growth Rate and Semicrystalline Morphology. *Colloid Polym. Sci.* **2015**, *293*, 2459–2467. [CrossRef]
12. Marubayashi, H.; Akaishi, S.; Akasaka, S.; Asai, S.; Sumita, M. Crystalline Structure and Morphology of Poly (L-Lactide) Formed under High-Pressure CO_2. *Macromolecules* **2008**, *41*, 9192–9203. [CrossRef]
13. Pan, P.; Zhu, B.; Kai, W.; Dong, T.; Inoue, Y. Effect of Crystallization Temperature on Crystal Modifications and Crystallization Kinetics of Poly (L-lactide). *J. Appl. Polym. Sci.* **2008**, *107*, 54–62. [CrossRef]
14. Cocca, M.; Di Lorenzo, M.L.; Malinconico, M.; Frezza, V. Influence of Crystal Polymorphism on Mechanical and Barrier Properties of Poly (l-Lactic Acid). *Eur. Polym. J.* **2011**, *47*, 1073–1080. [CrossRef]
15. Echeverría, C.; Limón, I.; Muñoz-Bonilla, A.; Fernández-García, M.; López, D. Development of Highly Crystalline Polylactic Acid with β-Crystalline Phase from the Induced Alignment of Electrospun Fibers. *Polymers* **2021**, *13*, 2860. [CrossRef]
16. Brizzolara, D.; Cantow, H.-J.; Diederichs, K.; Keller, E.; Domb, A.J. Mechanism of the Stereocomplex Formation between Enantiomeric Poly (Lactide) S. *Macromolecules* **1996**, *29*, 191–197. [CrossRef]
17. Wach, R.A.; Wolszczak, P.; Adamus-Wlodarczyk, A. Enhancement of Mechanical Properties of FDM-PLA Parts via Thermal Annealing. *Macromol. Mater. Eng.* **2018**, *303*, 1800169. [CrossRef]
18. Srinivas, V.; van Hooy-Corstjens, C.S.J.; Harings, J.A.W. Correlating Molecular and Crystallization Dynamics to Macroscopic Fusion and Thermodynamic Stability in Fused Deposition Modeling; a Model Study on Polylactides. *Polymer* **2018**, *142*, 348–355. [CrossRef]

19. Kitsakis, K.; Alabey, P.; Kechagias, J.; Vaxevanidis, N. A Study of the Dimensional Accuracy Obtained by Low Cost 3D Printing for Possible Application in Medicine. *IOP Conf. Ser. Mater. Sci. Eng.* **2016**, *161*, 12025. [CrossRef]
20. Aslani, K.-E.; Chaidas, D.; Kechagias, J.; Kyratsis, P.; Salonitis, K. Quality Performance Evaluation of Thin Walled PLA 3D Printed Parts Using the Taguchi Method and Grey Relational Analysis. *J. Manuf. Mater. Process.* **2020**, *4*, 47. [CrossRef]
21. Nathaphan, S.; Trutassanawin, W. Effects of Process Parameters on Compressive Property of FDM with ABS. *Rapid Prototyp. J.* **2021**, *27*, 905–917. [CrossRef]
22. Basavaraj, C.K.; Vishwas, M. Studies on Effect of Fused Deposition Modelling Process Parameters on Ultimate Tensile Strength and Dimensional Accuracy of Nylon. In Proceedings of the IOP Conference Series: Materials Science and Engineering, Bangalore, India, 14–16 July 2016; Volume 149, pp. 1–11.
23. Lalegani Dezaki, M.; Ariffin, M.K.; Serjouei, A.; Zolfagharian, A.; Hatami, S.; Bodaghi, M. Influence of Infill Patterns Generated by CAD and FDM 3D Printer on Surface Roughness and Tensile Strength Properties. *Appl. Sci.* **2021**, *11*, 7272. [CrossRef]
24. Padhi, S.K.; Sahu, R.K.; Mahapatra, S.S.; Das, H.C.; Sood, A.K.; Patro, B.; Mondal, A.K. Optimization of Fused Deposition Modeling Process Parameters Using a Fuzzy Inference System Coupled with Taguchi Philosophy. *Adv. Manuf.* **2017**, *5*, 231–242. [CrossRef]
25. Vahabli, E.; Rahmati, S. Improvement of FDM Parts' Surface Quality Using Optimized Neural Networks—Medical Case Studies. *Rapid Prototyp. J.* **2017**, *23*, 825–842. [CrossRef]
26. Sai, T.; Pathak, V.K.; Srivastava, A.K. Modeling and Optimization of Fused Deposition Modeling (FDM) Process through Printing PLA Implants Using Adaptive Neuro-Fuzzy Inference System (ANFIS) Model and Whale Optimization Algorithm. *J. Brazilian Soc. Mech. Sci. Eng.* **2020**, *42*, 617. [CrossRef]
27. Vyavahare, S.; Kumar, S.; Panghal, D. Experimental Study of Surface Roughness, Dimensional Accuracy and Time of Fabrication of Parts Produced by Fused Deposition Modelling. *Rapid Prototyp. J.* **2020**, *26*, 1535–1554. [CrossRef]
28. Camposeco-Negrete, C. Optimization of FDM Parameters for Improving Part Quality, Productivity and Sustainability of the Process Using Taguchi Methodology and Desirability Approach. *Prog. Addit. Manuf.* **2020**, *5*, 59–65. [CrossRef]
29. Mohamed, O.A.; Masood, S.H.; Bhowmik, J.L. Modeling, Analysis, and Optimization of Dimensional Accuracy of FDM-Fabricated Parts Using Definitive Screening Design and Deep Learning Feedforward Artificial Neural Network. *Adv. Manuf.* **2021**, *9*, 115–129. [CrossRef]
30. Mohanty, A.; Nag, K.S.; Bagal, D.K.; Barua, A.; Jeet, S.; Mahapatra, S.S.; Cherkia, H. Parametric Optimization of Parameters Affecting Dimension Precision of FDM Printed Part Using Hybrid Taguchi-MARCOS-Nature Inspired Heuristic Optimization Technique. *Mater. Today Proc.* **2021**, *50*, 893–903. [CrossRef]
31. Garg, A.; Bhattacharya, A.; Batish, A. On Surface Finish and Dimensional Accuracy of FDM Parts after Cold Vapor Treatment. *Mater. Manuf. Process.* **2016**, *31*, 522–529. [CrossRef]
32. Elkordy, A.G.E.-A.A. Application of Differential Scanning Calorimetry to the Characterization of Biopolymers. In *Applications of Calorimetry in a Wide Context-Differential Scanning Calorimetry, Isothermal Titration Calorimetry and Microcalorimetry*; IntechOpen: Rijeka, Croatia, 2013.
33. Jalali, A.; Huneault, M.A.; Elkoun, S. Effect of Thermal History on Nucleation and Crystallization of Poly (Lactic Acid). *J. Mater. Sci.* **2016**, *51*, 7768–7779. [CrossRef]
34. Šaparauskas, J.; Kazimieras Zavadskas, E.; Turskis, Z. Selection of Facade's Alternatives of Commercial and Public Buildings Based on Multiple Criteria. *Int. J. Strateg. Prop. Manag.* **2011**, *15*, 189–203. [CrossRef]
35. Bagal, D.K.; Giri, A.; Pattanaik, A.K.; Jeet, S.; Barua, A.; Panda, S.N. MCDM Optimization of Characteristics in Resistance Spot Welding for Dissimilar Materials Utilizing Advanced Hybrid Taguchi Method-Coupled CoCoSo, EDAS and WASPAS Method. In *Next Generation Materials and Processing Technologies*; Springer: Berlin/Heidelberg, Germany, 2021; pp. 475–490.
36. Chakraborty, S.; Zavadskas, E.K. Applications of WASPAS Method in Manufacturing Decision Making. *Informatica* **2014**, *25*, 1–20. [CrossRef]
37. Agarwal, K.M.; Shubham, P.; Bhatia, D.; Sharma, P.; Vaid, H.; Vajpeyi, R. Analyzing the Impact of Print Parameters on Dimensional Variation of ABS Specimens Printed Using Fused Deposition Modelling (FDM). *Sensors Int.* **2022**, *3*, 100149. [CrossRef]
38. Nancharaiah, T.; Raju, D.R.; Raju, V.R. An Experimental Investigation on Surface Quality and Dimensional Accuracy of FDM Components. *Int. J. Emerg. Technol.* **2010**, *1*, 106–111.
39. Wang, S.; Ma, Y.; Deng, Z.; Zhang, S.; Cai, J. Effects of Fused Deposition Modeling Process Parameters on Tensile, Dynamic Mechanical Properties of 3D Printed Polylactic Acid Materials. *Polym. Test.* **2020**, *86*, 106483. [CrossRef]
40. Mohamed, O.A.; Masood, S.H.; Bhowmik, J.L.; Nikzad, M.; Azadmanjiri, J. Effect of Process Parameters on Dynamic Mechanical Performance of FDM PC / ABS Printed Parts Through Design of Experiment. *J. Mater. Eng. Perform.* **2016**, *25*, 2922–2935. [CrossRef]
41. Akande, S.O. Dimensional Accuracy and Surface Finish Optimization of Fused Deposition Modelling Parts Using Desirability Function Analysis. *Int. J. Eng. Res. Technol* **2015**, *4*, 196–202.
42. Es-Said, O.S.; Foyos, J.; Noorani, R.; Mendelson, M.; Marloth, R.; Pregger, B.A. Effect of Layer Orientation on Mechanical Properties of Rapid Prototyped Samples. *Mater. Manuf. Process.* **2000**, *15*, 107–122. [CrossRef]
43. Akbaş, O.E.; Hıra, O.; Hervan, S.Z.; Samankan, S.; Altınkaynak, A. Dimensional Accuracy of FDM-Printed Polymer Parts. *Rapid Prototyp. J.* **2020**, *26*, 288–298. [CrossRef]

44. Abas, M.; Salman, Q.; Khan, A.M.; Rahman, K. Direct Ink Writing of Flexible Electronic Circuits and Their Characterization. *J. Brazilian Soc. Mech. Sci. Eng.* **2019**, *41*, 563. [CrossRef]
45. Brydson, J.A. *Flow Properties of Polymer Melts*; Plastics Institute, University of Michigan: Ann Arbor, MI, USA, 1970.
46. Hsueh, M.-H.; Lai, C.-J.; Liu, K.-Y.; Chung, C.-F.; Wang, S.-H.; Pan, C.-Y.; Huang, W.-C.; Hsieh, C.-H.; Zeng, Y.-S. Effects of Printing Temperature and Filling Percentage on the Mechanical Behavior of Fused Deposition Molding Technology Components for 3D Printing. *Polymers* **2021**, *13*, 2910. [CrossRef] [PubMed]
47. Afonso, J.A.; Alves, J.L.; Caldas, G.; Gouveia, B.P.; Santana, L.; Belinha, J. Influence of 3D Printing Process Parameters on the Mechanical Properties and Mass of PLA Parts and Predictive Models. *Rapid Prototyp. J.* **2021**, *27*, 487–495. [CrossRef]
48. Forrest, J.A.; Dalnoki-Veress, K.; Dutcher, J.R. Interface and Chain Confinement Effects on the Glass Transition Temperature of Thin Polymer Films. *Phys. Rev. E* **1997**, *56*, 5705. [CrossRef]
49. Spoerk, M.; Gonzalez-Gutierrez, J.; Sapkota, J.; Schuschnigg, S.; Holzer, C. Effect of the Printing Bed Temperature on the Adhesion of Parts Produced by Fused Filament Fabrication. *Plast. Rubber Compos.* **2018**, *47*, 17–24. [CrossRef]
50. Abdelrhman, A.M.; Gan, W.W.; Kurniawan, D. Effect of Part Orientation on Dimensional Accuracy, Part Strength, and Surface Quality of Three Dimensional Printed Part. In Proceedings of the IOP Conference Series: Materials Science and Engineering, Jakarta, Indonesia, 9–10 October 2019; IOP Publishing: Bristol, UK, 2019; Volume 694, p. 12048.
51. Chaudhry, M.S.; Czekanski, A. Evaluating FDM Process Parameter Sensitive Mechanical Performance of Elastomers at Various Strain Rates of Loading. *Materials* **2020**, *13*, 3202. [CrossRef]
52. Sood, A.K.; Ohdar, R.K.; Mahapatra, S.S. Improving Dimensional Accuracy of Fused Deposition Modelling Processed Part Using Grey Taguchi Method. *Mater. Des.* **2009**, *30*, 4243–4252. [CrossRef]
53. Enemuoh, E.U.; Duginski, S.; Feyen, C.; Menta, V.G. Effect of Process Parameters on Energy Consumption, Physical, and Mechanical Properties of Fused Deposition Modeling. *Polymers* **2021**, *13*, 2406. [CrossRef] [PubMed]
54. Benwood, C.; Anstey, A.; Andrzejewski, J.; Misra, M.; Mohanty, A.K. Improving the Impact Strength and Heat Resistance of 3D Printed Models: Structure, Property, and Processing Correlationships during Fused Deposition Modeling (FDM) of Poly (Lactic Acid). *Acs Omega* **2018**, *3*, 4400–4411. [CrossRef]
55. Jin, M.; Giesa, R.; Neuber, C.; Schmidt, H. Filament Materials Screening for FDM 3D Printing by Means of Injection-molded Short Rods. *Macromol. Mater. Eng.* **2018**, *303*, 1800507. [CrossRef]
56. Camposeco-Negrete, C. Optimization of Printing Parameters in Fused Deposition Modeling for Improving Part Quality and Process Sustainability. *Int. J. Adv. Manuf. Technol.* **2020**, *108*, 2131–2147. [CrossRef]
57. Peng, A.; Xiao, X.; Yue, R. Process Parameter Optimization for Fused Deposition Modeling Using Response Surface Methodology Combined with Fuzzy Inference System. *Int. J. Adv. Manuf. Technol.* **2014**, *73*, 87–100. [CrossRef]

Article

Micromechanical Modeling of the Biaxial Deformation-Induced Phase Transformation in Polyethylene Terephthalate

Fateh Enouar Mamache [1], Amar Mesbah [1], Hanbing Bian [2] and Fahmi Zaïri [2,*]

[1] Laboratory of Advanced Mechanics, University of Sciences and Technology Houari Boumediene, Algiers 16111, Algeria; mamacheanouar1991@gmail.com (F.E.M.); ammar_mesbah@yahoo.fr (A.M.)
[2] Laboratoire de Génie Civil et géo-Environnement, Université de Lille, IMT Nord Europe, JUNIA, Université d'Artois, ULR 4515-LGCgE, F-59000 Lille, France; hanbing.bian@univ-lille.fr
* Correspondence: fahmi.zairi@polytech-lille.fr

Abstract: In this paper, a micromechanics-based constitutive representation of the deformation-induced phase transformation in polyethylene terephthalate is proposed and verified under biaxial loading paths. The model, formulated within the Eshelby inclusion theory and the micromechanics framework, considers the material system as a two-phase medium, in which the active interactions between the continuous amorphous phase and the discrete newly formed crystalline domains are explicitly considered. The Duvaut–Lions viscoplastic approach is employed in order to introduce the rate-dependency of the yielding behavior. The model parameters are identified from uniaxial data in terms of stress–strain curves and crystallization kinetics at two different strain rates and two different temperatures above glass transition temperature. Then, it is shown that the model predictions are in good agreement with available experimental results under equal biaxial and constant width conditions. The role of the crystallization on the intrinsic properties is emphasized thanks to the model considering the different loading parameters in terms of mechanical path, strain rate and temperature.

Keywords: crystallizable PET; micromechanical model; viscoplasticity; temperature effect; biaxial loading

1. Introduction

Polyethylene terephthalate (PET) is one of the most used engineering polymers in applications where lightweight, chemical resistance, mechanical strength and thermal resistance are required. PET is a well-known example of a material system exhibiting a crystallization induced by mechanical loading. The phase transformation, highly dependent on the loading path, loading rate and temperature, implies a modification of the material response at the macroscale. Establishment of the structure–property relationship is a prerequisite for a reliable assessment of the material design whether under in-service or the manufacturing (e.g., hot-drawing near glass transition) process. Over the years, rheologically based models were formulated to reproduce the mechanical response of PET [1–12] and other crystallizable thermoplastics [13,14]. From the rheological viewpoint, these models combine different resistances to represent intermolecular and molecular network micromechanisms. The strain stiffening due to the appearance of the newly formed crystalline phase is introduced either implicitly via threshold conditions or explicitly by considering the crystallization kinetics via an Avrami expression. In the latter approach, the presence of crystallites is introduced thanks to the concept of volume fraction in which the polymer is seen as a two-phase composite. The concept was also extended to semicrystalline polymers [15–19]. Nonetheless, the active interactions between the continuous amorphous phase and the discrete crystalline domains are not considered, which constitutes a weakness from the physical viewpoint. Only micromechanical homogenization methods allow to incorporate local interactions and then lead to a physically convenient way to constitutively represent the heterogeneous problem. The latter is the basis of two main approaches to predict the effective mechanical properties of semicrystalline media, with different idealizations of the microstructure to derive the constitutive relations. In a first approach, the material can be defined as an aggregate of layered two-phase composite inclusions, each one represented by an amorphous layer and a crystalline lamella. The approach has been used in a viscoplastic framework [20,21], an elasto-viscoplastic framework [22–25] or a purely elastic framework [26,27]. In a second approach, the material system is treated as an Eshelby inclusion problem by seeing the crystals as reinforcing ellipsoidal inclusions embedded into a continuous amorphous matrix. The Eshelby-type inclusion approach was mainly used for the initial elastic behavior [26,28–33], the linear viscoelastic behavior [34], the initial yield behavior [35] and the postyield behavior [36].

The main objective of this paper is to explore the relevance of a micromechanics-based elasto-viscoplastic modeling to represent the deformation-induced phase transformation in PET. The model arises from the Eshelby inclusion theory and the micromechanics framework. The intrinsic viscosity of the amorphous phase is considered using the Duvaut–Lions viscoplastic approach, and the crystallization kinetics is governed by an Avrami equation. The model parameters are calibrated using available uniaxial data of a PET stretched under two different strain rates at two different temperatures above glass transition temperature. The model capacities to predict the mechanical response along with the phase transformation are verified by comparison with the available experimental data of a PET stretched under equal biaxial and constant width conditions.

The outline of the present paper is as follows. The micromechanical model is presented in Section 2. Section 3 is devoted to the comparison of the model simulations with experimental observations. Concluding remarks are given in Section 4.

2. Model

The constitutive representation of the PET system is treated as an Eshelby-type inclusion problem in which the material volume element consists of an amorphous phase as the continuous phase and discrete newly born crystalline domains. The two constitutive phases are supposed to be isotropic and homogeneous media with elastic stiffness tensors $\mathbf{C}_{am}$ and $\mathbf{C}_{cry}$.

2.1. Micromechanics-Based Theory for Deformation-Induced Phase Transformation

The constitutive relation between the macrostress tensor $\overline{\sigma}$ and the elastic part $\overline{\varepsilon}^e$ of the macrostrain tensor $\overline{\varepsilon}$ is given by:

$$\overline{\sigma} = \overline{\mathbf{C}} : \overline{\varepsilon}^e \qquad (1)$$

in which $\overline{\mathbf{C}}$ is the macroscopic elastic stiffness tensor of the semicrystalline material expressed as [37]:

$$\overline{\mathbf{C}} = \mathbf{C}_{am} \cdot \left[\mathbf{I} - \mathbf{Y} \cdot (\mathbf{S} \cdot \mathbf{Y} + \mathbf{I})^{-1} \right] \qquad (2)$$

where $\mathbf{I}$ is the identity tensor, $\mathbf{S}$ is the Eshelby tensor and $\mathbf{Y}$ is a tensor expressed as:

$$\mathbf{Y} = -\phi_{cry} \left[\mathbf{S} + (\mathbf{C}_{cry} - \mathbf{C}_{am})^{-1} \cdot \mathbf{C}_{am} \right]^{-1} \qquad (3)$$

where ϕ_{cry} is the volume fraction of the strain-induced crystalline phase for which the kinetics is governed by an Avrami-type formula that is specified below.

The fourth-order isotropic elastic stiffness tensors, $\mathbf{C}_{am}$ and $\mathbf{C}_{cry}$, are expressed, in Cartesian components, as follows:

$$(\mathbf{C}_{am})_{ijkl} = \frac{E_{am}}{2(1 + \nu_{am})} \left[\left(\delta_{ik}\delta_{jl} + \delta_{il}\delta_{jk} \right) + \frac{2\nu_{am}}{1 - 2\nu_{am}} \delta_{ij}\delta_{kl} \right] \qquad (4)$$

$$(C_{cry})_{ijkl} = \frac{E_{cry}}{2(1+\nu_{cry})}\left[\left(\delta_{ik}\delta_{jl}+\delta_{il}\delta_{jk}\right)+\frac{2\nu_{cry}}{1-2\nu_{cry}}\delta_{ij}\delta_{kl}\right] \quad (5)$$

in which E_{am} and E_{cry} are the Young's moduli, and ν_{am} and ν_{cry} are the Poisson's ratios. The term δ_{ij} denotes the Kronecker delta symbol.

The plastic yielding is considered from the continuum plasticity theory. Regarding the associative plastic flow rule, the macroscopic plastic strain rate $\dot{\bar{\varepsilon}}^p$ is expressed as:

$$\dot{\bar{\varepsilon}}^p = \dot{\lambda}\frac{\partial \overline{F}}{\partial \overline{\sigma}} = (1-\phi_{cry})\dot{\lambda}\frac{\overline{T}:\overline{\sigma}}{\sqrt{\overline{\sigma}:\overline{T}:\overline{\sigma}}} \quad (6)$$

where $\dot{\lambda}$ is the plastic multiplier and $\overline{F}$ is the macroscopic yield function. Considering the von Mises yield criterion with isotropic plastic hardening for the continuous amorphous phase, the yield function $\overline{F}$ can be expressed as [38]:

$$\overline{F} = (1-\phi_{cry})\sqrt{\overline{\sigma}:\overline{T}:\overline{\sigma}} - \sqrt{\frac{2}{3}}[\sigma_y + h(\bar{e}^p)^q] \leq 0 \quad (7)$$

in which $\bar{e}^p$ is the macroscopic equivalent plastic strain, σ_y is the initial yield strength of the amorphous phase and, h and q are the hardening parameters of the amorphous phase. The term $\overline{T}$ can be given by:

$$\overline{T}_{ijkl} = \overline{T}_1 \delta_{ij}\delta_{kl} + \overline{T}_2\left(\delta_{ik}\delta_{jl}+\delta_{il}\delta_{jk}\right) \quad (8)$$

where $\overline{T}_1$ and $\overline{T}_2$ are provided in Appendix A.

The volume fraction of the strain-induced crystalline phase ϕ_{cry} is given by:

$$\phi_{cry} = \phi_{\infty\_cry}\kappa \quad (9)$$

in which ϕ_{∞_cry} is the maximum crystal degree and κ is the total degree of transformation following the Avrami-type expression [4] modified by Ahzi et al. [5]:

$$\dot{\kappa} = \frac{\dot{\varepsilon}}{\dot{\varepsilon}_{ref}}\alpha_A K_{av}(-\ln(1-\kappa))^{\frac{\alpha_A-1}{\alpha_A}}(1-\kappa) \quad (10)$$

where α_A is the Avrami exponent, K_{av} is the phase transformation rate function, $\dot{\varepsilon}$ is the applied strain rate and $\dot{\varepsilon}_{ref}$ is the reference strain rate.

The transformation rate function K_{av} takes the empirical form defined as follows:

$$K_{av} = 1.47\times 10^{-3}\left(\frac{4\pi Nu}{3\phi_{\infty\_cry}}\right)^{1/3}\exp\left(-\left(\frac{\theta-141}{47.33}\right)^2\right) \quad (11)$$

in which Nu is the number density of nuclei in the amorphous phase.

This empirical formula first emerged for the study of spherulitic growth in thermally-induced crystallization and may not be an optimized choice for all kinetics of newly formed crystals due to differences in morphology and in size. Nonetheless, it was also employed in previous studies in the context of strain-induced crystallization in PET [5,6] and in PLA [13,14] as a phenomenological description of the evolution of a newly formed phase.

2.2. Model Implementation

The model allows to relate elasto-viscoplastic macrobehavior to microstructure variations depending on the loading parameters in terms of mechanical path, strain rate and temperature. It is identified using uniaxial (UA) data and its predictability is verified using equal biaxial (EB) and constant width (CW) data.

The components of the macrostress tensor $\overline{\sigma}$ under a general biaxial stretching are:

$$\overline{\sigma}_{11} > 0, \ \overline{\sigma}_{22} = R\overline{\sigma}_{11} \text{ and } \overline{\sigma}_{ij} = 0 \text{ for all other components} \tag{12}$$

where $R = \overline{\sigma}_{22}/\overline{\sigma}_{11}$ is the stress biaxial ratio:

$$\begin{array}{l} \text{UA}: \quad R = 0 \\ \text{EB}: \quad R = 1 \\ \text{CW}: 0 < R < 1 \end{array}, \tag{13}$$

The CW condition is a particular biaxial loading in which a stretching is performed in one direction while the transversal one is kept constant. In this regard, the stress biaxial ratio is not constant but changes iteratively during the loading.

The terms $\overline{\mathbf{T}} : \overline{\sigma}$ and $\overline{\sigma} : \overline{\mathbf{T}} : \overline{\sigma}$ in Equation (6) write:

$$\overline{\mathbf{T}} : \overline{\sigma} = \overline{\sigma}_{11} \text{diag}(\overline{T}_1 + R\overline{T}_1 + 2\overline{T}_2, \overline{T}_1 + R\overline{T}_1 + 2R\overline{T}_2) \tag{14}$$

$$\overline{\sigma} : \overline{\mathbf{T}} : \overline{\sigma} = \overline{\sigma}_{11}^2 \left[\overline{T}_1(1+R)^2 + 2\overline{T}_2(1+R)^2 \right] = \overline{\sigma}_{11}^2 \Phi(R) \tag{15}$$

The macroscopic plastic strain rate $\dot{\overline{\varepsilon}}^p$ becomes:

$$\dot{\overline{\varepsilon}}^p = (1 - \phi_{cry}) \frac{\dot{\lambda}}{\sqrt{\Phi(R)}} \text{diag}(\overline{T}_1 + R\overline{T}_1 + 2\overline{T}_2, \overline{T}_1 + R\overline{T}_1 + 2R\overline{T}_2) \tag{16}$$

The plastic multiplier $\dot{\lambda}$ was computed from the plastic consistency condition: $\dot{\lambda}\langle \dot{\overline{F}} \rangle = 0$, the yield condition being formulated in a Kuhn–Tucker form by: $\dot{\lambda} \geq 0$, $\langle \overline{F} \rangle \leq 0, \dot{\lambda}\langle \overline{F} \rangle = 0$. The model was coded in Matlab software using the flowchart provided in Figure 1.

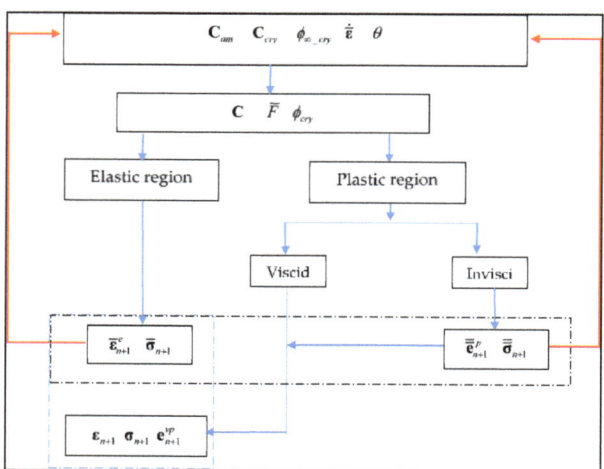

Figure 1. Algorithm of the model implementation.

The model inputs are the material constants, in particular the components of the amorphous elastic stiffness tensor, the components of the crystalline elastic stiffness tensor and the maximum crystal degree. The loading parameters, in terms of strain rate and temperature, are also specified. Both stiffness and yield function are updated iteratively while, once the threshold is reached, the crystallization amount increases with the mechanical loading. The rate-dependency of the yielding behavior is also taken into account by using the Duvaut–Lions viscoplastic approach [39,40]:

$$\dot{\bar{\varepsilon}}^{vp} = \frac{1}{\eta} \mathbf{C}_{am}^{-1} : (\overline{\sigma} - \overline{\overline{\sigma}}) \tag{17}$$

$$\dot{\bar{\mathbf{e}}}^{vp} = \frac{1}{\eta} \left(\overline{\mathbf{e}}^{vp} - \overline{\overline{\mathbf{e}}}^{p} \right) \tag{18}$$

where η is a viscosity parameter, $\overline{\sigma}$ and $\overline{\overline{\sigma}}$ are the total average viscoplastic stress tensor and the overall inviscid plastic stress tensor, respectively, and $\overline{\mathbf{e}}^{vp}$ and $\overline{\overline{\mathbf{e}}}^{p}$ are the viscoplastic strain tensor and the inviscid plastic strain tensor, respectively. The inviscid solution, in terms of the actual stress tensor $\overline{\overline{\sigma}}_{n+1}$ and the internal variable $\overline{\overline{\mathbf{e}}}^{p}_{n+1}$, is updated at each increment allowing the calculation of the new stress $\overline{\sigma}_{n+1}$ and the viscoplastic strain $\overline{\mathbf{e}}^{vp}_{n+1}$ by integrating the two previous equations using a backward Euler algorithm:

$$\overline{\sigma}_{n+1} = \frac{(\overline{\sigma}_n + \mathbf{C}_{am} : \Delta \overline{\varepsilon}_{n+1}) + \frac{\Delta t_{n+1}}{\eta} \overline{\overline{\sigma}}_{n+1}}{1 + \frac{\Delta t_{n+1}}{\eta}} \tag{19}$$

$$\overline{\mathbf{e}}^{vp}_{n+1} = \frac{\overline{\mathbf{e}}^{vp}_n + \frac{\Delta t_{n+1}}{\eta} \overline{\overline{\mathbf{e}}}^{p}_{n+1}}{1 + \frac{\Delta t_{n+1}}{\eta}} \tag{20}$$

where Δt_{n+1} is the time step. When $\Delta t_{n+1}/\eta \to \infty$, the inviscid solution is recovered, and when $\Delta t_{n+1}/\eta \to 0$, the elastic solution is achieved.

3. Results and Discussion

In what follows, the model is quantitatively compared with experimental results of PET loaded at different loading conditions. The model parameters for the two constitutive phases are listed in Tables 1 and 2.

Table 1. Model constants for the crystalline phase.

Parameter	Significance	Value
E_{cry}	Modulus	118 GPa
ν_{cry}	Poisson's ratio	0.3
α	Aspect ratio	2
$\dot{\varepsilon}_{ref}$	Reference strain rate	2.1/s
ϕ_{∞_cry}	Maximum crystal degree	0.3
α_A	Avrami exponent	3
Nu	Number density of nuclei	10^8

Table 2. Model constants for the amorphous phase.

Parameter	Significance	Value
E_g	Glassy modulus	2.1 GPa
E_r	Rubbery modulus	18 MPa
θ_g	Glass transition temperature	77 °C
$\Delta \theta$	Temperature shift	10 °C
X_g	Transition slope	−0.04 MPa/°C
ν_g	Glassy Poisson's ratio	0.32
ν_r	Rubbery Poisson's ratio	0.49
σ_y	Initial yield strength	3 MPa
h	Hardening	7 MPa
q	Hardening	0.9
η	Viscosity	0.01

3.1. Model Identification

3.1.1. Continuous Amorphous Phase Properties

Figure 2a shows the variation with temperature θ of the amorphous stiffness E_{am} around the glass transition temperature θ_g. Like all amorphous polymers, the PET elastic modulus exhibits a plateau within the glassy state, a drastic drop within the glass transition region and a continued decrease within the rubbery region. This behavior is represented by the following function [8]:

$$E_{am}(\theta) = \frac{1}{2}(E_g + E_r) - \frac{1}{2}(E_g - E_r)\tanh\left(\frac{5}{\Delta\theta}(\theta - \theta_g)\right) + X_g(\theta - \theta_g) \tag{21}$$

where E_g is the amorphous modulus in the glassy region, E_r is the amorphous modulus in the rubbery region, $\Delta\theta$ is the interval of the temperature range across which the glass transition occurs and X_g is the slope outside the glass transition region. The variation with temperature θ of the amorphous Poisson's ratio ν_{am} around θ_g is given by the following function:

$$\nu_{am}(\theta) = \nu_g + (\nu_r - \nu_g)\exp\left(\frac{\theta - (2(\theta_g + \Delta\theta - \theta))^2}{\theta_g + \Delta\theta}\right) \text{ for } \theta < \theta_g, \tag{22}$$

$$\nu_{am}(\theta) = \nu_r \text{ for } \theta \geq \theta_g, \tag{23}$$

where ν_g is the amorphous Poisson's ratio in the glassy region and ν_r is the amorphous Poisson's ratio in the rubbery region.

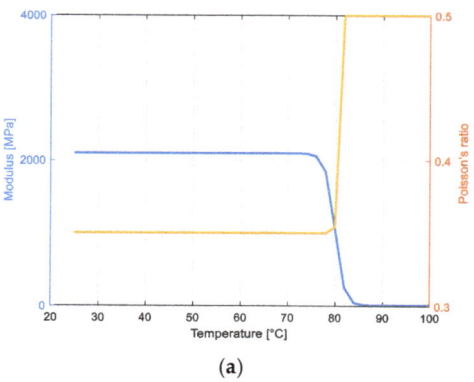
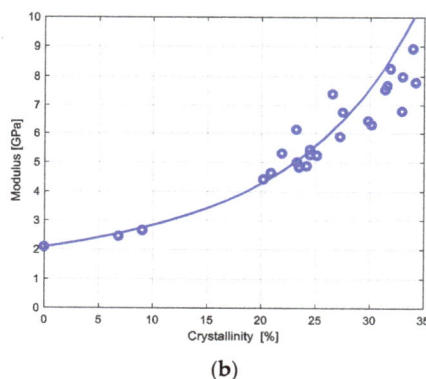

(a)   (b)

Figure 2. Model results of the linear elastic response: (**a**) amorphous elastic constants as a function of temperature, (**b**) overall modulus as a function of crystal content; solid lines: model; symbols: experimental data of Cosson et al. [11].

3.1.2. Discrete Crystalline Phase Properties

In the identification exercise, the crystal features, in terms of shape factor and elastic properties, are determined using the overall elastic properties at ambient temperature taken from the work of Cosson et al. [11] on previously crystallized PET. The theoretical and experimental stiffening is shown in Figure 2b as a function of crystal amount. Note that the amorphous elastic properties are considered to be independent of crystal amount.

3.1.3. Overall Response

The other model parameters were identified using the UA data of Salem [41] in terms of stress–strain and crystallization curves. The identification exercise includes the rate effect using two available strain rates: 0.42/s and 2.1/s. Only the straining temperature of 90 °C is used for the identification. Figure 3 presents the model results in comparison with the experimental stress–strain and crystallization curves. The solid lines represent the model results while the symbols designate the experimental data.

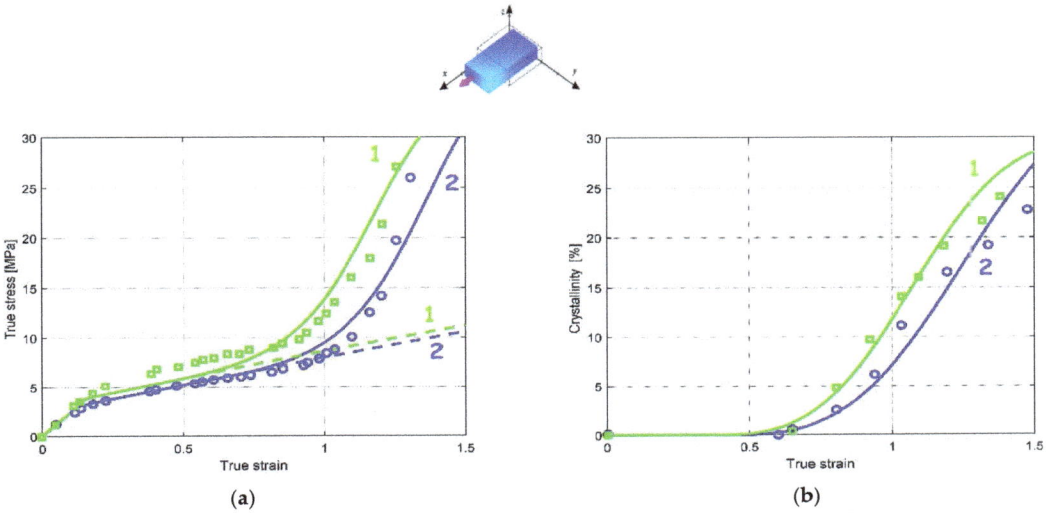

Figure 3. Model results under UA loading at a loading temperature of 90 °C and different strain rates (1: 2.1/s, 2: 0.42/s): (a) stress–strain response, (b) strain-induced crystallization; solid lines: model; dashed lines: model with no crystallization; symbols: experimental data of Salem [41].

It can be observed that the model is able to reproduce adequately the strain rate dependency on the overall UA response along with the crystallization. The crystallization kinetics and the intrinsic viscosity of the amorphous matrix are the two rate-dependent factors introduced in the model and affecting the overall response. The increase in strain rate slightly affects the initial yield region but significantly influences the progressive strain hardening and the dramatic strain hardening occurring at a strain of approximately 1. Moreover, the model is able to capture the acceleration of both the crystallization onset and the crystallization amount with the strain rate.

3.2. Model Prediction

3.2.1. Comparison between Model and Experiments

Figures 4–6 present the model predictions under EB and CW loadings using the model parameters identified under UA loading. The data extracted from the works of Buckley et al. [1] and Adams et al. [2] are reported in the form of symbols, while the solid lines represent the model predictions. The crystallization prediction accompanies the stress–strain curves. Note that the papers of Buckley et al. [1] and Adams et al. [2] do not present crystallization data, except at a temperature of 86 °C under EB condition.

A global view at these results shows that the model predictions are favorably compared with the experimental stress–strain curves for the two loading modes. Figures 4a and 5a show that the model adequately predicts the significant effect of the straining temperature on the stress–strain response including stiffness, yield strength and strain hardening region. The predicted crystallization is also presented in Figures 4b and 5b. Inversely to strain rate effects, the increase in loading temperature delays the onset of crystallization, which in turns affects the onset of strain hardening. The higher the loading temperature, the higher the strain level for which the dramatic strain hardening occurs. Figure 6 shows the theoretical and experimental CW stress–strain behavior at two strain rates: 1/s and 4/s.

It can be seen that the model is able to capture the highly nonlinear mechanical response, including the dramatic strain hardening occurring at a strain of approximately 1. The higher the strain rate, the higher the onset of crystallization and the amount of crystallization.

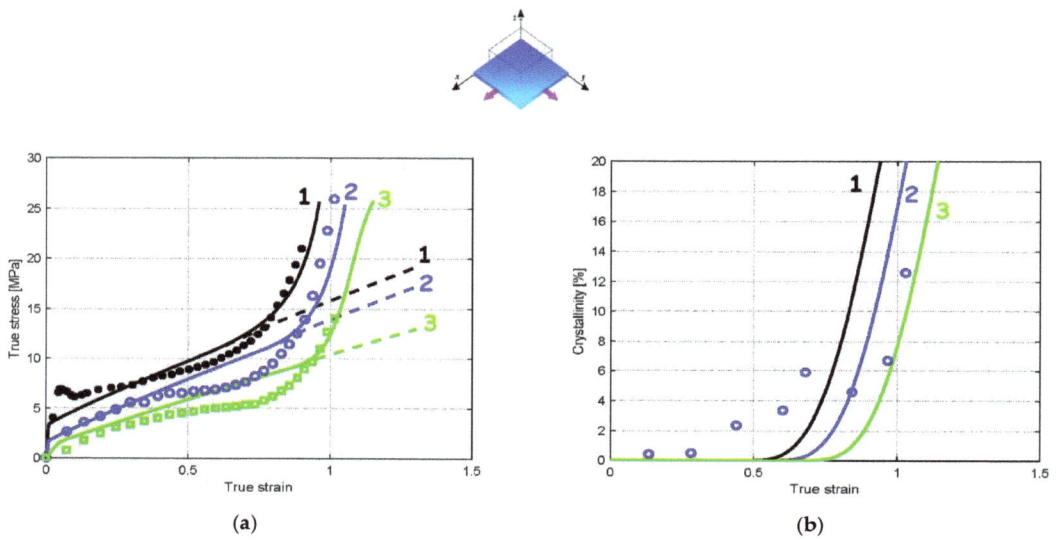

Figure 4. Model results under EB loading at a strain rate of 1/s and different loading temperatures (1: 83 °C, 2: 89 °C, 3: 94 °C): (**a**) stress–strain response, (**b**) strain-induced crystallization; solid lines: model; dashed lines: model with no crystallization; symbols: experimental data of Adams et al. [2].

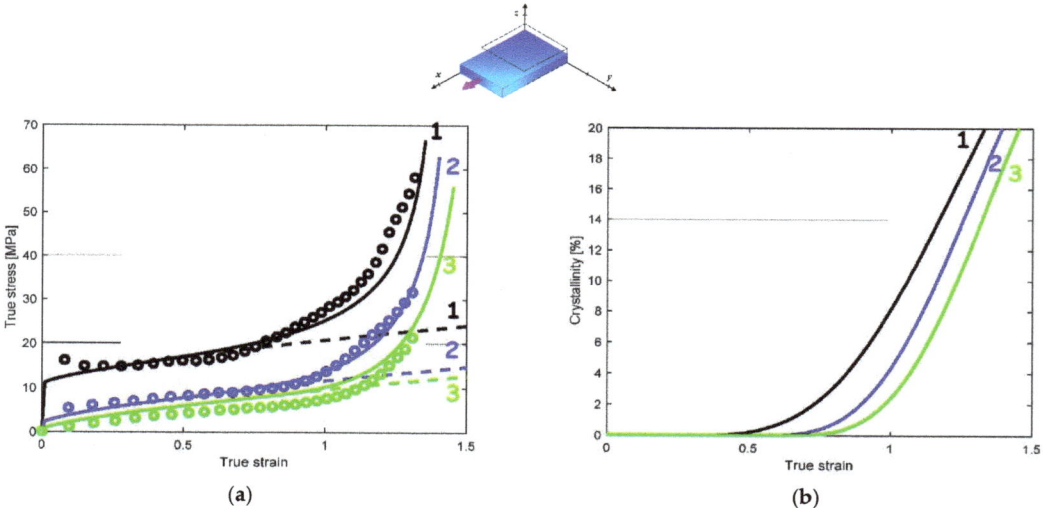

Figure 5. Model results under CW loading at a strain rate of 1 /s and different loading temperatures (1: 80 °C, 2: 86 °C, 3: 92 °C): (**a**) stress–strain response, (**b**) strain-induced crystallization; solid lines: model; dashed lines: model with no crystallization; symbols: experimental data of Adams et al. [2].

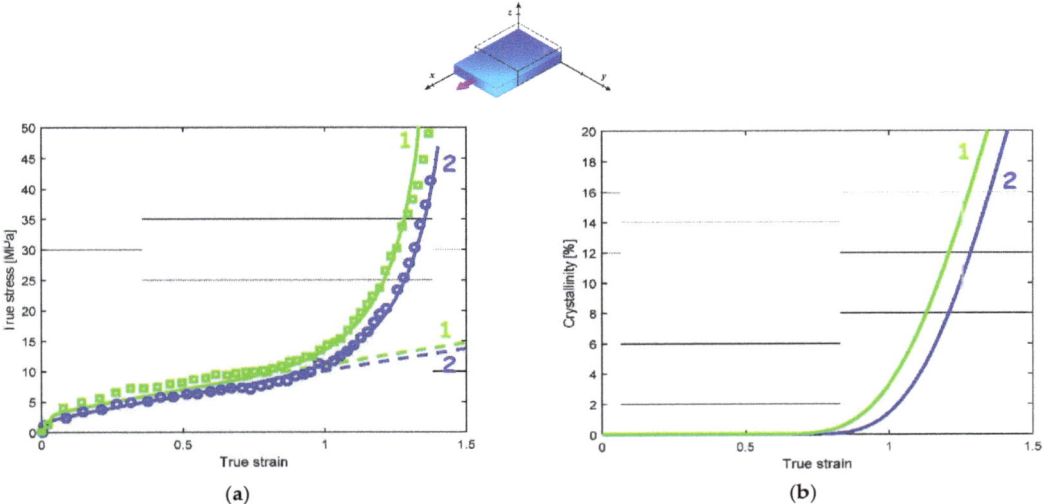

Figure 6. Model results under CW loading at a loading temperature of 87 °C and different strain rates (1: 1/s, 2: 4/s): (**a**) stress–strain response, (**b**) strain-induced crystallization; solid lines: model; dashed lines: model with no crystallization; symbols: experimental data of Buckley et al. [1].

3.2.2. Implication of the Phase Changes

The phase transformation effect on the overall response can be analyzed thanks to the model considering the different loading parameters in terms of loading path, loading rate and loading temperature. When the presence of crystals is neglected in the model (dashed lines in Figures 3a, 4a, 5a and 6a), no dramatic strain hardening occurs and the rate-dependency becomes weaker. The amorphous phase stretching may also play a role in the dramatic strain-hardening region [13]. The actual model introduces the occurrence of the crystallization as the origin of the strain hardening but, acknowledging its effective contribution [42], the supplementary effect of the amorphous phase stretching may be considered by an adequate modification of the amorphous plastic hardening in the yield function [36].

4. Concluding Remarks

In this work, a micromechanics-based elasto-viscoplastic constitutive model is presented. Formulated within the Eshelby inclusion theory and the micromechanics framework, the constitutive representation considers the material system as a two-phase medium in which the crystal is the reinforcement element of a viscoplastic amorphous medium which increases during stretching. The model provides a quantitative relation between deformation-induced phase transformation and mechanical response under different loading conditions in terms of loading path, loading rate and loading temperature. The model capacities are verified using available experimental observations under uniaxial, equal biaxial and constant width conditions.

Although the model can quite well capture the elasto-viscoplastic response of PET along with the strain-induced crystallization, some improvements are still necessary. It is indeed necessary to extend the mathematical description of the polymer deformation behavior to finite deformation. Furthermore, both isotropic and kinematic plastic hardening deserve to be considered in the local yield function [43] in order to bring a better description of the different steps of the overall mechanical response.

Author Contributions: Software, formal analysis and visualization, F.E.M.; software and project administration, A.M.; formal analysis and visualization, H.B.; conceptualization, methodology and project administration, F.Z. All authors have read and agreed to the published version of the manuscript.

Funding: This work was financially supported by the University of Science and Technology Houari Boumediene and the Algerian Ministry of Higher Education and Scientific Research within the international scientific cooperation program.

Institutional Review Board Statement: Not applicable.

Informed Consent Statement: Not applicable.

Data Availability Statement: The data presented in this study are available on request from the corresponding author.

Conflicts of Interest: The authors declare no conflict of interest.

Appendix A

The terms $\overline{T}_1$ and $\overline{T}_2$ of Equation (8) are expressed as:

$$\begin{aligned}\overline{T}_1 &= (3P_1 + 2P_2)^2 T_1 + 2P_1(3P_1 + 4P_2) T_2 \\ \overline{T}_2 &= 4P_2^2 T_2\end{aligned} \quad (A1)$$

where

$$\begin{aligned}T_1 &= \tfrac{1}{15}\left(T_{11}^{(1)} + 4T_{12}^{(1)} + 4T_{21}^{(1)} + 6T_{22}^{(1)} + 2T_{11}^{(2)} - 4T_{12}^{(2)} + 2T_{22}^{(2)}\right) \\ T_2 &= \tfrac{1}{15}\left(T_{11}^{(1)} - T_{12}^{(1)} - T_{21}^{(1)} + T_{22}^{(1)} + 2T_{11}^{(2)} + 6T_{12}^{(2)} + 7T_{22}^{(2)}\right)\end{aligned} \quad (A2)$$

$$\begin{aligned}T_{IK}^{(1)} = -\tfrac{1}{3} &+ \tfrac{\phi^{am}}{9450(1-\nu_{cry})^2 (Z_2+S_{II}^{(2)})(Z_2+S_{KK}^{(2)})} [1575(1-2\nu_{cry})^2 \Gamma_{II}\Gamma_{KK} \\ &+ 21(25\nu_{cry}-23)(1-2\nu_{cry})(\Gamma_{II}\Delta_K + \Gamma_{KK}\Delta_I) + 21(25\nu_{cry}-2) \\ &(1-2\nu_{cry})(\Gamma_{II}+\Gamma_{KK}) + 3(35\nu_{cry}^2 - 70\nu_{cry} + 36)\Delta_{IK} \\ &+ 7(50\nu_{cry}^2 - 59\nu_{cry} + 8)(\Delta_I + \Delta_K) - 2(175\nu_{cry}^2 - 343\nu_{cry} + 103)]\end{aligned} \quad (A3)$$

$$\begin{aligned}T_{IJ}^{(2)} = \tfrac{1}{2} &+ \tfrac{\phi^{am}}{6300(1-\nu_{cry})^2 (Z_2+S_{IJ}^{(2)})(Z_2+S_{IJ}^{(2)})}[(70\nu_{cry}^2 - 140\nu_{cry} - 72)\Delta_{IJ} \\ &- (175\nu_{cry}^2 - 266\nu_{cry} + 75)\tfrac{\Delta_I+\Delta_J}{2} + 350\nu_{cry}^2 - 476\nu_{cry} + 164]\end{aligned}$$

$$\begin{aligned}\Delta_1 &= \tfrac{3(1-\alpha^4 f(\alpha^2))}{1-\alpha^4}, \\ \Delta_2 &= \Delta_3 = \tfrac{1}{2}(3-\Delta_1), \\ \Delta_{11} &= \tfrac{5(2+\alpha^4-3\alpha^4 f(\alpha^2))}{2(1-\alpha^4)^2}, \\ \Delta_{12} &= \Delta_{21} = \Delta_{13} = \Delta_{31} = \tfrac{15\alpha^4[-3+(1+2\alpha^4)f(\alpha^2)]}{4(1-\alpha^4)^2}, \\ \Delta_{22} &= \Delta_{23} = \Delta_{32} = \Delta_{33} = \tfrac{15\alpha^4[1+2\alpha^4+(1-4\alpha^4)f(\alpha^2)]}{16(1-\alpha^4)^2}\end{aligned} \quad (A4)$$

$$f(\alpha) = \begin{cases} \tfrac{\cosh^{-1}\alpha}{\alpha\sqrt{\alpha^2-1}} & \text{if } \alpha > 1 \\ \tfrac{\cos^{-1}\alpha}{\alpha\sqrt{1-\alpha^2}} & \text{if } \alpha < 1 \end{cases} \quad (A5)$$

$$\begin{aligned}P_1 &= \tfrac{\phi^{am}(\Lambda_1 - \Omega_1)}{[1+2\phi^{am}(\Omega_2-\Lambda_2)][1+\phi^{am}(3\Omega_1+2\Omega_2-3\Lambda_1-2\Lambda_2)]} \\ P_2 &= \tfrac{1}{2+4\phi^{am}(\Omega_2-\Lambda_2)}\end{aligned} \quad (A6)$$

$$\begin{aligned}\Omega_1 &= \tfrac{1-\Gamma_{11}-4\Gamma_{12}}{30(Z_2+S_{11}^{(2)})} - \tfrac{1}{15(Z_2+S_{12}^{(2)})} + \tfrac{1-4\Gamma_{21}-6\Gamma_{22}}{30(Z_2+S_{22}^{(2)})} \\ \Omega_2 &= \tfrac{1-\Gamma_{11}+\Gamma_{12}}{30(Z_2+S_{11}^{(2)})} - \tfrac{1}{10(Z_2+S_{12}^{(2)})} + \tfrac{7+\Gamma_{21}-\Gamma_{22}}{60(Z_2+S_{22}^{(2)})}\end{aligned} \quad (A7)$$

$$\Lambda_1 = \frac{\left(S_{11}^{(1)}+4S_{21}^{(1)}+2S_{11}^{(2)}\right)(1-\Gamma_{11}-4\Gamma_{12})+10S_{11}^{(1)}\Gamma_{12}}{30\left(Z_2+S_{11}^{(2)}\right)} - \frac{2S_{12}^{(2)}}{15\left(Z_2+S_{12}^{(2)}\right)}$$
$$+\frac{\left(3S_{22}^{(1)}+2S_{12}^{(1)}+3S_{22}^{(2)}\right)(3-4\Gamma_{21}-6\Gamma_{22})-6S_{22}^{(2)}+5S_{12}^{(1)}\Gamma_{21}}{45\left(Z_2+S_{22}^{(2)}\right)}$$

$$\Lambda_2 = \frac{\left(S_{11}^{(1)}-S_{21}^{(1)}+2S_{11}^{(2)}\right)(1-\Gamma_{11}+\Gamma_{12})}{30\left(Z_2+S_{11}^{(2)}\right)} + \frac{S_{12}^{(2)}}{5\left(Z_2+S_{12}^{(2)}\right)}$$
$$+\frac{\left(S_{22}^{(1)}-S_{12}^{(1)}+S_{22}^{(2)}\right)(1+2\Gamma_{21}-2\Gamma_{22})+6S_{22}^{(1)}}{30\left(Z_2+S_{22}^{(2)}\right)}$$
(A8)

$$\Gamma_{I1} = \frac{\left(Z_1+Z_2+S_{22}^{(1)}+S_{22}^{(2)}\right)\left(Z_1+S_{11}^{(1)}\right)-\left(Z_1+S_{21}^{(1)}\right)\left(Z_1+S_{12}^{(1)}\right)}{\left(Z_1+Z_2+S_{22}^{(1)}+S_{22}^{(2)}\right)\left(Z_1+2Z_2+S_{11}^{(1)}+2S_{11}^{(2)}\right)-\left(Z_1+S_{12}^{(1)}\right)\left(Z_1+S_{21}^{(1)}\right)}$$
$$\Gamma_{I2} = \Gamma_{I3} = \frac{\left(Z_1+2Z_2+S_{11}^{(1)}+2S_{11}^{(2)}\right)\left(Z_1+S_{12}^{(1)}\right)-\left(Z_1+S_{12}^{(1)}\right)\left(Z_1+S_{11}^{(1)}\right)}{2\left(Z_1+Z_2+S_{22}^{(1)}+S_{22}^{(2)}\right)\left(Z_1+2Z_2+S_{11}^{(1)}+2S_{11}^{(2)}\right)-2\left(Z_1+S_{12}^{(1)}\right)\left(Z_1+S_{21}^{(1)}\right)}$$
(A9)

$$Z_1 = \frac{\lambda_{cry}\mu_{am}-\lambda_{am}\mu_{cry}}{\left(\mu_{am}-\mu_{cry}\right)\left[2\left(\mu_{am}-\mu_{cry}\right)+3\left(\lambda_{am}-\lambda_{cry}\right)\right]}$$
$$Z_2 = \frac{\mu_{cry}}{2\left(\mu_{am}-\mu_{cry}\right)}$$
(A10)

$$\lambda_{cry} = \frac{E_{cry}\nu_{cry}}{(1+\nu_{cry})(1-2\nu_{cry})}, \quad \mu_{cry} = \frac{E_{cry}}{2(1+\nu_{cry})}$$
$$\lambda_{am} = \frac{E_{am}\nu_{am}}{(1+\nu_{am})(1-2\nu_{am})}, \quad \mu_{am} = \frac{E_{am}}{2(1+\nu_{am})}$$
(A11)

$$S_{11}^{(1)} = \left(4\nu_{cry}+\frac{2}{\alpha^2-1}\right)g(\alpha)+4\nu_{cry}+\frac{4}{3(\alpha^2-1)}$$
$$S_{12}^{(1)} = S_{13}^{(1)} = \left(4\nu_{cry}+\frac{2\alpha^2+1}{\alpha^2-1}\right)g(\alpha)+4\nu_{cry}-\frac{2\alpha^2}{\alpha^2-1}$$
$$S_{21}^{(1)} = S_{31}^{(1)} = \left(-2\nu_{cry}-\frac{2\alpha^2+1}{\alpha^2-1}\right)g(\alpha)-\frac{2\alpha^2}{\alpha^2-1}$$
$$S_{22}^{(1)} = S_{23}^{(1)} = S_{32}^{(1)} = S_{33}^{(1)} = \left(-2\nu_{cry}+\frac{4\alpha^2-1}{4(\alpha^2-1)}\right)g(\alpha)+\frac{\alpha^2}{2(\alpha^2-1)}$$
$$S_{11}^{(2)} = \left(-4\nu_{cry}+\frac{4\alpha^2-2}{\alpha^2-1}\right)g(\alpha)-4\nu_{cry}+\frac{12\alpha^2-8}{3(\alpha^2-1)}$$
$$S_{12}^{(2)} = S_{13}^{(2)} = S_{21}^{(2)} = S_{31}^{(2)} = \left(-\nu_{cry}-\frac{\alpha^2+2}{\alpha^2-1}\right)g(\alpha)-2\nu_{cry}-\frac{2}{\alpha^2-1}$$
$$S_{22}^{(2)} = S_{23}^{(2)} = S_{32}^{(2)} = S_{33}^{(2)} = \left(2\nu_{cry}-\frac{4\alpha^2-7}{4(\alpha^2-1)}\right)g(\alpha)+\frac{\alpha^2}{2(\alpha^2-1)}$$
(A12)

$$g(\alpha) = \begin{cases} \frac{\alpha}{(\alpha^2-1)^{3/2}}\left[\cosh^{-1}\alpha-\alpha\left(\alpha^2-1\right)^{1/2}\right] & \text{if } \alpha>1 \\ \frac{\alpha}{(1-\alpha^2)^{3/2}}\left[\alpha\left(1-\alpha^2\right)^{1/2}-\cos^{-1}\alpha\right] & \text{if } \alpha<1 \end{cases}$$
(A13)

References

1. Buckley, C.P.; Jones, D.C.; Jones, D.P. Hot-drawing of poly(ethylene terephthalate) under biaxial stress: Application of a three-dimensional glass-rubber constitutive model. *Polymer* **1996**, *37*, 2403–2414.
2. Adams, A.M.; Buckley, C.P.; Jones, D.P. Biaxial hot drawing of poly(ethylene terephthalate): Measurements and modelling of strain-stiffening. *Polymer* **2000**, *41*, 771–786.
3. Boyce, M.C.; Socrate, S.; Llana, P.G. Constitutive model for the finite deformation stress-strain behavior of poly(ethylene terephthalate) above the glass transition. *Polymer* **2000**, *41*, 2183–2201.
4. Doufas, A.K.; McHugh, A.J.; Miller, C. Simulation of melt spinning including flow-induced crystallization: Part I. Model development and predictions. *J. Non-Newton. Fluid Mech.* **2000**, *92*, 27–66.
5. Ahzi, S.; Makradi, A.; Gregory, R.V.; Edie, D.D. Modeling of deformation behavior and strain-induced crystallization in poly(ethylene terephthalate) above the glass transition temperature. *Mech. Mater.* **2003**, *35*, 1139–1148.
6. Makradi, A.; Ahzi, S.; Gregory, R.V.; Edie, D.D. A two-phase self-consistent model for the deformation and phase transformation behavior of polymers above the glass transition temperature: Application to PET. *Int. J. Plast.* **2005**, *21*, 741–758.
7. Dupaix, R.B.; Krishnan, D. A constitutive model for strain-induced crystallization in poly(ethylene terephthalate) (PET) during finite strain load-hold simulations. *J. Eng. Mater. Technol.* **2006**, *128*, 28–33.
8. Dupaix, R.B.; Boyce, M.C. Constitutive modeling of the finite strain behavior of amorphous polymers in and above the glass transition. *Mech. Mater.* **2007**, *39*, 39–52.
9. Figiel, L.; Buckley, C.P. On the modelling of highly elastic flows of amorphous thermoplastics. *Int. J. Non-Linear Mech.* **2009**, *44*, 389–395.

10. Chevalier, L.; Luo, Y.M.; Monteiro, E.; Menary, G.H. On visco-elastic modelling of polyethylene terephthalate behaviour during multiaxial elongations slightly over the glass transition temperature. *Mech. Mater.* **2012**, *52*, 103–116.
11. Cosson, B.; Chevalier, L.; Régnier, G. Simulation of the stretch blow moulding process: From the modelling of the microstructure evolution to the end-use elastic properties of polyethylene terephthalate bottles. *Int. J. Mater. Form.* **2012**, *5*, 39–53.
12. Menary, G.H.; Tan, C.W.; Harkin-Jones, E.M.A.; Armstrong, C.G.; Martin, P.J. Biaxial deformation and experimental study of PET at conditions applicable to stretch blow molding. *Polym. Eng. Sci.* **2012**, *52*, 671–688.
13. Mahjoubi, H.; Zaïri, F.; Tourki, Z. A micro-macro constitutive model for strain-induced molecular ordering in biopolymers: Application to polylactide over a wide range of temperatures. *Int. J. Plast.* **2019**, *123*, 38–55.
14. Mahjoubi, H.; Zaïri, F.; Tourki, Z. Strain-induced phase transformation in poly(lacticacid) across the glass transition: Constitutive model and identification. *Int. J. Non-Linear Mech.* **2020**, *118*, 103241.
15. Ayoub, G.; Zaïri, F.; Naït-Abdelaziz, M.; Gloaguen, J.M. Modelling large deformation behaviour under loading-unloading of semicrystalline polymers: Application to a high density polyethylene. *Int. J. Plast.* **2010**, *26*, 329–347.
16. Ayoub, G.; Zaïri, F.; Fréderix, C.; Gloaguen, J.M.; Naït-Abdelaziz, M.; Seguela, R.; Lefebvre, J.M. Effects of crystal content on the mechanical behaviour of polyethylene under finite strains: Experiments and constitutive modelling. *Int. J. Plast.* **2011**, *27*, 492–511.
17. Abdul-Hameed, H.; Messager, T.; Zaïri, F.; Naït-Abdelaziz, M. Large-strain viscoelastic-viscoplastic constitutive modeling of semi-crystalline polymers and model identification by deterministic/evolutionary approach. *Comput. Mater. Sci.* **2014**, *90*, 241–252.
18. Makki, M.; Ayoub, G.; Abdul-Hameed, H.; Zaïri, F.; Mansoor, B.; Naït-Abdelaziz, M.; Ouederni, M.; Zaïri, F. Mullins effect in polyethylene and its dependency on crystal content: A network alteration model. *J. Mech. Behav. Biomed. Mater.* **2017**, *75*, 442–454.
19. Bernard, C.A.; Lame, O.; Deplancke, T.; Cavaillé, J.Y.; Ogawa, K. From rheological to original three-dimensional mechanical modelling of semi-crystalline polymers: Application to a wide strain rate range and large deformation of Ultra-High Molecular Weight PolyEthylene. *Mech. Mater.* **2020**, *151*, 103640.
20. Lee, B.J.; Parks, D.M.; Ahzi, S. Micromechanical modeling of large plastic deformation and texture evolution in semi-crystalline polymers. *J. Mech. Phys. Solids* **1993**, *41*, 651–687.
21. Lee, B.J.; Argon, A.S.; Parks, D.M.; Ahzi, S.; Bartczak, Z. Simulation of large strain plastic deformation and texture evolution in high density polyethylene. *Polymer* **1993**, *34*, 3555–3575.
22. Nikolov, S.; Doghri, I.; Pierard, O.; Zealouk, L.; Goldberg, A. Multi-scale constitutive modeling of the small deformations of semi-crystalline polymers. *J. Mech. Phys. Solids* **2002**, *50*, 2275–2302.
23. Van Dommelen, J.A.W.; Parks, D.M.; Boyce, M.C.; Brekelmans, W.A.M.; Baaijens, F.P.T. Micromechanical modeling of the elasto-viscoplastic behavior of semi-crystalline polymers. *J. Mech. Phys. Solids* **2003**, *51*, 519–541.
24. Agoras, M.; Ponte Castaneda, P. Multi-scale homogenization-based modeling of semi-crystalline polymers. *Philos. Mag.* **2012**, *92*, 925–958.
25. Poluektov, M.; van Dommelen, J.A.W.; Govaert, L.E.; MacKerron, D.H.; Geers, M.G.D. Micromechanical modeling of roll-to-roll processing of oriented polyethylene terephthalate films. *J. Appl. Polym. Sci.* **2016**, *133*, 43384.
26. Bedoui, F.; Diani, J.; Régnier, G. Micromechanical modeling of elastic properties in polyolefins. *Polymer* **2004**, *45*, 2433–2442.
27. Sedighiamiri, A.; Van Erp, T.B.; Peters, G.W.M.; Govaert, L.E.; van Dommelen, J.A.W. Micromechanical modeling of the elastic properties of semicrystalline polymers: A three-phase approach. *J. Polym. Sci. Part B Polym. Phys.* **2010**, *48*, 2173–2184.
28. Bedoui, F.; Diani, J.; Régnier, G.; Seiler, W. Micromechanical modeling of isotropic elastic behavior of semicrystalline polymers. *Acta Mater.* **2006**, *54*, 1513–1523.
29. Guan, X.; Pitchumani, R. A micromechanical model for the elastic properties of semicrystalline thermoplastic polymers. *Polym. Eng. Sci.* **2004**, *44*, 433–451.
30. Ahzi, S.; Bahlouli, N.; Makradi, A.; Belouettar, S. Composite modeling for the effective elastic properties of semicrystalline polymers. *J. Mech. Mater. Struct.* **2007**, *2*, 1–21.
31. Gueguen, O.; Ahzi, S.; Makradi, A.; Belouettar, S. A new three-phase model to estimate the effective elastic properties of semi-crystalline polymers: Application to PET. *Mech. Mater.* **2010**, *42*, 1–10.
32. Anoukou, K.; Zaïri, F.; Naït-Abdelaziz, M.; Zaoui, A.; Qu, Z.; Gloaguen, J.M.; Lefebvre, J.M. A micromechanical model taking into account the contribution of α- and γ-crystalline phases in the stiffening of polyamide 6-clay nanocomposites: A closed-formulation including the crystal symmetry. *Compos. Part B Eng.* **2014**, *64*, 84–96.
33. Yao, S.; Hu, D.; Xi, Z.; Liu, T.; Xu, Z.; Zhao, L. Effect of crystallization on tensile mechanical properties of PET foam: Experiment and model prediction. *Polym. Test.* **2020**, *90*, 106649.
34. Bedoui, F.; Guigon, M. Linear viscoelastic behavior of poly(ethylene terephtalate) above T_g amorphous viscoelastic properties Vs crystallinity: Experimental and micromechanical modeling. *Polymer* **2010**, *51*, 5229–5235.
35. Hachour, K.; Zaïri, F.; Naït-Abdelaziz, M.; Gloaguen, J.M.; Aberkane, M.; Lefebvre, J.M. Experiments and modeling of high-crystalline polyethylene yielding under different stress states. *Int. J. Plast.* **2014**, *54*, 1–18.
36. Mesbah, A.; Elmeguenni, M.; Yan, Z.; Zaïri, F.; Ding, N.; Gloaguen, J.M. How stress triaxiality affects cavitation damage in high-density polyethylene: Experiments and constitutive modeling. *Polym. Test.* **2021**, *100*, 107248.
37. Liu, H.T.; Sun, L.Z. Multi-scale modeling of elastoplastic deformation and strengthening mechanisms in aluminium-based amorphous nanocomposites. *Acta Mater.* **2005**, *53*, 2693–2701.

38. Ju, J.W.; Sun, L.Z. Effective elastoplastic behavior of metal matrix composites containing randomly located aligned spheroidal inhomogeneities. Part I: Micromechanics-based formulation. *Int. J. Solids Struct.* **2001**, *38*, 183–201.
39. Simo, J.C.; Kennedy, J.G.; Govindjee, S. Non-smooth multisurface plasticity and viscoplasticity. Loading/unloading conditions and numerical algorithms. *Int. J. Numer. Methods Eng.* **1988**, *26*, 2161–2185.
40. Ju, J.W.; Zhang, X.D. Effective elastoplastic behavior of ductile matrix composites containing randomly located aligned circular fibers. *Int. J. Solids Struct.* **2001**, *38*, 4045–4069.
41. Salem, D.R. Development of crystalline order during hot-drawing of poly(ethylene terephthalate) film: Influence of strain rate. *Polymer* **1992**, *33*, 3182–3188.
42. Yan, Z.; Guo, Q.; Zaïri, F.; Zaoui, A.; Jiang, Q.; Liu, X. Continuum-based modeling large-strain plastic deformation of semi-crystalline polyethylene systems: Implication of texturing and amorphicity. *Mech. Mater.* **2021**, *162*, 104060.
43. Basiri, A.; Zaïri, F.; Azadi, M.; Ghasemi-Ghalebahman, A. Micromechanical constitutive modeling of tensile and cyclic behaviors of nano-clay reinforced metal matrix nanocomposites. *Mech. Mater.* **2022**, *168*, 104280.

Article
Numerical Investigation of the Infill Rate upon Mechanical Proprieties of 3D-Printed Materials

Laszlo Racz and Mircea Cristian Dudescu *

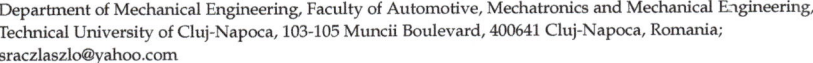

Department of Mechanical Engineering, Faculty of Automotive, Mechatronics and Mechanical Engineering, Technical University of Cluj-Napoca, 103-105 Muncii Boulevard, 400641 Cluj-Napoca, Romania; sraczlaszlo@yahoo.com
* Correspondence: mircea.dudescu@rezi.utcluj.ro; Tel.: +40-264401663

Abstract: The paper proposes a novel method of numerical simulation of the fused deposition molding 3Dprinted parts. The single filaments are modeled by a script using the G-code of the 3D printer. Based on experimental evaluation of the cross-sectional geometry of a printed tensile specimen, the connection between the filaments is determined and the flattening effect of the filaments can be counted. Finite element (FE) simulations considering different element lengths were validated by experimental tests. The methodology allows, on one hand, numerical estimation of the true cross-sectional area of a specimen and correction of the experimental stress-strain curves and, on the other hand, accurate determination of the E-modulus of a printed tensile specimen with different deposition densities (20%, 40%, 60%, 80% and 100% infill rate). If the right method to connect the single filaments is established and validated for a 3D printer, the mechanical properties of the 3D specimens can be predicted without physical tensile test, only using FE method, which will allow the designers to print out the parts with variable infill rate and tunable stiffness only after the FE result are suitable for their needs, saving considerably materials and time.

Keywords: fused deposition modeling; finite element analysis; tensile testing; infill rate

1. Introduction

Additive manufacturing or three-dimensional printing (3D) technologies demonstrated over the last few years a great potential to produce parts from a digital model cost efficiently, without the need of additional tooling and assembly. Three-dimensional printing successfully integrates the design and manufacturing process with an efficient use of material and ability to create parts with highly complex geometries. A very popular 3D printing method for creating prototypes or functional parts out of thermoplastics is Fused Deposition Modeling (FDM) [1], a type of material extrusion additive manufacturing technique, also known as Fused Filament Fabrication (FFF). The FDM technique is able to construct physical parts out of a range of thermoplastic materials such as acrylonitrile butadiene styrene (ABS), polylactic acid (PLA), polycarbonate, polyether-ether-ketone (PEEK) or fiber reinforced thermoplastics. The FDM process consists of the deposition of thermoplastic filaments in a semi-molten state through a heated deposition nozzle onto the build platform contributing layer by layer to the constructed part.

Despite all these advantages, parts created with FDM technique present inferior mechanical properties, due to additional porosity and anisotropy caused by the nature of the manufacturing process. In this regard, both porosity and mechanical anisotropy strongly depend on the printing parameters [2–9]. Therefore, the influence of printing parameters may be used to customize the mechanical properties of the printed components. FDM technique has the potential to produce parts with locally controlled properties [10–13] by changing the deposition density (infill ratio) and orientation (infill pattern).

Strength, toughness, and geometric accuracy of the manufactured parts depend on various process parameters such as infill ratio, infill pattern, layer thickness, layer height

and machine settings [14–18]. Using the optimum process-parameter settings can greatly improve the mechanical strength, surface quality and geometric accuracy by a considerable amount [10,15,19–22]. Modeling of parts by FDM technique implies also analytical models; some techniques being employed in the simulation models are presented in current work [23,24]. An important role to a reliable model and strength evaluation of the parts manufactured by fused deposition modeling is the proper estimation of bonding between the filaments within a layer (intra-layer bonding) and bonds formed between the filaments of the two successive layers (termed as 'inter-layer bonding'). The bonding quality among filaments in FDM parts is an important factor in determining mechanical properties of the parts [24,25]. Meso-structure and void density analysis are usually necessary for theoretical calculation of strength and E-modulus of the printed structures. Also, finite element analysis can be used to predict the mechanical behavior of the FDM prototypes [14,26–28]. The analysis method proposed by Garg et al. [14] uses an FE method wherein the modeling process is accurately replicating the real inter-layer and intra-layer necking of the filaments during the diffusion of the raster layers throughout the printing process. Three different layer heights and three different raster angles were analyzed by FEM using the ABAQUS platform to simulate the elastic-plastic deformation under a uniaxial tensile load. Experimental studies using fractographic analysis are also performed to validate the results.

The present work presents a novel method to simulate the 3D-printed parts manufactured by FDM technique. The method is based on original G-code of the 3D printer to generate the single filaments. Based on experimental evaluation of the cross-sectional geometry of a printed tensile specimen, the connection between the filaments is determined and also the flattening effect of the filaments can be counted. Uniaxial tensile tests are carried out to investigate the mechanical property of the 3D printed material. Stress calculation is undertaken based on the outside dimensions of the tensile specimen, not taking into account the material/airgap ratio from a cross-sectional area. The resulting E modulus will not correspond to reality, but the tensile strain could serve as a reference to validate the proposed geometrical models and the corresponding numerical simulations. The methodology was applied for specimens with different infill ratios (20%, 40%, 60%, 80 and 100%) and the variation of E-modulus and tensile strength were determined. Specific characteristics of a simulation such as element length effect on the precision and simulation time were also addressed. The paper ends with the conclusions underlying the advantage of the proposed methodology that allows, if the proper connection between filaments is identified, accurate prediction of mechanical behavior of a part only using FE analysis.

2. Materials and Methods

2.1. Geometrical Mesostructure Analysis of the of FDM Printed Specimens

After carefully studying the method developed by Garg et al. [14], a disadvantage was identified regarding its real-life applicability; the method does not take into consideration the shell section of the printed part. In real-life application, it is not possible to print parts without the shell section having a low percentage of infill rate, for example, 20%. The analysis method proposed in [14] would not be feasible to determine the true cross-section area of the printed parts because it is a simplified model, applicable only if the infill percentage is high, for example, close to 100%, and does not consider the effect of the shell section on the mechanical behavior of the printed parts.

Given the lack of specific testing standard for 3D printed materials, selection of dog bone specimens has been undertaken considering their wide use and acceptance for 3D-printed specimens. A cross-section of a 3D printed ISO_527 1A tensile specimen is presented in Figure 1 where both infill and shell areas can be distinguished. Each layer of the specimen is built in the x-y plane by a series of lines parallel to x-axis through nozzle movement. The cross-section is lying in the y-z plane, where x, y and z axes are defined by ISO/ASTM 52900:2015.

In order to determine the real area of material in a printed specimen's cross-section, a new, improved model is necessary. In this study, the efficacy of a geometrical method was analyzed, and designed to be able to build up a very realistic geometrical model of the printed parts, where the material and void ratio is clearly identifiable in a virtual model. This geometrical model should also be suitable to undergo a quick meshing process, where the printed mesostructure will be transformed into finite elements, thus facilitating the prediction of the mechanical behavior of the printed parts using FEM analysis. This method allows us to closely examine the printed parts *Shell* section and *Infill* section as well. The cross-section of the printed tensile specimen is presented in Figure 1.

Figure 1. Infill section (**1**) and Shell section (**2**) of a printed ISO_5271A tensile specimen.

To determine the real cross-sectional area of the specimens with different infill rates, a geometrical method has been employed. In order to create a realistic 3D model of the specimens, G-codes of the printed parts were used to build up the geometrical model. Based on the G-code, which was used for printing the specimens, a geometrical model was constructed in ANSA 17.1.2 (BetaCae, Kato Scholari, Greece). A plug-in script was created in Beta Scripting language, which reads the information from the G-code and transforms the movement of the tool into geometrical lines. The dimensions of the geometrical lines—representing the printed fibers—are identical with the primary layer height of 0.2 mm; the width is considered the extruder width at 0.35mm, as presented in Figure 2, and these dimensions are used to determine the real cross-sectional area of the specimens, using cutting planes transversal to the longitudinal direction of the printed fiber, shown in Figure 3.

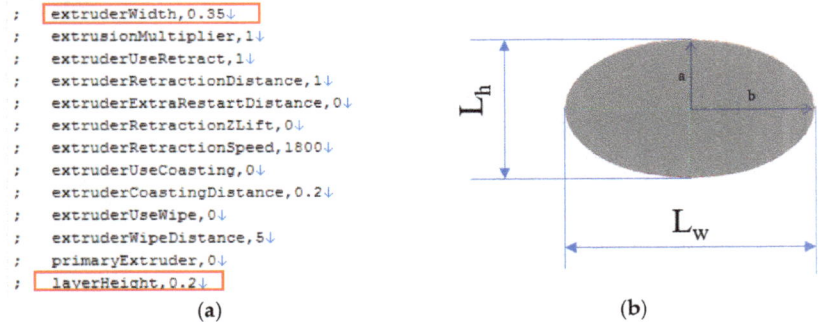

Figure 2. Printed filaments geometry: (**a**) Segment form the printed G-code; (**b**) Theoretical dimensions of a fiber (L_w and L_h correspond to extruder Width and layer Height).

The height of the fiber is considered the primary layer height (L_h = 2a); the width is considered equal with the extruder width (L_w = 2b) values of the printing parameter, according to G-code. The theoretical area of a single filament can be calculated now, using the formula $A = \pi ab$ (a and b from Figure 2). For extracting the area of the cross-section of a printed specimen, a geometric method is used. The virtual (CAD) tensile specimen is constructed in ANSA according to the process described above. Initially five cases of standard ISO_527 1A tensile specimens were studied, where the infill rate ranged from 20% to 100%, the printing direction was kept the same, 0-degree according to the longitudinal axes of the specimen. In all of the five cases the extraction of the cross-section was undertaken in the same place, which is 75.708 mm from the edge of the tensile specimen, as shown in Figure 3.

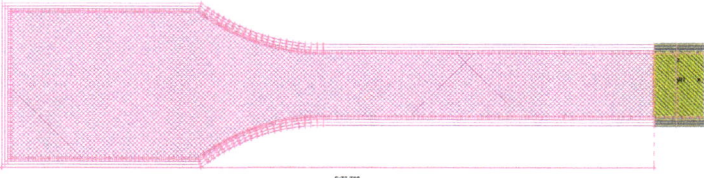

Figure 3. Cross-section distance from the edge of the tensile specimen.

The extraction of the cross-section area for the tensile specimens was undertaken in ANSA; the geometry of the tensile specimen is presented in Figure 4a. The geometrical representation of the filaments is shown in Figure 4b. The first step was to define a different property for the face, which was used to intersect the sample; it was named Cutting Plane. The File-Intersect function was used in ANSA to intersect the sample with the Cutting Plane, and the resulting cross-section was assigned a different property named C_P_CrossSection, as represented in Figure 4c. The surface area of the property C_P_CrossSection can be measured using the D.Util function; the result is shown in Figure 4d. The filament representation as a perfect eclipse is a theoretic assumption.

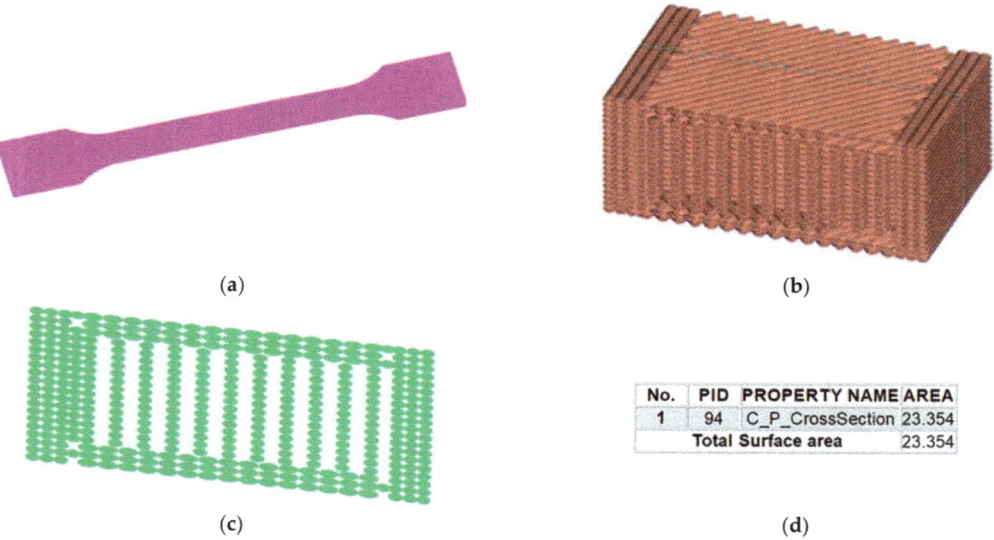

Figure 4. Geometrical modeling and cross-section extraction in ANSA software: (**a**) geometry of the tensile specimen; (**b**) geometrical representation of the filaments; (**c**) cross-section of the geometrical model; (**d**) the result of the area calculation for the cross-section.

In reality, the molted material is slightly compacted, and the cross-sections of the filaments are irregular. To verify the area determination method using a CAD model in ANSA, microscopic study of printed specimens is necessary. Five set of tensile specimens were printed using the same G-code's as in the CAD model construction in ANSA, and the cross-section is analyzed under a microscope to determine the real shape and size of the filaments.

2.2. Estimation of Contact Areas of the Printed Specimens

The cross-section, meso-structure and infill pattern of the printed specimens was analyzed under a microscope, with 50× magnifying factor, to determine the real shape and structure of the fibers. The result of the cross-section extraction from the geometrical model was compared to the microscopic pictures of the printed specimens. Because the extruded filament is in semi-molten state, both its top and bottom would flatten slightly when deposited onto the previous layer, as indicated in the microscopic photo Figure 5a. Correlation of the geometrical area is necessary, because the cross-section of the fibers are represented as a perfect ellipsoid in the geometrical model; however, in reality, the cross-section of the fibers has a bigger contact patch, due to the fact that their cross-section is not perfectly ellipsoid shaped, as presented in Figure 5b, and pointed out in [13,23].

Figure 5. Cross-section comparison of the fiber structure between a printed tensile specimen and the geometrical model in ANSA: (**a**) printed specimen under microscope; (**b**) geometrical model.

Therefore, a modified calculation method is required when considering the flattening effect that can be measured experimentally. This flattening effect leads to the formation of a bigger contact area between the filaments, known as intra- and inter-layer necking, as presented in Figure 6.

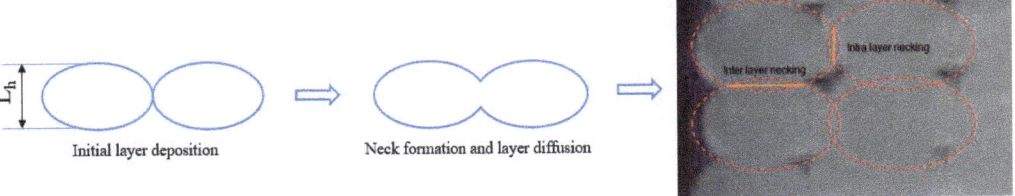

Figure 6. Necking formation between adjacent raster layers.

In the case of using only the printing direction or infill pattern to build up a part or a sample, determination of the real cross-section area is fairly easy, considering that the areas of inter-layer and intra-layer necking can be determined experimentally by optical measurements. The void and material ratio are practically an adjusted area of the printed filament layer multiplied with the number of layers. In the geometrical model, the cross-section of the fibers will be considered perfectly ellipsoidal, and the relationship between the printing parameters and settings can be established. For example, if the relationship between the printing parameters (primary layer height, extruder width) and the proportion of the flattening effect is found, the real area of the fibers can be accurately predicted, and the information implemented in the script, which calculates the fiber cross-section area. A scale factor has to be established to create the correct assumption of the cross-section area of the individual fiber, which will be used to correct the geometrical model area calculation errors. The scale factor is determined by optical analysis of the cross-sections of different specimens under microscope. To determine the real size of the contact patch between the layers (inter-layer necking and intra-layer necking), according to infill rate and infill pattern, microscopic pictures were analyzed with Digimizer software (MedCalc Software Ltd., Ostend, Belgium), where the length of the contact areas, both inter-layer and intra-layer necking was measured, as presented in Figure 7.

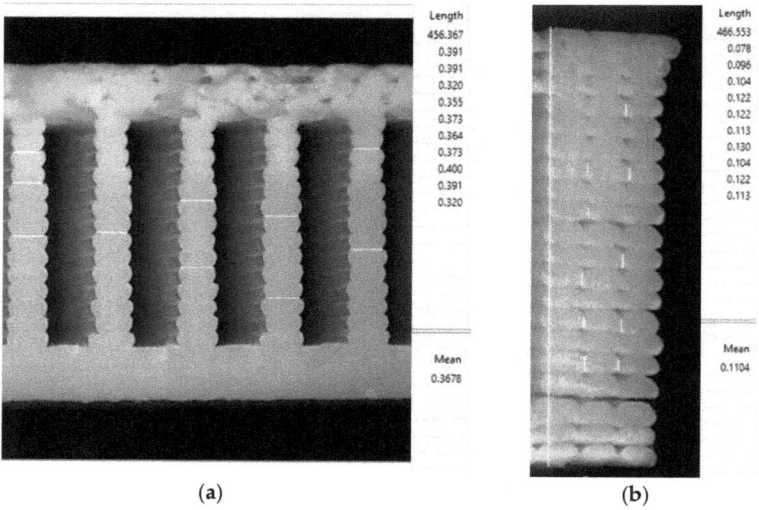

Figure 7. Determination of the contact area between the filaments (**a**) inter layer necking; (**b**) intra layer necking.

Taking into account the values of the horizontal (inter-layer necking) and vertical overlapping (intra-layer necking) determined experimentally, an updated geometrical model was created wherein the filaments are still represented as a perfect ellipsoid, but the contact area between the layers was modified to fit the measured intra- and inter-layer necking, as shown in the Figure 8.

It can be observed that the inter-layer necking is close to the extruder width (extruder with 0.35 mm, inter layer necking ~0.36 mm) and the intra-layer necking is about half of the primary layer height (layer height 0.2 mm, intra-layer necking ~0.11 mm). Taking into account these observations, the geometric model was updated in order to better approximate the geometric model to reality, thus facilitating a more realistic cross-section area determination. Furthermore, if the filament area is accurately determined for different printing parameters, the air-gap material ratio can be found to determine the density of the printed specimens. To validate the method, real-life tensile tests are necessary, where the results of the tensile test, especially tensile strain will be compared to the result of numeric simulation, therefore validating the geometrical model. Another important fact should be taken into consideration, i.e., when determining the area of the cross-section by geometric method in ANSA, the area of the cross-section is dependent on the settings in ANSA, especially "Perimeter length" and "Resolution tolerance". Perimeter length defines the node density on the edges. A high node density (shorter distance between the nodes) allows a finer mesh generation, which better approximates the mesh to the edges of the geometry, thus minimizing the loss of material (cross-section area in our case). Resolution tolerance displays the geometrical details in high or lower resolution. A fine tolerance displays all the features in high detail, and also allows a low perimeter length, but uses a lot of computing and graphic resources, which can make a model difficult to work with.

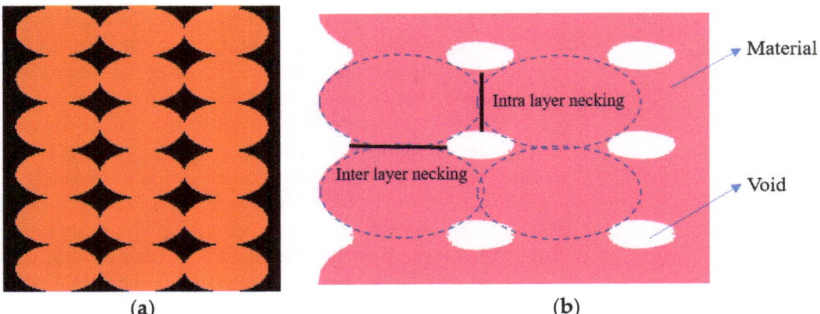

Figure 8. Representation of the filament: (**a**) initial geometry; (**b**) adjusted to represent the intra and inter layer necking.

2.3. Finite Element Analysis of FDM 3D Printed Specimen

Finite element analysis is conducted on the geometrical models, described in Section 2.1. In order to validate the geometrical model proposed, the result of the finite element analysis and experimental data should match. To verify the cross-section of the specimens, numerical simulation will be performed on the geometrical models where the resultant tensile strain will be compared with the experimentally obtained values. If they are comparable, the area of the cross-section is correct and the FEA1 and FEA2 tensile moduli are considered correct. After the tensile moduli are determined for different infill rates, they can be used for FE simulations utilizing a homogenous mesh.

To obtain accurate results in Finite Element simulations the model creation part is especially important. The model should be built up as precisely as it can be, and all the boundary conditions should be represented and applied as realistically as possible. Choosing the right representation of the modeling approach, element type, element length and quality file is also crucial. Conventional Finite Element modeling techniques are not suitable for analyzing 3D-printed parts; due to the air gaps inside the parts, they cannot be considered either simple shell or volume parts. The modeling method used in this study is utilizing the experience of Garg et al. [14], where the authors used a microstructure modeling approach, wherein each filament line is modeled with Tetra elements and they are connected according to previously established intra-layer and inter-layer necking. As pointed out earlier in this document, this approach presents some limitations in terms of applicability in case of functional parts, where the printed parts are more complex and are built up with a shell section around the infill area. In order to eliminate this problem, the shell section should also be accurately modeled, and the modeling time should be reduced and if possible automatized. The method proposed desires to meet this requirement, where the modeling can be at least partially automatized in order to reduce modeling time, and capability to build up complex models with shell section ca be included.

The process used in this study starts with the initial geometrical model, built up in a conventional design tool. The CAD model will be processed in a "slicer tool", wherein the printing parameters are defined. The result of this process will be a G-code, which is readable by 3D printers. The G-code contains all the information that the 3D printer requires, such as temperature of the extruder head, layer height and width, printing speed and, most importantly, all the coordinates of the extruder heads movements. This coordinates and printer settings can be used to create a very realistic micro-structure model, which is suitable for later Finite Element analysis.

The process can be described as follows: the geometrical models were established in ANSA 19.1.2 using a script that translates the movement of the extruder head based on the printed G-code and transforms the movements of the extruder of the 3D printer to geometrical lines. The geometrical lines are then used to extrude an ellipsoid section of the infill along the lines laid down in the previous step. The shape of the individual layers is determined by the primary setting of the printing parameters, for example, primary layer height, extruder diameter, etc., which the script reads from the G-code and creates the primary ellipsoid accordingly. The ellipsoid section is adjusted with the overlapping necking areas, (inter-layer and intra-layer necking) in order to adjust the perfect geometrical model to a more realistic model, wherein the printed layers do not have a regular ellipsoid shape, rather a deformed quad-like shape. In order to extract the cross-section area of the tensile specimens, the created geometry has to be meshed. In the software ANSA, there is a possibility to obtain information about surface areas, but the surface representation in term of accuracy is connected to "perimeter length". Perimeter length finally defines the length of the mesh created on the surfaces. To extract the real cross-section area as precisely as possible, a very low perimeter length was used: 0.02, as shown in Figure 9.

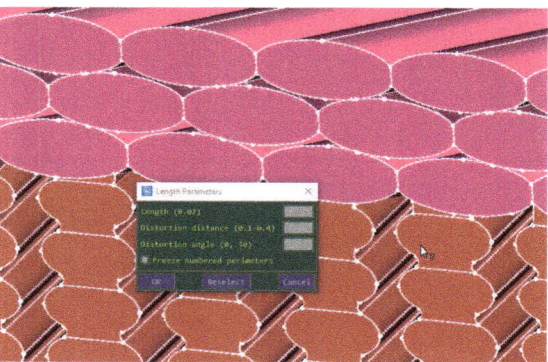

Figure 9. Setting the Perimeter Length to a value that allows a precise representation of the layers.

After the determination of the cross-section area of the tensile specimen, Finite Element analysis was conducted on a portion of the parallel part of the tensile specimen. In order to decrease the simulation time and to fit the processing power of a generic workstations, only a segment of the tensile specimen was analyzed. The length of the segment is 5.43 mm, elected randomly with the consideration to have at least four repetitive elements in case of complex infill patterns such as Honeycomb or Triangle, as shown in Figure 10.

Several simulations were conducted in order to determine the optimal modeling procedure. The factors that influence the result are element type, element order, element length and quality of the elements. The first five sets of models (with infill rate from 20%, 40%, 60%, 80%, 100%, direction: 0° according to the longitudinal axis of the sample) were build up using Tetra, first order elements, with a generic element length of 0.05 mm. The models were loaded with 160 N on one end, and fixed on the other with a rigid body *RBE2* element connected at the center of gravity to a single point of constrain (*SPC*), which forbids all movements and rotations.

Linear static simulations, assuming isotropic material, were conducted in Epilysis, which is a solver provided by BETA CAE, and it is integrated in the pre-processing tool ANSA. Epilysis is using a Nastran source code, and the results are comparable with the results obtained in Nastran. In fact, the first model was simulated in Nastran and in Epilysis too, and the output was compared, resulting in displacement difference of 4.3% (from 0.0203476 to 0.02127), as presented in the Figure 11.

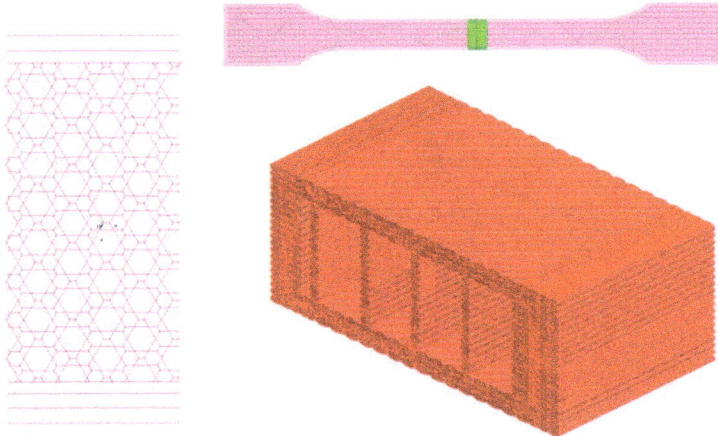

Figure 10. FEM representation of the filaments. (Segment from a standard tensile test specimen).

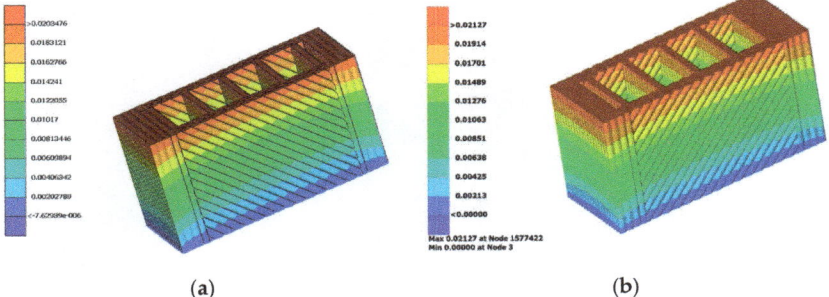

Figure 11. Result of simulation in: (**a**) Epilysis and in (**b**) Nastran.

Although the Nastran delivers a more precise result than Epilysis, the further simulations were performed in Epilysis because of the availability of licenses, and because the average result of the simulations performed with Epilysis are off by 5.2%, (compared to experimental results), which can be considered acceptable. The post-processing of the data was undertaken in Animator_v2.0.8 for the result of Nastran, and in Metapost 19.1.2 for the results of the simulations performed in Epilysis. The two post-processing softwares were compared also, and no significant difference was found in reading the displacements and stresses from the resulted.op2 files.

A model with Tetra 2nd order element was built up and simulated using the exact same boundary conditions as the 1st order model. In case of Tetra 2nd order model, the total number of elements are the same, but the number of nodes are increased from 373,777 to 2,738,200 (in case of 100% infill) due to the introduction of the middle nodes on each element edge. The results are showing a highest deformation in case of 2nd order element, for example, in the case of 40% infill rate where the experimentally measured displacement was 0.00352 mm, the simulation result was 0.00369105 mm, which represents a 4.63% increase in displacement. After analyzing the results, we can notice, that the simulation model built up using tetra 1st order elements is always showing a deformation less than the experimentally obtained values, which is expected due to the fact that Tetra 1st order elements are stiffer than Tetra 2nd order elements.

As expected, the simulation time was gradually increasing with the complexity and size of the models varying from 12 min for model with the lowest infill rate 20%, up to

45 min for the highest infill rate 100%. With the increase in infill rate from 20%- to 100%, the models increased from (element length 0.05) 3,740,121 tetra element to 11,940,382 elements, which can be expressed as a 219.25% increase. In this case, a regular personal computer is unable to process the amount of information, therefore, professional calculation/simulation servers are recommended.

In order to analyze the effect of the element length on the precision and simulation time, another model setup was tested. The element length was doubled from 0.05 to 0.1 mm. The simulation time of the models with 0.1 mm element length varied between minimum 1 min for 20% infill rate up to 8 min for the more complex model of 100% infill rate. Comparing the two models in terms of the computing time: using a larger element length (0.1 mm) reduced the time with a factor of 5.6, while delivering an average precision of results with an error of 7.5% compared to the experimental results and 2.3% compared to finer mesh (0.05). The differences are caused by the loss of details in geometry due to the representation of the layers in a simplified way, as shown in Figure 12.

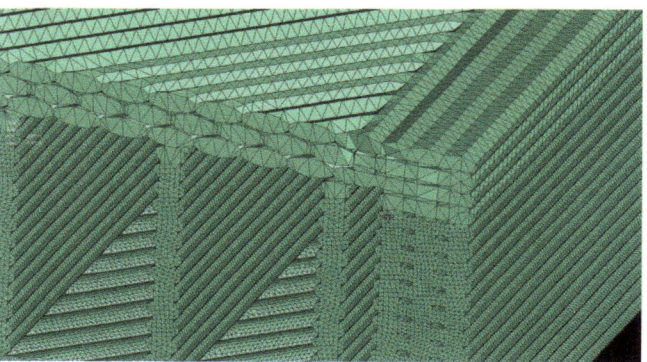

Figure 12. Influence of the higher target length on the representation of the individual filaments.

After reviewing the effects of the FE model build parameters, the simulation and post-processing tools, it can be concluded that the "Best Practice" for our investigation concerning FE analysis is as follows: the Tetra 1st order element model, with a global element length of 0.1 mm. The pre-processor will remain Ansa, solver: Epilysis and post-processing tool: Metapost. This approach to modeling seems to be a well-balanced solution between accuracy of results, computing time and the need for processing power. Another conclusion can be drawn from analyzing the result of the simulations, i.e., that using a bigger element length greatly reduces the computation time, while delivering the results with an acceptable precision.

2.4. Tensile Test

The geometric models of 3D-printed specimens were realized according to ISO 527-2-2012 (International standard, plastics Determination of tensile properties Part 2: Test conditions for molding and extrusion plastics). Uniaxial tensile tests are carried out to investigate the mechanical property of the 3D-printed material. The specimens are tested with a universal testing machine type INSTRON 3366, 10 kN capacity. The loading speed of this machine is 1 mm/min and the test stops once the specimens are broken. A uniaxial extensometer was used to measure the tensile strain. For each infill rate, a number of five specimens were tested. The material used for specimen's preparation was acrylonitrile butadiene styrene ABS filament (Plasty Mladeč, Czech Republic). Its mechanical properties, according to the producer, are as follows: tensile modulus E_f = 2140 MPa, tensile stress σ_f = 43 MPa and tensile strain ε_f = 2.7%.

Stress calculation is undertaken for all infill rates based on the cross-sectional dimensions of the tensile specimen (4 mm thickness, 10 mm width; not taking in account of the material/airgap ratio, results in a cross-section area of 40 mm^2). The resulting tensile stress will not correspond to reality, but the tensile strain could serve as a reference to validate the geometric model with numeric simulations. If the experimental tensile strain is comparable to the one obtained by FE simulation, the area of the cross-section is validated.

A tensile test simulation was conducted on the FE model using Epilysis solver. Previously, the same tensile specimen was tested experimentally. The same boundary conditions were applied to the FE model as for the real tensile specimen during the tensile test. The objective of the simulation was to find out if the elongation of the FE model corresponds with the experimental results. The results can be summarized by Table 1 where the experimentally measured strain is compared to the results of the FEM simulations performed with a finer (0.05 mm) and a bigger element length (0.01 mm) and the deviation between them is expressed as percentage. The results are selected from the elastic domain of the material at 160 N load.

Table 1. Comparison of the displacements from experimental tensile test (EXP) to simulation results with two different element lengths (FEA1 & FEA2).

Infill Rate (%)	Strain EXP (mm/mm)	Strain FEA1 (mm/mm)	Strain FEA2 (mm/mm)	Relative Dev. FEA1-EXP (%)	Relative Dev. FEA2-EXP (%)	Relative Dev. FEA1-FEA2 (%)
20%	0.004040	0.003917	0.003796	3.14	6.43	3.09
40%	0.003520	0.003300	0.003250	6.65	8.31	1.53
60%	0.003040	0.002907	0.002866	4.57	6.06	1.40
80%	0.002740	0.002584	0.002561	6.04	6.97	0.87
100%	0.002510	0.002323	0.002310	8.06	8.66	0.56

Ten simulations were conducted to determine the effect of the element length on the precision of the result. The basis of comparison is the displacement obtained by real experiment (tensile test) in relation to the FE results using two different element lengths, a finer mesh 0.05 and course mesh of 0.1 mm. The results showed a range of minimum 3.13% and maximum 8.05% difference in the case of FE model meshed with 0.05 mm element length (FEA1) compared to the experimental result (EXP). The FE model meshed with 0.1 mm element length (FEA2) delivers results with a slightly higher error, where the range is minimum 6.42% and maximum is 8.66%. Comparing the two FE models, the range is between 0.55% and 3.09% difference, which is considered a good result factoring in the savings in terms of processing power and simulation time, as mentioned in the previous paragraph. A convergence is noticeable when comparing the two FE models (Table 1—Relative Dev. FEA1- FEA2), with the increasing of the infill rate.

The obtained results are consistent with the measured data that proved that our estimations of contact areas and filament shape of the printed specimens and the proposed finite element models are accurate, and it can be successfully further employed.

3. Results and Discussions

In Figure 13, the cross-section of the tensile specimens is presented (as they are extracted from the geometrical model from ANSA) from 20% to 100% infill rate.

The first set of cross-section values are representing the areas of the samples printed with different infill rates, determined with a perimeter length (PL) of 0.05 mm. The value 0.05 mm is the mesh length used for FE simulation (Perimeter length = Mesh length) to validate the geometrical model based on tensile strain comparison. As previously presented, two different meshing lengths have been used to simulate the specimens, a finer mesh: 0.05 mm (FEA1) and a coarser mesh: 0.1 mm (FEA2).

In the case of FEA1 and FEA2, the area of the cross-section was extracted from the geometrical model from ANSA, as presented in Table 2, and introduced into the testing machine software allowing for a more accurate calculation of the tensile stress.

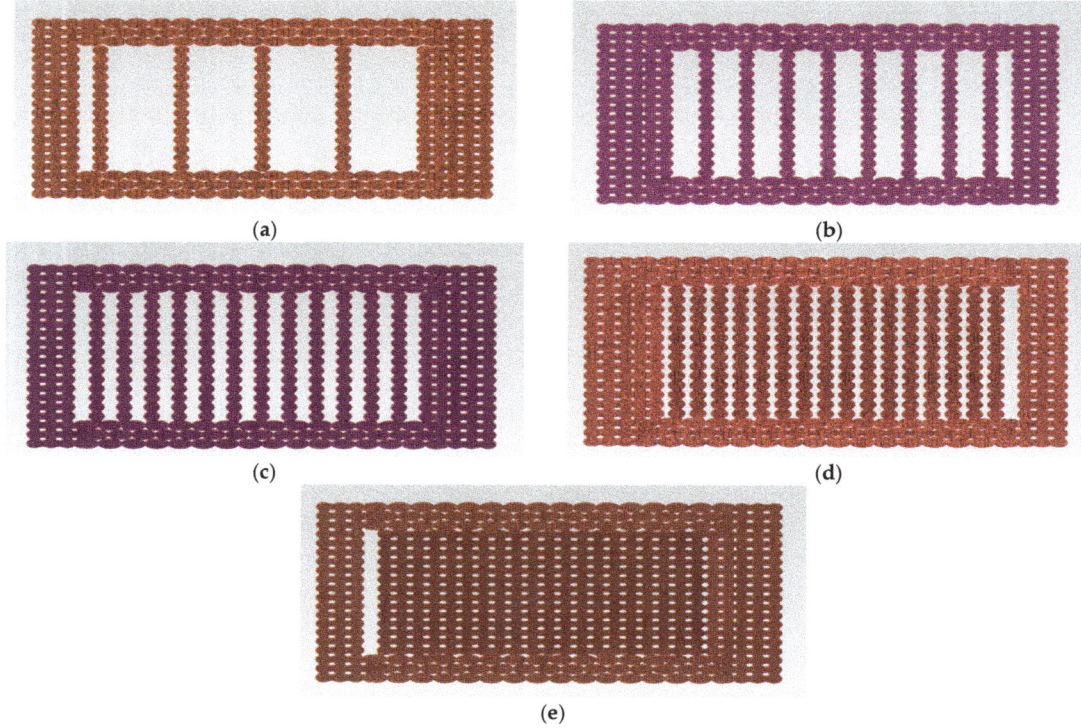

Figure 13. Cross-sections patterns according to infill rate: (**a**) 20%; (**b**) 40%; (**c**) 60%; (**d**) 80%; (**e**) 100%.

Table 2. Estimated cross-sectional areas according to infill rate and mesh size.

Infill Rate	Cross-Sectional Area of the Specimen (mm^2)	
	FEA1	FEA2
20%	20.50	20.30
40%	23.85	23.35
60%	27.05	26.52
80%	30.23	29.66
100%	34.72	34.19

The increase in the cross-section area FEA1 and FEA2 is, on average, 13%, so the 20% increase in "Infill rate" does not result in a similar increase step in cross-section. Examining the quantity of raw material (filament length and material weight) added to specimens with different infill rate (Table 3), values delivered by the 3D-printer software, a decrease from 20% to 13% can be observed. A possible explanation is that a change in the infill rate determines not only an increase in the quantity of the deposited filament, but it is also constrained by the disposal (pattern) of the filaments according to the part geometry.

Table 3. Filament length and specimen weight variation with the infill rate.

Infill Rate Change	Filament Length (%)	Material Weight (%)
20–40%	19.07	18.99
40–60%	16.04	15.96
60–80%	13.83	13.88
80–100%	13.57	13.57

Comparative strain-stress curves for all analyzed infill rates are presented in Figure 14. The initial experimental curve (EXP) is based on a constant cross-sectional area of 40 mm² given by the outside dimensions of the specimens. The other two curves (FEA1 & FEA2) presented in Figure 14 are recalculated experimental strain-stress curves considering numerically estimated cross-sectional areas. It can be observed that utilizing the full cross-section of the 3D-printed specimen, even for a 100% infill rate, will not deliver a result within an expectable error range.

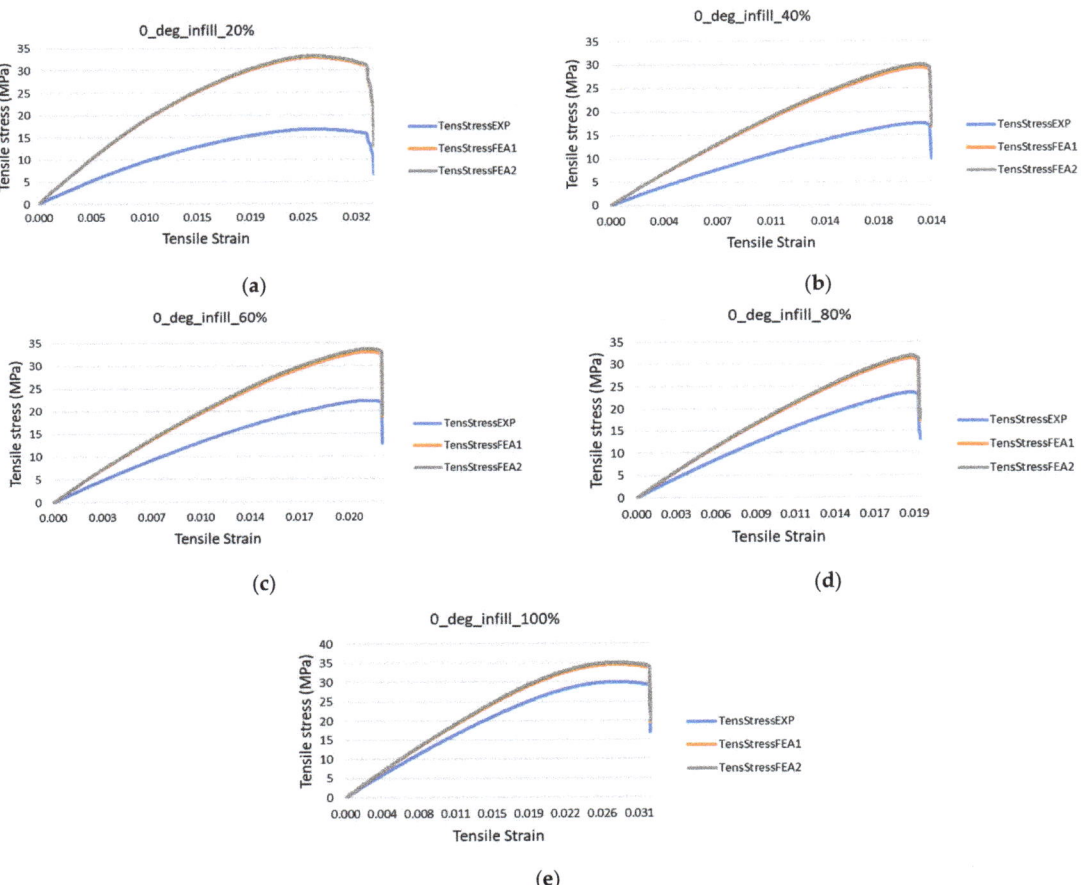

Figure 14. Stress-strain curves recalculated based on numerically estimated cross-sectional areas specimens with the infill rates: (**a**) 20%; (**b**) 40%; (**c**) 60%; (**d**) 80%; (**e**) 100%.

In order to predict the tensile modulus of the 3D-printed specimen with different infill rate, the cross-section extracted from the geometric model was reintroduced into the testing machine software and the results were recalculated according to the two set of values.

In Table 4, the result of the tensile modulus is presented where: E modulus EXP is representing the tensile modulus of the specimen—experimentally determined—utilizing the full cross-section of the sample; E modulus FEA1 and FEA2 are representing the recalculated tensile moduli of the specimens utilizing the area extracted from the geometric model employing the finer mesh (0.05 mm) and the course mesh (0.1 mm) definition, respectively.

Table 4. Tensile modulus of the printed specimens with geometrically determined cross-section.

Infill Rate	E-Modulus (MPa)		
	FEA1	FEA2	EXP
20%	1916.0	1934.9	982.0
40%	1974.8	2017.1	1177.5
60%	2046.7	2087.7	1384.1
80%	1983.4	2021.5	1498.9
100%	1703.4	1729.8	1521.8

The influences of the perimeter length on the area of the cross-section (which translates to the values of the E modulus) are presented in Figure 15. It can be observed that there is an inverse correlation between the numerically and experimentally determined E modulus in relation to the infill rate. In case of a low infill rate of 20%, the difference between the E modulus EXP and E Modulus FEA1 is 53%, while for an infill rate of 100%, the difference between E modulus EXP and FEA1 is reduced to a still significant 36%. It can be concluded that for tensile modulus, utilizing the full cross-section of the 3D printed specimen, even for a 100% infill rate, will not deliver a result within an expectable error range.

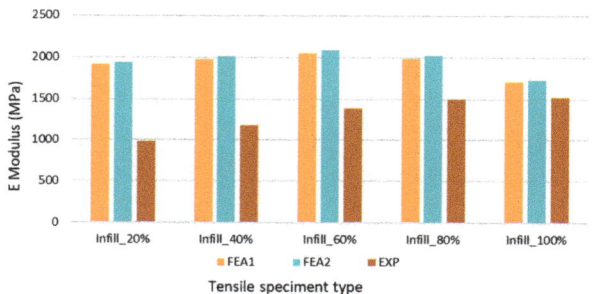

Figure 15. E-moduli value for different infill rates recalculated based on numerically estimated cross-sectional areas.

Comparing the deviation between E-modulus FEA1 and FEA2, we can observe an average of 1.72% difference, which is coming from the mesh size difference. As expected, for higher infill rates, the values of FEA1 and FEA2 are converging to the experimentally obtained (EXP) values, but contrary to our initial expectation, not in a proportional manner. The results are clearly showing that the increase in 20% in infill rate does not directly translate to a 20% change in E modulus. In the case of the experimental results, the difference from EXP Infill_20% to EXP Infill_40% is a 19.9% step, but this shrinks to 2.1% when comparing EXP Infill_80% to EXP Infill_100%. Notably, the values of FEA1 and FEA2 are increasing, up to Infill_60%, then slightly decreasing to Infill_100%.

Variation of E-modulus with respect to the infill rate is depicted fin Figure 16. A quadratic polynomial fit delivers the best results.

Analyzing the obtained equations of the fitting curve:

$$\text{E-modulus/MPa} = 2085 - 18.6 \text{ Infill rate}/\% + 0.6307 \text{ Infill rate}^2/\%^2 - 0.006569 \text{ Infill rate}^3/\%^3 + 1.74 \times 10^{-5} \text{ Infill rate}^4/\%^4 \quad (1)$$

it can be observed that the initial term (2085) is very close to the value of measured E-modulus of the filament (E_f = 2140 MPa), and has a slight increase up to 60% infill rate and then a decrease.

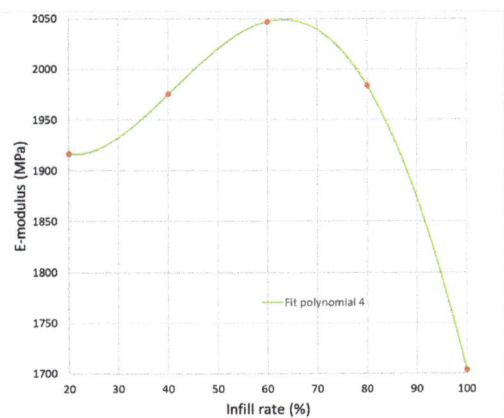

Figure 16. Polynomial fit of the infill rate vs. E-modulus of tensile specimens.

In order to fully understand and explain this behavior, a closer analysis on the cross-sections is necessary (Figure 13, Tables 2 and 3). Inconsistencies can be observed in orienting the infill pattern into the shell section of the printed part, where the shell section has a wall that consists of three rows of filament on one side and five rows of filament on the other, shown in Figure 11. This effect can be especially important at high infill rate (Infill 80%, 100%) and can influence the way the infill is connected to the shell part of the specimen, which can result in different bonding forces between filaments, on one hand, and inner filaments and outer shell, on the other hand. A stiffness difference between the left and the right sides can be also produced. Our results indicate that besides strength reported in [24], the E-modulus of the FDM specimen is not only dependent on the infill rates, but an important contribution is made by intra-layer bonding, inter-layer bonding and neck growth between filaments. As long as the printer is not respecting a symmetrical disposal of the rows of filaments and outer shell construction correlated with the infill rate such behavior can occur and can be estimated only by applying a methodology as those presented in this paper.

4. Conclusions

In this paper 3D-printed tensile specimens were analyzed in order to understand the effect of different infill rates on the mechanical behavior. A novel approach to analyze 3D-printed FDM models was presented, which is utilizing a geometrical model, constructed based on the printer generated G-code, which facilitates Finite Element analysis. The issue of cross-section determination was addressed, and a geometrical model was developed to investigate the "air gap—material ratio" problem. A case study was conducted on finite element models in order to establish the best modeling method for optimal balance between accuracy and simulation running time. The tensile strain resulted when the two cases of simulations (FEA1 and FEA2) were compared to experimental result, which confirmed that the area of cross-section extracted from the geometric model is predicted with good accuracy.

In order to determine the E moduli of different infill rates, the cross-sections extracted from the geometric model were reintroduced into the tensile testing machine. After conducting twelve simulations, we can establish that the simulation method proposed in this study is a viable option to predict the behavior of 3D printed parts even before they are being printed, only by running an analysis on the G-code generated by the "slicer" tool. The obtained E moduli for different infill rates can be used for FE simulation where the microstructure no longer has to be modeled; a simple volume mesh should be enough because the E modulus also contains the correct airgap-material ratio.

The findings presented in this paper allow the following overall conclusions to be drawn:
- The proposed approach for constructing a complex finite element model based on the printer generated G-code is a reliable methodology to predict the behavior of the FDM printed parts but adjustments to represent the intra- and inter-layer necking are necessary for accurate results.
- Cross-sectional area of a tensile specimen extracted from the numerical model is predicted with good accuracy and allows estimation of strain-stress curves and E-moduli closer to reality.
- For higher infill rates the values of tensile stress and E-modulus of the specimens are converging to the experimentally obtained values, but not in a proportional manner.
- The results are clearly showing that the increase in the infill rate does not directly translate to a corresponding change in E modulus, disposal of the rows of filaments influencing the bonding forces between them and the outer shell.

The main purpose of this modeling method is to being able to model complex 3D-printed parts with variable infill rate and tunable stiffness. If the right method to connect the single filaments is established for a specific printer (accounting printing speed, resolution, nozzle diameter, temperature, material behavior during deposition, etc.), and tested to provide a reliable result, the mechanical properties of the 3D specimens can be predicted without physical tensile test, which will allow the designers to print out the parts with variable infill rate only after the FE result are suitable for their needs, saving considerable materials and time. If the computation power allows, we consider that the methodology can be extended to complex parts. Having information about the mechanical behavior of the parts before the actual printing could be important in the design strategy of functional prints or in applications with a reduced number of filaments such as 3D-printed houses.

Author Contributions: Conceptualization, L.R. and M.C.D.; methodology, M.C.D.; software, L.R.; validation, L.R. and M.C.D.; formal analysis, M.C.D.; investigation, L.R.; resources, L.R.; data curation, L.R.; writing—original draft preparation, L.R.; writing—review and editing, M.C.D.; supervision, M.C.D. All authors have read and agreed to the published version of the manuscript.

Funding: This research received no external funding.

Institutional Review Board Statement: Not applicable.

Informed Consent Statement: Not applicable.

Data Availability Statement: Not applicable.

Conflicts of Interest: The authors declare no conflict of interest.

References

1. Dezaki, M.L.; Ariffin, M.K.A.M.; Hatami, S. An overview of fused deposition modelling (FDM): Research, development and process optimisation. *Rapid Prototyp. J.* **2021**, *27*, 562–582. [CrossRef]
2. Dudescu, C.; Racz, L. Effects of raster orientation, infill rate and infill pattern on the mechanical properties of 3D printed materials. *ACTA Univ. Cibiniensis* **2017**, *69*, 23–30. [CrossRef]
3. Lalegani Dezaki, M.; Mohd Ariffin, M.K.A. The effects of combined infill patterns on mechanical properties in fdm process. *Polymers* **2020**, *12*, 2792. [CrossRef] [PubMed]
4. Popescu, D.; Zapciu, A.; Amza, C.; Baciu, F.; Marinescu, R. FDM process parameters influence over the mechanical properties of polymer specimens: A review. *Polym. Test.* **2018**, *69*, 157–166. [CrossRef]
5. Onwubolu, G.C.; Rayegani, F. Characterization and optimization of mechanical properties of ABS parts manufactured by the fused deposition modelling process. *Int. J. Manuf. Eng.* **2014**, *2014*, 598531. [CrossRef]
6. Rankouhi, B.; Javadpour, S.; Delfanian, F.; Letcher, T. Failure analysis and mechanical characterization of 3D printed ABS with respect to layer thickness and orientation. *J. Fail. Anal. Prev.* **2016**, *16*, 467–481. [CrossRef]
7. Ćwikła, G.; Grabowik, C.; Kalinowski, K.; Paprocka, I.; Ociepka, P. The influence of printing parameters on selected mechanical properties of FDM/FFF 3D-printed parts. In *IOP Conference Series: Materials Science and Engineering*; IOP Publishing: Bristol, UK, 2017; Volume 227, p. 012033.
8. Dave, H.K.; Patadiya, N.H.; Prajapati, A.R.; Rajpurohit, S.R. Effect of infill pattern and infill density at varying part orientation on tensile properties of fused deposition modeling-printed poly-lactic acid part. *Proc. Inst. Mech. Eng. Part C J. Mech. Eng. Sci.* **2021**, *235*, 1811–1827. [CrossRef]

9. Samykano, M.; Selvamani, S.K.; Kadirgama, K.; Ngui, W.K.; Kanagaraj, G.; Sudhakar, K. Mechanical property of FDM printed ABS: Influence of printing parameters. *Int. J. Adv. Manuf. Technol.* **2019**, *102*, 2779–2796. [CrossRef]
10. Racz, L.; Dudescu, M.C. Mechanical behavior of beams with variable stiffness obtained by 3D printing. In Proceedings of the 10th International Conference on Manufacturing Science and Education—MSE 2021, Sibiu, Romania, 2–4 June 2021; Volume 343.
11. Gebisa, A.W.; Lemu, H.G. Investigating effects of fused-deposition modeling (FDM) processing parameters on flexural properties of ULTEM 9085 using designed experiment. *Materials* **2018**, *11*, 500. [CrossRef]
12. Sheth, S.; Taylor, R.M. Numerical investigation of stiffness properties of FDM parts as a function of raster orientation. In Proceedings of the 2017 International Solid Freeform Fabrication Symposium, Austin, TX, USA, 7–9 August 2017.
13. Li, L.; Sun, Q.; Bellehumeur, C.; Gu, P. Composite modeling and analysis for fabrication of FDM prototypes with locally controlled properties. *J. Manuf. Process.* **2002**, *4*, 129–141. [CrossRef]
14. Garg, A.; Bhattacharya, A. An insight to the failure of FDM parts under tensile loading: Finite element analysis and experimental study. *Int. J. Mech. Sci.* **2017**, *120*, 225–236. [CrossRef]
15. Mohamed, O.A.; Masood, S.H.; Bhowmik, J.L. Optimization of fused deposition modeling process parameters: A review of current research and future prospects. *Adv. Manuf.* **2015**, *3*, 42–53. [CrossRef]
16. Bakır, A.A.; Atik, R.; Özerinç, S. Mechanical properties of thermoplastic parts produced by fused deposition modeling: A review. *Rapid Prototyp. J.* **2021**, *27*, 537–561. [CrossRef]
17. Teraiya, S.; Vyavahare, S.; Kumar, S. Experimental investigation on influence of process parameters on mechanical properties of PETG parts made by fused deposition modelling. In *Advances in Manufacturing Processes*; Springer: Singapore, 2021; pp. 283–293.
18. Khan, I.; Kumar, N. Fused deposition modelling process parameters influence on the mechanical properties of ABS: A review. *Mater. Today Proc.* **2021**, *44*, 4004–4008. [CrossRef]
19. Sood, A.K.; Ohdar, R.K.; Mahapatra, S.S. Parametric appraisal of mechanical property of fused deposition modelling processed parts. *Mater. Des.* **2010**, *31*, 287–295. [CrossRef]
20. Özen, A.; Auhl, D.; Völlmecke, C.; Kiendl, J.; Abali, B.E. Optimization of manufacturing parameters and tensile specimen geometry for fused deposition modeling (FDM) 3D-printed PETG. *Materials* **2021**, *14*, 2556. [CrossRef]
21. Kiendl, J.; Gao, C. Controlling toughness and strength of FDM 3D-printed PLA components through the raster layup. *Compos. Part B Eng.* **2020**, *180*, 107562. [CrossRef]
22. Shojib Hossain, M.; Espalin, D.; Ramos, J.; Perez, M.; Wicker, R. Improved mechanical properties of fused deposition modeling-manufactured parts through build parameter modifications. *J. Manuf. Sci. Eng.* **2014**, *136*, 061002. [CrossRef]
23. Croccolo, D.; De Agostinis, M.; Olmi, G. Experimental characterization and analytical modelling of the mechanical behaviour of fused deposition processed parts made of ABS-M30. *Comput. Mater. Sci.* **2013**, *79*, 506–518. [CrossRef]
24. Gurrala, P.K.; Regalla, S.P. Part strength evolution with bonding between filaments in fused deposition modelling: This paper studies how coalescence of filaments contributes to the strength of final FDM part. *Virtual Phys. Prototyp.* **2014**, *9*, 141–149. [CrossRef]
25. Sun, Q.; Rizvi, G.M.; Bellehumeur, C.T.; Gu, P. Effect of processing conditions on the bonding quality of FDM polymer filaments. *Rapid Prototyp. J.* **2008**, *14*, 72–80. [CrossRef]
26. Domingo-Espin, M.; Puigoriol-Forcada, J.M.; Garcia-Granada, A.A.; Llumà, J.; Borros, S.; Reyes, G. Mechanical property characterization and simulation of fused deposition modeling Polycarbonate parts. *Mater. Des.* **2015**, *83*, 670–677. [CrossRef]
27. Paul, S. Finite element analysis in fused deposition modeling research: A literature review. *Measurement* **2021**, *178*, 109320. [CrossRef]
28. Górski, F.; Kuczko, W.; Wichniarek, R.; Hamrol, A. Computation of mechanical properties of parts manufactured by fused deposition modeling using finite element method. In Proceedings of the 10th International Conference on Soft Computing Models in Industrial and Environmental Applications, Burgos, Spain, 15–17 June 2015; pp. 403–413.

Article

Adaptive Mechanism for Designing a Personalized Cranial Implant and Its 3D Printing Using PEEK

Syed Hammad Mian [1,*], Khaja Moiduddin [1], Sherif Mohammed Elseufy [1] and Hisham Alkhalefah [1]

Department of Industrial Engineering, College of Engineering, King Saud University, P.O. Box 800, Riyadh 11421, Saudi Arabia; khussain1@ksu.edu.sa (K.M.); sseufy@ksu.edu.sa (S.M.E.); halkhalefah@ksu.edu.sa (H.A.)
* Correspondence: smien@ksu.edu.sa

Abstract: The rehabilitation of the skull's bones is a difficult process that poses a challenge to the surgical team. Due to the range of design methods and the availability of materials, the main concerns are the implant design and material selection. Mirror-image reconstruction is one of the widely used implant reconstruction techniques, but it is not a feasible option in asymmetrical regions. The ideal design approach and material should result in an implant outcome that is compact, easy to fit, resilient, and provides the perfect aesthetic and functional outcomes irrespective of the location. The design technique for the making of the personalized implant must be easy to use and independent of the defect's position on the skull. As a result, this article proposes a hybrid system that incorporates computer tomography acquisition, an adaptive design (or modeling) scheme, computational analysis, and accuracy assessment. The newly developed hybrid approach aims to obtain ideal cranial implants that are unique to each patient and defect. Polyetheretherketone (PEEK) is chosen to fabricate the implant because it is a viable alternative to titanium implants for personalized implants, and because it is simpler to use, lighter, and sturdy enough to shield the brain. The aesthetic result or the fitting accuracy is adequate, with a maximum deviation of 0.59 mm in the outside direction. The results of the biomechanical analysis demonstrate that the maximum Von Mises stress (8.15 MPa), Von Mises strain (0.002), and deformation (0.18 mm) are all extremely low, and the factor of safety is reasonably high, highlighting the implant's load resistance potential and safety under high loading. Moreover, the time it takes to develop an implant model for any cranial defect using the proposed modeling scheme is very fast, at around one hour. This study illustrates that the utilized 3D reconstruction method and PEEK material would minimize time-consuming alterations while also improving the implant's fit, stability, and strength.

Keywords: 3D reconstruction; customized implant; PEEK; accuracy evaluation; finite element analysis; 3D printing

Citation: Mian, S.H.; Moiduddin, K.; Elseufy, S.M.; Alkhalefah, H. Adaptive Mechanism for Designing a Personalized Cranial Implant and Its 3D Printing Using PEEK. *Polymers* **2022**, *14*, 1266. https://doi.org/10.3390/polym14061266

Academic Editor: Mohammadali Shirinbayan

Received: 16 February 2022
Accepted: 16 March 2022
Published: 21 March 2022

Publisher's Note: MDPI stays neutral with regard to jurisdictional claims in published maps and institutional affiliations.

Copyright: © 2022 by the authors. Licensee MDPI, Basel, Switzerland. This article is an open access article distributed under the terms and conditions of the Creative Commons Attribution (CC BY) license (https://creativecommons.org/licenses/by/4.0/).

1. Introduction

Skull or cranial deformities are becoming more common as a result of increased traffic accidents, tumors, and disasters, thus increasing the demand for skull reconstruction [1]. Cranial defect reconstruction is one of the most arduous tasks confronted by surgeons because of the unique shape of the skull. It is one of the most challenging surgical procedures owing to the complication of the skull's form and the differences in the physiology of the skull. The very first phase in addressing cranial problems is to detach abnormally connected skull bones and replace them with accurate implants in order to restore their functionality [2]. According to Park et al. [3], customized implants for skull skeletal augmentation should be fitted properly to the cranial deficiency, reducing the gap between the implant and the bone, as well as matching the tapered borders with the neighboring bone boundaries. Personalized implants, when properly designed and manufactured, can significantly shorten the operation time, while also improving the correctness of the implant's shape and geometry in relation to the patient's anatomy. The fundamental idea behind these design strategies is to employ a cranium template to create an implant with an exterior shape that will adapt well to the skull while also distributing mechanical force effectively in the situation of an accident. Two primary design approaches that are considered for cranial defect reconstruction (depending on the intricacy of the cranial injury) are discussed below.

- Mirrored reconstruction design technique: This is based on the symmetrical nature of bones, and is more suitable for the treatment of unilateral skull damage [4–7]. The unhealthy or tumorous region is removed and replaced by the healthy opposite side in this technique. The precise execution of the mirroring approach is not simple, as it necessitates many manual operations, including the determination of the symmetry plane, the segregation of the healthier portion relating to the lesion, and the tweaking of the generated patch on the deficient region of the skull [8]. One of the major drawbacks of mirror-image reconstruction technique is that it cannot be used in asymmetrical body regions. If there is tumor in the center of the skull, the mirror-image design technique cannot provide the healthy opposite side.
- Anatomical reconstruction design approach: This is a curve-based surface manipulation and modification method in which two end curves of the resected regions are chosen and a guiding curve is employed for anatomical bone regeneration [9]. The fundamental benefit of the anatomical reconstruction technique is that it could be employed in both symmetrical and asymmetrical regions; nevertheless, in highly contoured parts, it requires technical competence. According to Moiduddin et al. [10], in mandible reconstruction, the anatomical reconstruction technique is more efficacious than the mirroring technique, resulting in less variation from the reference bone.

The introduction of Computer-Aided Design (CAD) and Computer-Aided Manufacturing (CAM) technologies supplanted the preceding processes that relied on manual shaping [11] and casting [12,13], allowing novel materials to be used in implant manufacturing, resulting in superior quality and enhanced postoperative results. Until recently, autogenous bone transplantation has been the most popular method of fixing skull abnormalities because it has fewer problems such as infection, aggressive foreign body reaction, soft tissue, and skin damage, etc. [14]. However, the utilization of autologous bone repair in the case of large and complicated lesions is constrained due to the scarcity of donors. As a result, there is a drive for implants made of alternative materials. Other developments in implant industries are being leveraged by recent advances in CAD systems and three-dimensional (3D) printing technologies. Patient-specific implants can now be 3D printed from a range of polymer, ceramic, or metal components [15].

Three-dimensional printing has revolutionized the manufacturing business—particularly medical, aerospace, and automobile manufacturing, construction, and so on—with its groundbreaking material deposition technique [16,17]. The possibilities of 3D printing in the healthcare industry have substantially improved the ability to build components with complex geometries using medical imaging data, which would be unachievable through traditional approaches [18]. Furthermore, 3D printing reduces the patient's surgery time and discomfort by avoiding surgery revisions. However, there are some limits when it comes to the employment of 3D printing to fabricate bespoke implants. For instance, it is currently unclear which material produces the greatest results in cranioplasty [19]. The neurosurgeon's choice of materials has frequently been influenced by availability, regulation, expense, and experience [15,20]. A variety of biomaterials have been employed in the fabrication of cranioplasty implants. For instance, polymethyl methacrylate (PMMA) was commonly employed because of its biocompatibility and inexpensive cost; however, it created heat during polymerization and did not chemically bind to adjacent tissue [21]. Calcium phosphate and hydroxyapatite are osteoconductive, and facilitate osteointegration, but their brittleness made them prone to breaking [22–24]. The most frequent biomaterials for implant reconstructions have been titanium (Ti) alloys. They offer numerous advantages, such as superior biocompatibility, great mechanical qualities, osseointegration abilities, high corrosion and wear resistance, and so on [25,26]. However, Ti and its alloys have certain disadvantages, including the possibility of metal ion release and subsequent osteolysis, metal corrosion, and inadequate compliance with contemporary imaging technologies [27]. Ti has a substantially higher modulus of elasticity (110 GPa) when compared to bone (14 GPa) (see Table 1). This significant disparity between the two frequently leads to implant malfunction due to stress shielding, bone resorption, and implant rupture [28,29]. When metallic implants are exposed to irradiation, they emit scattering rays that are detrimental to tissues. Economic reasons connected to manufacturing procedures (especially Electron Beam Melting, EBM, which is the most prevalent process for the production of Ti implants) have led researchers to hunt for new implant materials [30]. Researchers have switched to the examination of potential Ti substitutes in order to address the above-mentioned restrictions and reduce biological issues after implant insertion.

Table 1. Properties of various materials [31–40].

Material	Elastic Modulus (GPA)	Density (g/cm^3)	Thermal Conductivity (W/m.K)	Biocompatible
Titanium alloy (Ti6Al4V)	110	4.5	7.1	Yes
Cobalt-Chromium	180–210	10	9.4	Yes
Zirconia	210	5.68	1.7–2.7	Yes
Porcelain	68.9	2.3–2.4	1.5	Yes
PMMA	3–5	1.18	0.167–0.25	Yes
PEEK	3–4	1.3–1.32	0.25–0.93	Yes
CFR-PEEK	18	1.42	0.95	Yes
GFR-PEEK	12	1.55	0.35	Yes
Cortical Bone	14	1.6–2	0.68	
Cancellous bone	1.34	0.05–0.3	0.42	
Enamel	40–83	2.6–3	0.45–0.93	
Dentin	15–30	1.79–2.12	0.11–0.96	

An elevated thermoplastic polymer called polyetheretherketone (PEEK)—which is a partly crystalline, poly-aromatic, linear, thermoplastic material [41,42]—is by far the most plausible innovative substitute to Ti. It is an organic synthetic polymeric material with higher temperature stability above 300 °C, high mechanical strength, adaptable mass production, production possibility utilizing plastic techniques, natural radiolucency and MRI compatibility, nontoxicity, excellent chemical and sterilization tolerance, and a pure and quantifiable supply channel [41,43,44]. The relative density of PEEK is 1.3 to 1.5. The water absorption for PEEK after 24 h is 0.06 to 0.3%, whereas its water absorption at saturation was found to be 0.22 to 0.5% [45]. PEEK, unlike metals and alloys, offers great strength with a low Young's modulus (3.6 GPa in its pure form, and 12 GPA in glass fiber-reinforced PEEK (GFR-PEEK)), which is nearer to that of human bone than Ti [41,44,46]. This characteristic may reduce stress by dispersing it in a most healthy manner, promoting bone growth and minimizing osteolysis surrounding the implant. The PEEK composite material has indeed been utilized extensively in the disciplines of orthopedics [47–50], dentistry [51–54], and other domains as a proven implant alternative in load-bearing body parts. However, there is a scarcity of information for PEEK as a cranial implant material, with only a few pieces of research emphasizing its application. Cranioplasty is a complicated procedure that may result in a number of serious side effects, prompting the need for revision surgery. Apparently, as mentioned in [55–57], the price of tailored PEEK implants is regarded to be excessively costly. Practitioners must be aware that the material is costly, but the cost of the material does not account for a significant amount of the total cost. For example, PEEK can be sterilized frequently and at a low cost. Other materials are less expensive, but sterilizing is exorbitant [58]. As a result, PEEK performs extremely fairly and provides favorable value for cranial implants. Likewise, the higher the rate of failure, the more changes and reoperations are required. Additional revisions imply greater expenses and much more mental anguish for the patient. Thien et al. [59] observed 12.5% cranioplasty failures with PEEK compared to 25% with Ti, indicating that PEEK is a more cost-effective option for cranial implants.

The idea of customized implant design employing medical modeling software and fabrication using PEEK must be adopted in order to perfectly accommodate bone shapes and give a better aesthetic outcome. The purpose of this study is to investigate the fabrication of a cranial implant made of PEEK, as well as its mechanical characteristics and accuracy. The cranial implant is designed using a new procedure based on anatomical reconstruction. The method is discussed in detail, from computer tomography (CT) scans through to the final design and subsequent material extrusion-based 3D printing. This research also evaluates the performance of a 3D-printed PEEK implant in terms of strength using Finite Element Analysis (FEA), and in terms of accuracy through 3D Comparison analysis.

2. Proposed Methodology

Customized 3D cranial implants are essential, and they employ contemporary production methods, particularly 3D printing or additive manufacturing (AM). These approaches should incorporate the use of multiple design and analysis tools at the same time. As a result, this research develops a hybrid framework (Figure 1) that combines a CT system (as a reverse engineering or data acquisition mechanism), a design (or modeling) scheme, a computational analysis system, a fabrication process, and an accuracy assessment tool. The aim of the use of the newly established hybrid approach is to obtain an impeccable customized cranial implant irrespective of the tumor location. The designed model of the cranial implant is always well-suited to the boundary of the missing fragment in the skull, which is a unique attribute of this procedure.

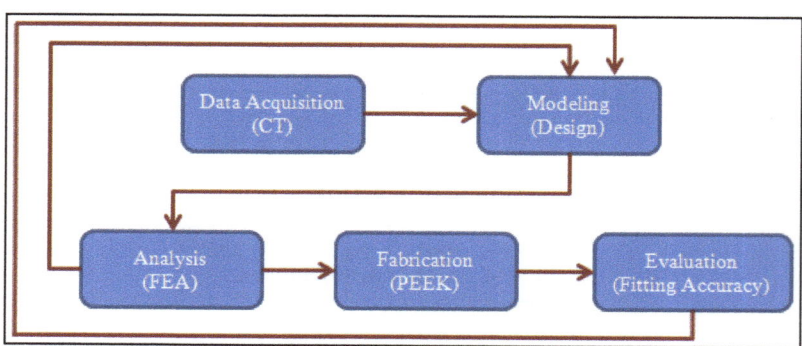

Figure 1. Deriving a cranial implant using a hybrid system.

2.1. Design and Modeling

The input component to the modeling process of the skull implant is the two-dimensional (2D) images obtained from the CT scan, as illustrated in Figure 2a. The 2D images are processed using medical modeling software, such as Mimics® (Materialise, Leuven, Belgium), which converts them into a 3D structure. The hard and soft tissues are differentiated using a grayscale metric, while region growing segmentation is used to exclude unnecessary data and label the 3D model into distinct areas. Following segmentation, the region of interest is extracted and saved as a Standard Tessellation Language (STL) file for the implant's design (Figure 2b). An experimental segmental defect (Figure 2c) is marked on a healthy skull, which acts as a template for implant design, as shown in Figure 2d. The skull template with a resected tumor is crafted to evade the inconvenience of obtaining patient and ethical committee approvals. The ultimate purpose is to communicate how to model the implant design precisely, and to replicate the characteristics of a complex injury, tumor or wound on the skull regardless of the location. Furthermore, using this approach, the healthy STL skull model can be used as a reference in order to compare the implanted skull's fitting accuracy, which is explained in the following sections.

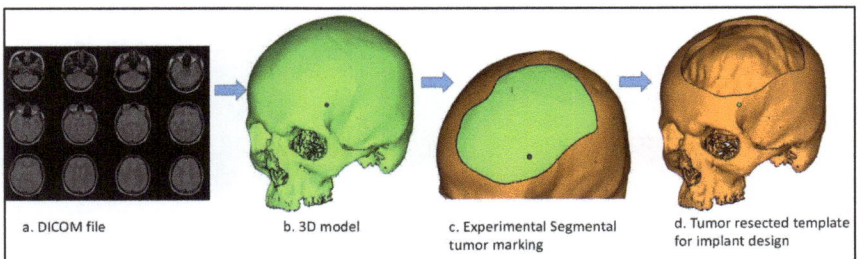

Figure 2. Steps involved in the reconstruction of a tumor-resected implant template. (**a**) DICOM file. (**b**) 3D model. (**c**) Experimental Segmental tumor making. (**d**) Tumor resected template for implant design.

The next task is to design a custom cranial implant model based on the asymmetrical skull regional defect using the skull template (Figure 3a). The application of interpolation spline curves is the foundation for the construction of an implant model that is adapted to a skull bone with an incomplete segment [60]. The entire modeling process is accomplished by combining different modules such as the Digitized Shape Editor, Generative Shape Design, and Quick Surface Reconstruction in Catia V5. The approach proposed is universal, and can be used regardless of the location where the skull bone loss occurs.

The distorted region (Figure 3b) is identified for further processing using the Activate tool in the Digitized Shape Editor during the first stage of the modeling process. This stage can be skipped; however, working on a relatively smaller volume is less time-consuming and easier. As shown in Figure 3c, spline curves tangent to the surface of the skull model are then defined. The Planar Sections option is used for this reason. It computes curves by cutting a cloud of points or a mesh using planes. Because the curves are disconnected due to the resected skull hole fragment, the shape of the curves is interpolated between the existing neighboring nodes, culminating in a spline curve that passes seamlessly through all of the appropriate points (Figure 3d). For this, the Connect Curve option (in Generative Shape Design) is used, which produces a connecting curve between two curves. As a consequence, the obtained spline curves (Figure 3e) are utilized to facilitate surface patching. The surface generation process is carried out in the CATIA V5 system's Quick Surface Reconstruction environment. The multi-section surface mechanism is chosen to build the surface patch along the curvature of the spline curves. The established surface patch is much larger than the resected skull tumor region, as depicted in Figure 3f. The redundant (or excess) surface is split into the shape of the opening on the skull by Boolean operation in order to achieve the desired form of the implant surface (Figure 3g). The transformation of the implant surface model into the part model, as shown in Figure 3h, is the final stage of the process. As can be seen in Figure 3i, the implant is precisely positioned on the defect region of the skull.

2.2. Computational Model and Analysis

The 3D model is corrected for small geometrical errors prior to the FEA study, such as the deletion of intersecting and overlapping triangles on both the implant and skull portion. The FEA model was chosen to examine the biomechanical stability of the custom cranial implant that has been developed [1]. The purpose is to objectively assess and simulate the behavior of the designed PEEK implant in actual working scenarios. The FEA can forecast the possibility of cracks, vulnerable spots, and malfunctions of the point of implant–bone contact in a digital setting. The ANSYS (Version 19.1, Canonsburg, Pennsylvania, United States) and Hypermesh program (Version 14.0, Altaire Hyper works, Troy, Michigan, United States) are used to perform the pre-processing, post-processing, and execution of the built FEA model. The computational model constructed consists of three elements: the Skull, the PEEK implant and the fixation screws. Table 2 lists the properties of the materials assigned to the FEA model. Different regions of the skull are given different material properties [61]. The cranium is given cortical bone characteristics, while the personalized cranial implant is assigned PEEK properties [62]. Six screws of titanium are fastened at selected reference points.

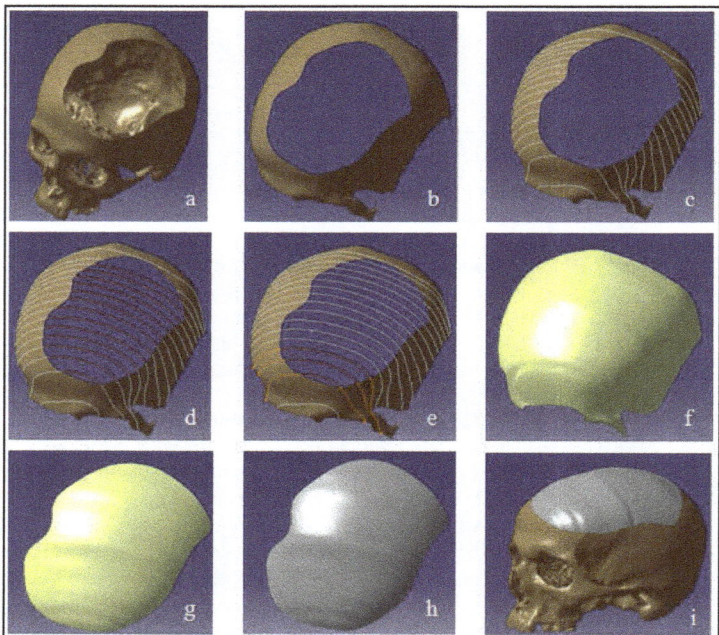

Figure 3. Modeling of the customized implant (**a**) asymmetrical skull defect; (**b**) identification of distorted region; (**c**) recognition of curves tangent to surface; (**d**) interpolation of curves; (**e**) surface generation; (**f**) surface acquired through curves; (**g**) splitting appropriate surface patch; (**h**) conversion of surface to part model; (**i**) implant placed on the skull.

Table 2. Material properties used in the computational model [61,62].

Materials	Youngs Modulus (MPa)	Poisson's Ratio	Yield Strength (MPa)
Cortical bone	13,700	0.3	122
PEEK implant	3738	0.4	99.9
Titanium screws	120,000	0.3	930

As illustrated in Figure 4a, the implant is fixed onto the skull model using six screws, whereby the contact between the plate–bone and the bone-screws is defined as bonded. The Hypermesh is applied (Figure 4b) in the establishment of the finite element (FE) model of the skull and the implant using the solid element of the tetra4 type, where the dimensions of the mesh range in size between 3 and 0.25 mm. A finer mesh is created to minimize mesh distortion and to optimize element quality, making a total of 828,683 elements and

213,805 nodes for the implant and bone portion. The mesh is then imported into ANSYS to provide the boundary and loading conditions. Figure 4c illustrates the cross-sectional view of the interface between the implant, the screw and the skull bone. A set of three circular patches of X, Y and Z are considered on the implant for loading (Figure 4d). The base of the skull is held constant for all of the configuration by fixing it at the bottom, and a static force of 50 N is exerted (Figure 4e) at each of the three regions of the implant over a small area of 225 mm^2 [63,64]. In cranial reconstruction, the implant–bone assembly fixture and its mechanical robustness is of extreme importance for the long-term clinical success. Several others research studies have demonstrated the FE model of skull implants with respect to different shapes, geometry, fixation devices and materials [64,65]. In this study, a computational model of a PEEK implant with a static load of 50 N is evaluated for the biomechanical responses of stress, strain, deformation, and for the factor of safety. The factor of safety is also a good measure for the acceptance of a design on the basis of its withstanding capacity for the intended load.

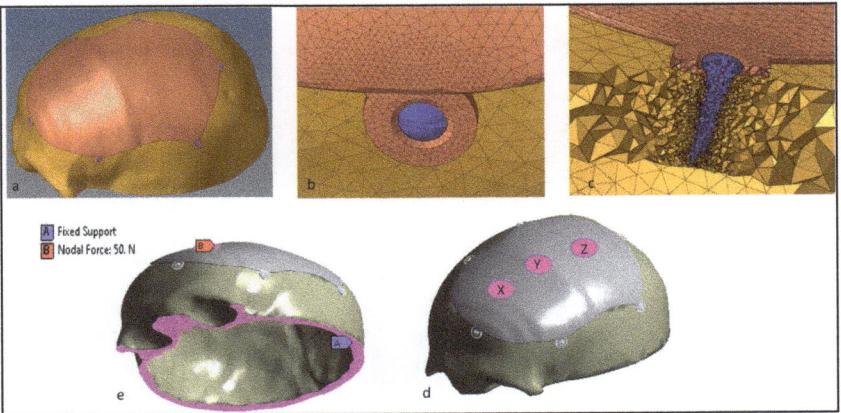

Figure 4. (**a**) Cranial PEEK implant on the skull model; (**b**) mesh generation; (**c**) cross-sectional view of the screw, implant and bone; (**d**) exertion of forces at the three implant regions; (**e**) representation of the fixed supports and the nodal force.

2.3. Fabrication of the Skull Model and Cranial Implant

The fabrication of the cranial implant is initiated by the correction of errors in the STL file. Magics® is employed to correct defects such as overlapping, intersecting triangles, poor corners, and other flaws. The processing of the STL file and the specification of the optimal location and orientation for fabrication are the goals of this step. Once the STL model is perfect, appropriate supports are created on the overhanging structures. In order to construct the prototype precisely and without any abnormality, the proper supports are needed. The support structures are also needed for efficient heat transfer, in order to prevent any distortion and to make it easier to fabricate overhanging sections. The slicing, orientation and support generation on the PEEK 3D model are performed using INTAMSUITE slicing software (Version 3.6.2, INTAMSYS Technology Co. Ltd, Shanghai, China) as illustrated in Figure 5. There are several build-plate adhesion types in the slicing software. In this model, a raft build-plate is utilized because it provides the added thick grid between the model and the build plate, and avoids warping effects, thus ensuring that the model better sticks to the build plate.

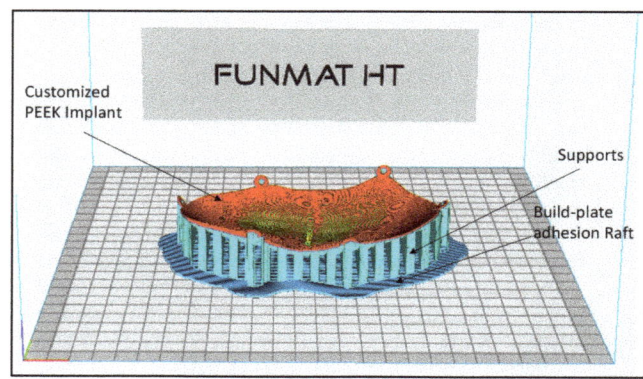

Figure 5. INTAMSUITE software for the orientation and support generation of the PEEK 3D model.

An INTAMSYS (Intelligent Additive Manufacturing Systems) FUNMAT HT 3D printer (INTAMSYS Technology Co. Ltd, Shanghai, China) is used for the manufacturing of the personalized PEEK cranial implant. The printer uses a fused filament fabrication (FFF) technique, which has the advantage of consuming less energy, producing more, and delivering higher tensile strength when the printing parameters are set appropriately [66,67]. FFF is amongst the most widespread and straightforward AM procedures [68]. The fabrication of 3D-printed objects is controlled by many parameters in this technology. As a result, the selection of the appropriate parameters for the fabrication of any component is essential. The process parameters used for the fabrication of the PEEK cranial implant are provided in Table 3. The printer is equipped with both the heated build plate (reaching up to 160 °C) and a build chamber (reaching up to 90 °C). The high-temperature extruder (450 °C) is used in the extrusion of high-temperature PEEK material. The infill parameter in the slicing software is set to 100% in order to provide a completely solid structure. The post-processing of the PEEK cranial implant includes the removal and pulling away of the support structures. The wall of the support structures is made less dense for easy removal. Cutting and gripping pliers are used with protective gloves to access the underneath of the supports, and to carefully bend them upwards to remove them. The time taken to remove the supports is approximately 30 to 40 min. Figure 6a, b illustrates the 3D-printed customized PEEK implant with supports (Figure 6c), where the supports are easily removed manually using plyers (Figure 6d) to obtain the final PEEK implant (Figure 6e).

Table 3. Process parameters in the fabrication of the PEEK implant using the FUNMAT HT printer.

Description	Parameters
Printing Technology	FFF
Extruder	Single
Extruder diameter (mm)	0.4
Layer thickness ((mm)	0.15
Print speed (mm/s)	50
Filament diameter (mm)	1.75
Build adhesion type	Raft
Nozzle temperature (°C)	420
Build temperature (°C)	130
Chamber temperature (°C)	90
Nozzle used	High-temperature nozzle set

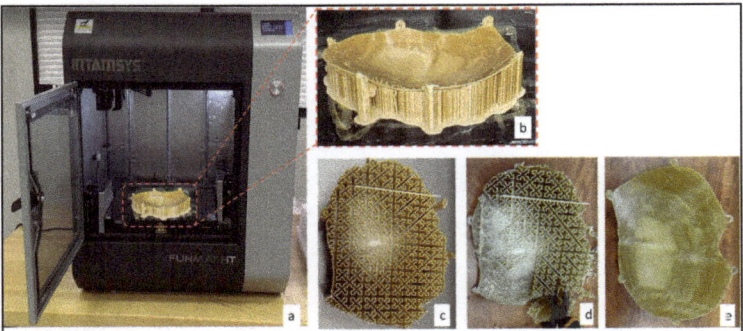

Figure 6. (**a**) Intamsys 3D Printer used in the fabrication of the customized (**b**) PEEK cranial Implant; (**c**) Support structures bottom view; (**d**) removal of the supports using pliers; (**e**) the PEEK cranial implant after the supports' removal.

Similarly, skull model is fabricated (Figure 7a) by employing the same INTAMSYS 3D Printer using acrylonitrile butadiene styrene (ABS) material for the purposes of evaluation and testing. The ABS skull model after the removal of supports (Figure 7b) and the final PEEK implant are tested for custom fitting and rehearsal evaluation, as shown in Figure 7c.

Figure 7. (**a**) 3D printed skull model of the ABS material with supports, (**b**) ABS skull model after the removal of the supports, and (**c**) the rehearsal and fitting evaluation of the PEEK cranial implant on the ABS skull model.

2.4. Accuracy Assessment

The Intamsys PEEK implant with a polymer skull structure is used for the planning and fitting evaluation. According to Wyleżoł et al. [60], an overall inspection of the implant shape should be performed by professionals prior to the fabrication and implanting. As a result, the implant's placement on the skull and cosmetic effects are evaluated using a visual analogue score (VAS) of 1 to 5 (1, bad; 2, average; 3, satisfactory; 4, fine; 5, excellent) [69]. Five health professionals and five research experts from the surgical unit are provided with the implanted skull. A PEEK implant model is produced for the specified design and delivered to the advisors for approval. Every analyst is encouraged to grade the implant independently by looking at cranial uniformity, connectivity, and visual attractiveness. The model assembly is then evaluated using a visual score ranging from 1 to 5 by all of the reviewers. The average aesthetic score is then calculated. In this study, the null hypothesis and alternative hypothesis are investigated for statistical analysis [70,71]. The null hypothesis asserts that the median aesthetic score is less than or equal to 3 (H_0: $\mu \leq 3$), whereas the alternative hypothesis asserts that it is larger than 3 (H_a: $\mu > 3$) [18]. Finally, the implant is validated using the FARO if the null hypothesis is rejected. Otherwise, the implant is redesigned and produced again if the null hypothesis is proven, as illustrated in Figure 8.

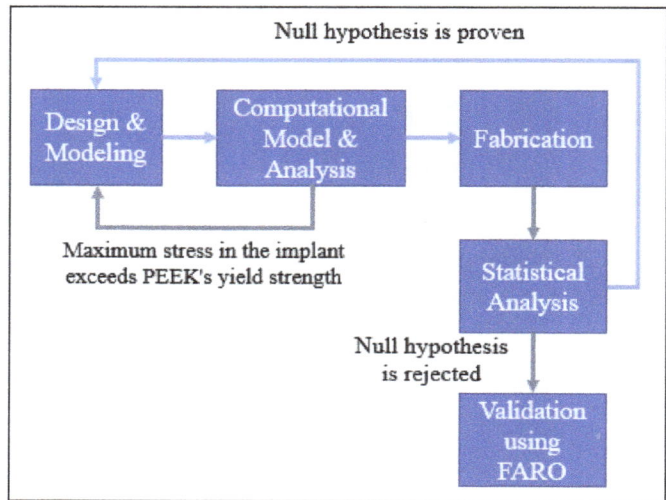

Figure 8. Scenarios for redesigning the implant.

The fitting precision of the PEEK implant is also quantified in order to ascertain deviation from the reference (or actual) skull shape. It is estimated utilizing a 3D comparison analysis in Geomagics Control® (Version 2014, 3D Systems Inc., Cary, NC, USA) [72]. It is considered to be amongst the most comprehensive and structured approaches for the interactive assessment of surface variations between the real object and the reference CAD model [73]. The test data is acquired after the custom-made cranial implant is built and fixed on the skull. The scanning is gathered with the laser scanner on the Faro Platinum arm (FARO, Lake Mary, FL, USA), as illustrated in Figure 9a. The captured point cloud data (shown in Figure 9b) is exploited as test data in the 3D comparison assessment.

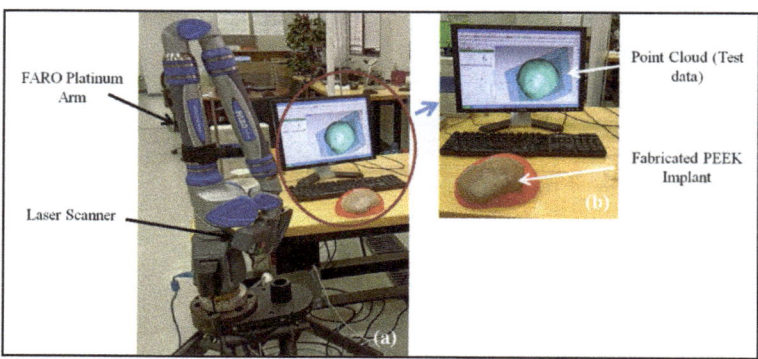

Figure 9. (**a**) Scanning system; (**b**) acquired point cloud data.

The initial stage in the 3D comparison analysis is to classify the test object and the reference CAD model. The outside surface of the reconstructed skull (or the skull with the implant) is digitized and imported as an STL model into Geomagics Control®. The outer surface (test data) of the skull is inspected because the personalized cranial implant is built depending on the outside contour of the skull. The CAD model of a healthy (or the original) skull is employed as a reference. The quality of the reconstructed skull is then examined using a 3D comparison analysis. The test data (collected point cloud) is aligned with the reference CAD model using the best fit alignment technique. The best-fit alignment approach is used to ensure that the test and reference entities are in the same coordinate system. The average deviation in the positive direction is used to assess the implant's accuracy. The average deviation statistic is chosen because it indicates the average deviation in outward direction, thus approximating the gap between the remodeled skull (or the tailored implant) and the original skull.

As shown in Figure 10, the accuracy evaluation is divided into two phases. The virtual model of the designed implant is compared to the original skull in phase one. This determines the amount of error (or accuracy) caused by the modeling method used. The fabricated PEEK implant is examined in phase 2 in relation to the virtual model of the designed implant. This assists in the quantification of the fabrication error. The cumulative error can be derived by adding the errors from Phases 1 and 2, as depicted in Figure 11. This also provides the implant's overall fitting accuracy.

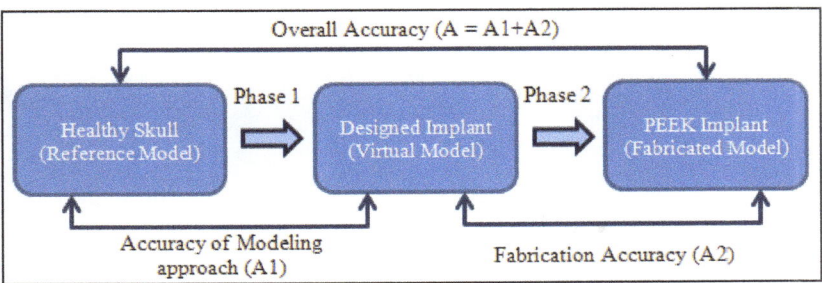

Figure 10. Procedure to estimate the fitting accuracy of the implant.

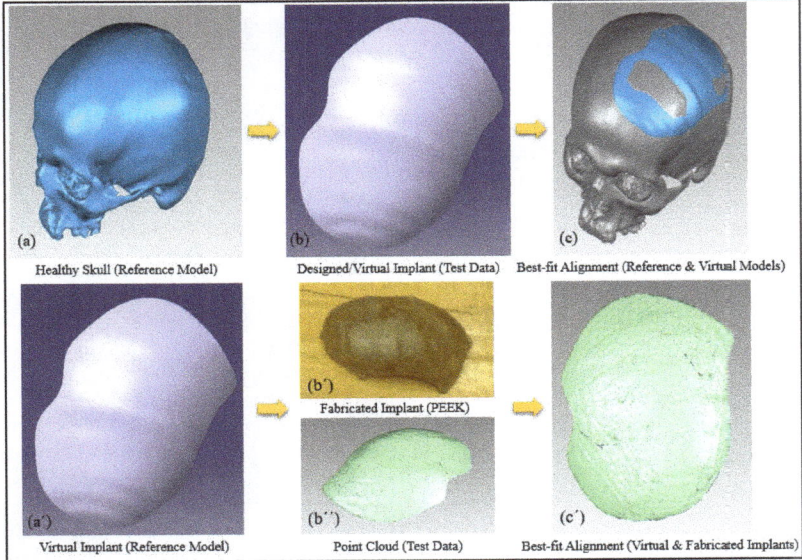

Figure 11. Computation of implant's overall fitting accuracy. (a–c) Phase 1: Accuracy of the modeling approach; (a′,b′,b″,c′) Phase 2: Fabrication accuracy.

3. Results and Discussions

The results of the computational model are illustrated in Figure 12, displaying the Von Mises stress, Von Mises strain, total deformation, and safety factor at three points: X, Y and Z. The ANSYS FE solver with a static framework was chosen. The time taken to complete the computational model was approximately 450 s using the HP Z800 workstation. The results of the FEA are also used to provide input for the modeling process. For example, if the maximum stress in the implant exceeds the PEEK's yield strength, the implant's thickness should be modified.

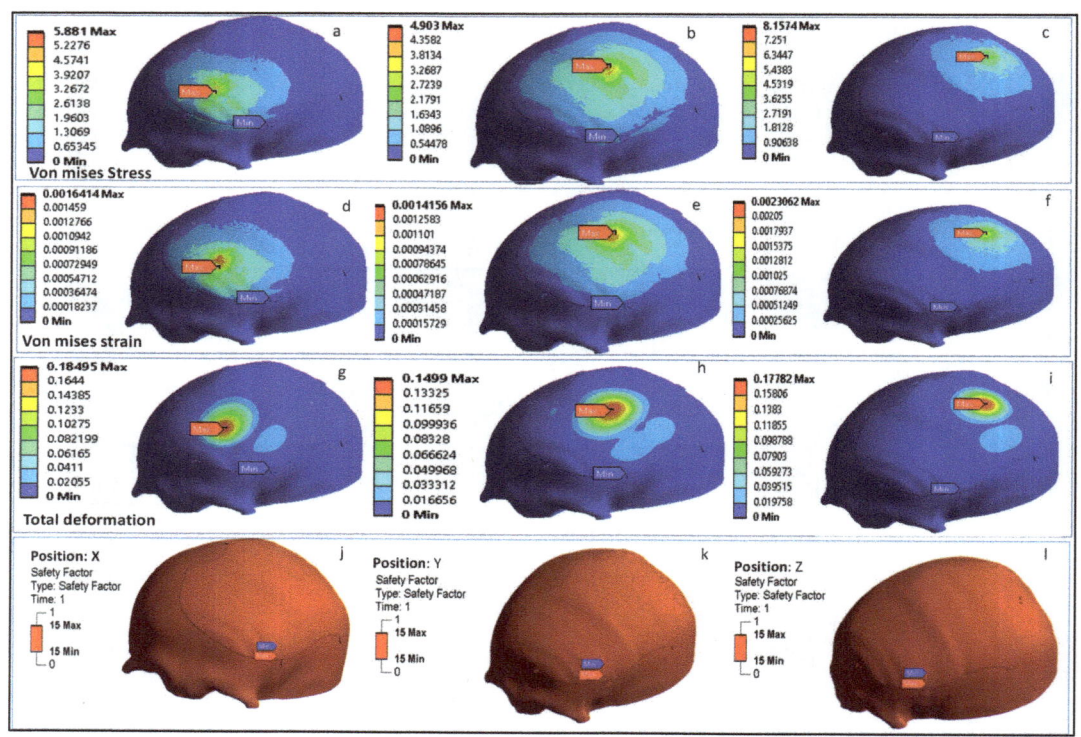

Figure 12. FEA results illustrating: (**a**–**c**) Von Mises stress at the point, (**d**–**f**) Von mises strain, (**g**–**i**) total deformation, and (**j**–**l**) the factor of safety at points X, Y and Z.

The analysis results showed the largest Von Mises stress of 8.15 MPa at the patch Z, and the lowest—4.90 MPa—at patch Y, which is well below the yield point (the failure criterion) of the material, which is 99.9 MPa. The maximum Von Mises strain was found to be 0.002 (2000 με) at patch Z, and lowest was found to be 0.0014 (1400 με) at patch Y, which is also well below the assumed mechanical strain limit of humans, i.e., 3000 με [74]. In addition, studies stated that strains of more than 3500 με lead to bone resorption, and strains over the limit of 4000 με lead to bone fracture [75,76]. The implant deformation—with a maximum of 0.18 mm at patch X and a minimum of 0.14 mm at patch Z—is also within the millimeter domain. The FEA results also indicate that the factor of safety for the PEEK implant is reasonably high, suggesting safe performance under high loading. The analysis results of our computational model represent favorable nominal values which are sufficient for optimality analysis and within the scope in comparison to other cranial studies [64]. In addition, the values obtained from the FEA results in this work match those of other research studies using similar kinds of PEEK cranial implant [77]. The computational model setup used in this study illustrates an effective and relatively simple way to evaluate cranial reconstruction while avoiding complex design setup and computational expenses.

In the implant fitting evaluation, the mean aesthetic score of 3.47 out of 5 ($n = 10$) was achieved, which signifies favorability in aesthetic achievement. With a *p*-value of less than 0.05, the result was statistically significant. The hypothesis testing was carried out using a one-sample *t*-test in Minitab Statistical Software (Minitab 21, MINITAB Ltd., United Kingdom). The null hypothesis was disproved because the *p*-value (=0.003) was less than the significance level (α) of 0.05, and it can be claimed that the average aesthetic score was larger than 3, showing expert contentment and assurance. This emphasizes the fact that the implant design allows for a good fit on the skull and a pleasing cosmetic appearance.

Figure 13a,b demonstrates the results of the 3D Comparison analysis for the healthy skull and virtual implant model, as well as the virtual implant model and the manufactured model. The modeling approach (A1) had an accuracy of 0.0687 mm, indicating that the anatomical design approach used is quite precise, and that its perfection may be improved with further expertise and experience. Furthermore, the fabrication accuracy (A2) was estimated to be 0.5232 mm, as evidenced by the outer-direction deviation. This indicates that the implant's overall accuracy (including both A1 and A2) is 0.5919 mm. It was rather more accurately fitted when compared to the Ti4Al6V implant, which had an overall deviation of 0.9529 mm, as reported in [18]. As a result, the repaired skull produced with the proposed design technique and PEEK material is acceptable as well as appealing. In addition to the deviation analysis, the gap analysis results are also provided in Figure 14. According to the gap analysis, the distance between the implant and the skull is reasonably low in both the X and Y directions. For instance, the average implant lengths in the X and Y directions are 97.11 and 124.97 mm, respectively. The average length of the cavity (formed after removing the flaw) is 98.02 and 125.69 mm, in the X- and Y-directions respectively. This means that there is a gap of 0.91 mm and 0.71 mm in the X- and Y-direction, respectively, which is less than 1 mm, thus implying a greater fit.

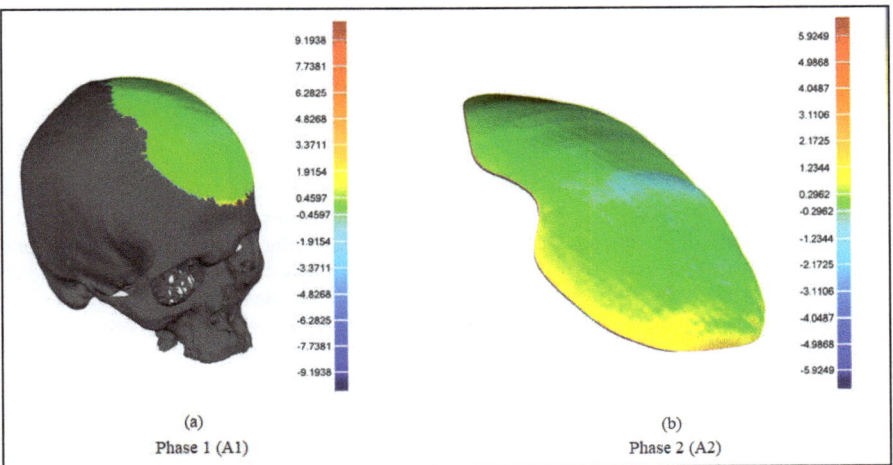

Figure 13. Deviation in the outward direction. (**a**) Healthy skull and the designed implant (virtual model); (**b**) designed implant (virtual model) and the fabricated implant.

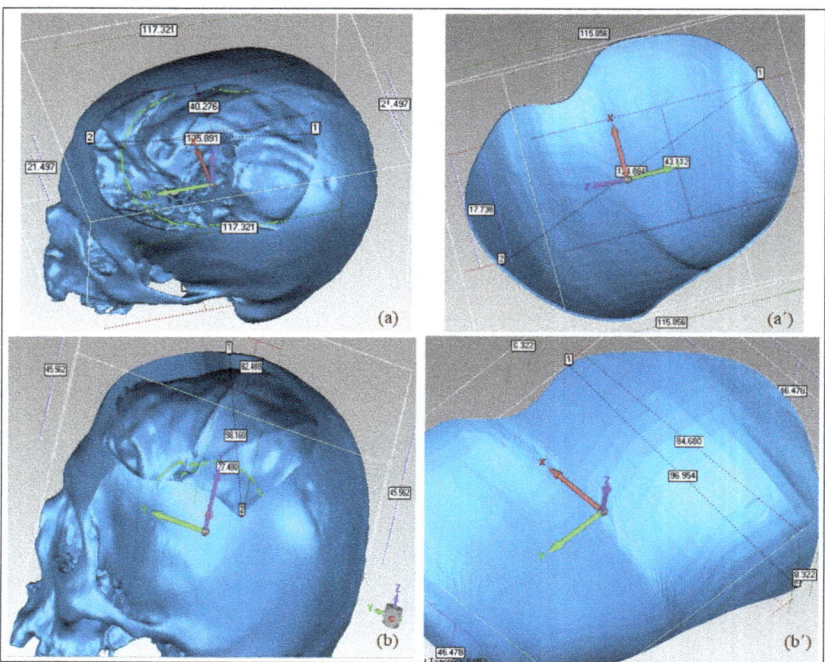

Figure 14. Gap Analysis between the implant and the skull. (**a**) Defect length in the Y-direction; (**a′**) implant length in the Y-direction; (**b**) defect width in the X-direction; (**b′**) implant width in the X-direction.

4. Conclusions

The surgical team faces a challenging task when it comes to the reconstruction of the skull's bones. The major aspects in cranial reconstruction are the implant design and the selection of a suitable material, owing to the variety of design methods and materials available. The optimum design strategy and material should result in a compact, easy-to-fit, robust implant that achieves the best cosmetic and functional results irrespective of the defective location. The customized implant design technique must be simple to implement and irrespective of the location of the lesion on the cranium.

The cranial implant—built using a revolutionary design technique based on anatomical reconstruction—delivered good precision, with a maximum deviation of 0.0687 mm. Similarly, the implant made of PEEK performed satisfactorily in terms of both gaps and deviations. The overall deviation of the PEEK implant was 0.5919 mm, with an average gap of less than 1 mm in both the X and Y axes, which is absolutely acceptable. Furthermore, the biomechanical analysis showed that the maximal Von Mises stress (8.15 MPa), Von Mises strain (0.002), and deformation (0.18 mm) were all remarkably low, emphasizing the implant's load-bearing capacity.

The adopted design technique can be used as a substitute for other existing design approaches because it is also quick and precise. PEEK material can also be used as an alternative to Ti implants because it has greater fitting accuracy and enhanced load bearing capabilities. The established design technique can also be applied to enhance the effectiveness of highly complex implant designs in both symmetrical and asymmetrical body regions. The PEEK material and the personalized implant design technique used in this study can also be applied in other areas of bone surgery where implants are used for rehabilitation. Additionally, more research is essential in order to confirm the biomechanical and biological viability of the FFF-based PEEK implants.

Author Contributions: Conceptualization, S.H.M. and K.M.; methodology, S.H.M. and K.M.; software, S.H.M., K.M. and S.M.E.; validation, S.H.M. and H.A.; formal analysis, S.H.M., K.M. and S.M.E.; investigation, H.A. and S.M.E.; resources, H.A.; data curation, S.H.M. and S.M.E.; writing—original draft preparation, S.H.M. and K.M.; writing—review and editing, S.H.M., K.M. and H.A.; visualization, S.H.M. and K.M.; supervision, H.A.; project administration, S.H.M. and H.A.; funding acquisition, H.A. All authors have read and agreed to the published version of the manuscript.

Funding: This research was funded through the Researchers Supporting Project number (RSP2022R499), King Saud University, Riyadh, Saudi Arabia.

Institutional Review Board Statement: Not Applicable.

Informed Consent Statement: Informed consent was obtained from the subject involved in the study.

Data Availability Statement: The data presented in this study are available in the article.

Acknowledgments: The authors extend their appreciation to King Saud University for funding this work through Researchers Supporting Project number (RSP2022R499), King Saud University, Riyadh, Saudi Arabia.

Conflicts of Interest: The authors declare no conflict of interest.

References

1. Ameen, W.; Al-Ahmari, A.; Mohammed, M.K.; Abdulhameed, O.; Umer, U.; Moiduddin, K. Design, Finite Element Analysis (FEA), and Fabrication of Custom Titanium Alloy Cranial Implant Using Electron Beam Melting Additive Manufacturing. *Adv. Prod. Eng. Manag.* **2018**, *13*, 267–278. [CrossRef]
2. Mandolini, M.; Brunzini, A.; Serrani, E.B.; Pagnoni, M.; Mazzoli, A.; Germani, M. Design of a Custom-Made Cranial Implant in Patients Suffering from Apert Syndrome. *Proc. Des. Soc. Int. Conf. Eng. Des.* **2019**, *1*, 709–718. [CrossRef]
3. Park, D.K.; Song, I.; Lee, J.H.; You, Y.J. Forehead Augmentation with a Methyl Methacrylate Onlay Implant Using an Injection-Molding Technique. *Arch. Plast Surg* **2013**, *40*, 597–602. [CrossRef]
4. Chen, X.; Xu, L.; Li, X.; Egger, J. Computer-Aided Implant Design for the Restoration of Cranial Defects. *Sci. Rep.* **2017**, *7*, 4199. [CrossRef] [PubMed]
5. Kung, W.-M.; Tzeng, I.-S.; Lin, M.-S. Three-Dimensional CAD in Skull Reconstruction: A Narrative Review with Focus on Cranioplasty and Its Potential Relevance to Brain Sciences. *Appl. Sci.* **2020**, *10*, 1847. [CrossRef]
6. Brunzini, A.; Mandolini, M.; Manieri, S.; Germani, M.; Mazzoli, A.; Pagnoni, M.; Iannetti, G.; Modugno, A. Orbital Wall Reconstruction by Selective Laser Sintered Mould. In Proceedings of the 2017 13th IASTED International Conference on Biomedical Engineering (BioMed), Innsbruck, Austria, 20–21 February 2017; pp. 260–264.
7. Igor, D.; Hren, N.; Strojnik, T.; Brajlih, T.; Valentan, B.; Pogačar, V.; Hartner, T. Applications of Rapid Prototyping in Cranio-Maxilofacial Surgery Procedures. *Int. J. Biol. Biomed. Eng.* **2008**, *1*, 29–38.
8. Buonamici, F.; Furferi, R.; Genitori, L.; Governi, L.; Marzola, A.; Mussa, F.; Volpe, Y. Reverse Engineering Techniques for Virtual Reconstruction of Defective Skulls: An Overview of Existing Approaches. *Comput. -Aided Des. Appl.* **2019**, *16*, 103–112. [CrossRef]
9. Harrysson, O.L.; Hosni, Y.A.; Nayfeh, J.F. Custom-Designed Orthopedic Implants Evaluated Using Finite Element Analysis of Patient-Specific Computed Tomography Data: Femoral-Component Case Study. *BMC Musculoskelet. Disord.* **2007**, *8*, 91. [CrossRef]
10. Moiduddin, K.; Al-Ahmari, A.; Nasr, E.S.A.; Mian, S.H.; Al Kindi, M. A Comparison Study on the Design of Mirror and Anatomy Reconstruction Technique in Maxillofacial Region. *Technol. Health Care* **2016**, *24*, 377–389. [CrossRef]
11. Chee Kai, C.; Siaw Meng, C.; Sin Ching, L.; Seng Teik, L.; Chit Aung, S. Facial Prosthetic Model Fabrication Using Rapid Prototyping Tools. *Integr. Mfg Syst.* **2000**, *11*, 42–53. [CrossRef]
12. Bhargava, D.; Bartlett, P.; Russell, J.; Liddington, M.; Tyagi, A.; Chumas, P. Construction of Titanium Cranioplasty Plate Using Craniectomy Bone Flap as Template. *Acta Neurochir* **2010**, *152*, 173–176. [CrossRef] [PubMed]
13. Dumbrigue, H.B.; Arcuri, M.R.; LaVelle, W.E.; Ceynar, K.J. Fabrication Procedure for Cranial Prostheses. *J. Prosthet. Dent.* **1998**, *79*, 229–231. [CrossRef]
14. Abdel-Haleem, A.K.; Nouby, R.; Taghian, M. The Use of the Rib Grafts in Head and Neck Reconstruction. *Egypt. J. Ear Nose Throat Allied Sci.* **2011**, *12*, 89–98. [CrossRef]
15. Bonda, D.J.; Manjila, S.; Selman, W.R.; Dean, D. The Recent Revolution in the Design and Manufacture of Cranial Implants: Modern Advancements and Future Directions. *Neurosurgery* **2015**, *77*, 814–824. [CrossRef]
16. Khorasani, A.; Gibson, I.; Veetil, J.K.; Ghasemi, A.H. A Review of Technological Improvements in Laser-Based Powder Bed Fusion of Metal Printers. *Int. J. Adv. Manuf Technol.* **2020**, *108*, 191–209. [CrossRef]
17. Ali, M.d.H.; Issayev, G.; Shehab, E.; Sarfraz, S. A Critical Review of 3D Printing and Digital Manufacturing in Construction Engineering. *Rapid Prototyp. J.* **2022**, ahead-of-print. [CrossRef]
18. Mian, S.H.; Moiduddin, K.; Abdo, B.M.A.; Sayeed, A.; Alkhalefah, H. Modelling and Evaluation of Meshed Implant for Cranial Reconstruction. *Int. J. Adv. Manuf Technol.* **2022**, *118*, 1967–1985. [CrossRef]

19. Rukskul, P.; Suvannapruk, W.; Suwanprateeb, J. Cranial Reconstruction Using Prefabricated Direct 3DP Porous Polyethylene. *Rapid Prototyp. J.* **2019**, *26*, 278–287. [CrossRef]
20. Yadla, S.; Campbell, P.G.; Chitale, R.; Maltenfort, M.G.; Jabbour, P.; Sharan, A.D. Effect of Early Surgery, Material, and Method of Flap Preservation on Cranioplasty Infections: A Systematic Review. *Neurosurgery* **2011**, *68*, 1124–1130. [CrossRef]
21. Marchac, D.; Greensmith, A. Long-Term Experience with Methylmethacrylate Cranioplasty in Craniofacial Surgery. *J. Plast. Reconstr. Aesthetic Surg.* **2008**, *61*, 744–752. [CrossRef]
22. Rupprecht, S.; Merten, H.-A.; Kessler, P.; Wiltfang, J. Hydroxyapatite Cement (BoneSourceTM) for Repair of Critical Sized Calvarian Defects—an Experimental Study. *J. Cranio-Maxillofac. Surg.* **2003**, *31*, 149–153. [CrossRef]
23. Pang, D.; Tse, H.H.; Zwienenberg-Lee, M.; Smith, M.; Zovickian, J. The Combined Use of Hydroxyapatite and Bioresorbable Plates to Repair Cranial Defects in Children. *J. Neurosurg.* **2005**, *102*, 36–43. [CrossRef] [PubMed]
24. Kanno, Y.; Nakatsuka, T.; Saijo, H.; Fujihara, Y.; Atsuhiko, H.; Chung, U.; Takato, T.; Hoshi, K. Computed Tomographic Evaluation of Novel Custom-Made Artificial Bones, "CT-Bone", Applied for Maxillofacial Reconstruction. *Regen. Ther.* **2016**, *5*, 1–8. [CrossRef]
25. De Viteri, V.S.; Fuentes, E. *Titanium and Titanium Alloys as Biomaterials*; InTechOpen: London, UK, 2013.
26. Trevisan, F.; Calignano, F.; Aversa, A.; Marchese, G.; Lombardi, M.; Biamino, S.; Ugues, D.; Manfredi, D. Additive Manufacturing of Titanium Alloys in the Biomedical Field: Processes, Properties and Applications. *J. Appl. Biomater. Funct. Mater.* **2018**, *16*, 57–67. [CrossRef] [PubMed]
27. Tengvall, P.; Lundström, I. Physico-Chemical Considerations of Titanium as a Biomaterial. *Clin. Mater.* **1992**, *9*, 115–134. [CrossRef]
28. Huiskes, R.; Ruimerman, R.; van Lenthe, G.H.; Janssen, J.D. Effects of Mechanical Forces on Maintenance and Adaptation of Form in Trabecular Bone. *Nature* **2000**, *405*, 704–706. [CrossRef] [PubMed]
29. Lee, W.-T.; Koak, J.-Y.; Lim, Y.-J.; Kim, S.-K.; Kwon, H.-B.; Kim, M.-J. Stress Shielding and Fatigue Limits of Poly-Ether-Ether-Ketone Dental Implants. *J. Biomed. Mater. Res. B Appl. Biomater.* **2012**, *100*, 1044–1052. [CrossRef]
30. Agapovichev, A.; Sotov, A.; Kokareva, V.; Smelov, V. Possibilities and Limitations of Titanium Alloy Additive Manufacturing. *MATEC Web Conf.* **2018**, *224*, 01064. [CrossRef]
31. Rahmitasari, F.; Ishida, Y.; Kurahashi, K.; Matsuda, T.; Watanabe, M.; Ichikawa, T. PEEK with Reinforced Materials and Modifications for Dental Implant Applications. *Dent. J.* **2017**, *5*, 35. [CrossRef]
32. Dondi, M.; Ercolani, G.; Marsigli, M.; Melandri, C.; Mingazzini, C. The Chemical Composition of Porcelain Stoneware Tiles and Its Influence on Microstructure and Mechanical Properties. *InterCeram: Int. Ceram. Rev.* **1999**, *48*, 75–83.
33. Brockett, C.L.; Carbone, S.; Fisher, J.; Jennings, L.M. PEEK and CFR-PEEK as Alternative Bearing Materials to UHMWPE in a Fixed Bearing Total Knee Replacement: An Experimental Wear Study. *Wear* **2017**, *374–375*, 86–91. [CrossRef] [PubMed]
34. Grimal, Q.; Laugier, P. Quantitative Ultrasound Assessment of Cortical Bone Properties Beyond Bone Mineral Density. *IRBM* **2019**, *40*, 16–24. [CrossRef]
35. Zioupos, P.; Cook, R.B.; Hutchinson, J.R. Some Basic Relationships between Density Values in Cancellous and Cortical Bone. *J. Biomech.* **2008**, *41*, 1961–1968. [CrossRef] [PubMed]
36. Gradl, R.; Zanette, I.; Ruiz-Yaniz, M.; Dierolf, M.; Rack, A.; Zaslansky, P.; Pfeiffer, F. Mass Density Measurement of Mineralized Tissue with Grating-Based X-Ray Phase Tomography. *PLoS ONE* **2016**, *11*, e0167797. [CrossRef]
37. Lancaster, P.; Brettle, D.; Carmichael, F.; Clerehugh, V. In-Vitro Thermal Maps to Characterize Human Dental Enamel and Dentin. *Front. Physiol.* **2017**, *8*, 461. [CrossRef]
38. Feldmann, A.; Wili, P.; Maquer, G.; Zysset, P. The Thermal Conductivity of Cortical and Cancellous Bone. *Eur Cell Mater.* **2018**, *35*, 25–33. [CrossRef]
39. Rivière, L.; Caussé, N.; Lonjon, A.; Dantras, É.; Lacabanne, C. Specific Heat Capacity and Thermal Conductivity of PEEK/Ag Nanoparticles Composites Determined by Modulated-Temperature Differential Scanning Calorimetry. *Polym. Degrad. Stab.* **2016**, *127*, 98–104. [CrossRef]
40. Properties: Zirconia—ZrO2, Zirconium Dioxide. Available online: https://www.azom.com/properties.aspx?ArticleID=133 (accessed on 24 January 2022).
41. Gowda, E.M.; Iyer, S.R.; Verma, K.; Murali Mohan, S. Evaluation of PEEK Composite Dental Implants: A Comparison of Two Different Loading Protocols. *J. Dent. Res. Rep.* **2018**, *1*. [CrossRef]
42. Sharma, N.; Aghlmandi, S.; Dalcanale, F.; Seiler, D.; Zeilhofer, H.-F.; Honigmann, P.; Thieringer, F.M. Quantitative Assessment of Point-of-Care 3D-Printed Patient-Specific Polyetheretherketone (PEEK) Cranial Implants. *Int. J. Mol. Sci.* **2021**, *22*, 8521. [CrossRef]
43. Skirbutis, G.; Dzingutė, A.; Masiliūnaitė, V.; Šulcaitė, G.; Žilinskas, J. A Review of PEEK Polymer's Properties and Its Use in Prosthodontics. *Stomatologija* **2017**, *19*, 19–23.
44. Mishra, S.; Chowdhary, R. PEEK Materials as an Alternative to Titanium in Dental Implants: A Systematic Review. *Clin. Implant. Dent. Relat. Res.* **2019**, *21*, 208–222. [CrossRef] [PubMed]
45. Polyetheretherketone (PEEK): MakeItFrom.com. Available online: https://www.makeitfrom.com/material-properties/Polyetheretherketone-PEEK (accessed on 15 February 2022).
46. Papathanasiou, I.; Kamposiora, P.; Papavasiliou, G.; Ferrari, M. The Use of PEEK in Digital Prosthodontics: A Narrative Review. *BMC Oral Health* **2020**, *20*, 217. [CrossRef] [PubMed]
47. Kurtz, S.M.; Devine, J.N. PEEK Biomaterials in Trauma, Orthopedic, and Spinal Implants. *Biomaterials* **2007**, *28*, 4845–4869. [CrossRef]

48. Ma, H.; Suonan, A.; Zhou, J.; Yuan, Q.; Liu, L.; Zhao, X.; Lou, X.; Yang, C.; Li, D.; Zhang, Y. PEEK (Polyether-Ether-Ketone) and Its Composite Materials in Orthopedic Implantation. *Arab. J. Chem.* **2021**, *14*, 102977. [CrossRef]
49. Garrido, B.; Albaladejo-Fuentes, V.; Cano, I.G.; Dosta, S. Development of Bioglass/PEEK Composite Coating by Cold Gas Spray for Orthopedic Implants. *J. Spray Tech.* **2022**, *31*, 186–196. [CrossRef]
50. Yadav, D.; Garg, R.K.; Ahlawat, A.; Chhabra, D. 3D Printable Biomaterials for Orthopedic Implants: Solution for Sustainable and Circular Economy. *Resour. Policy* **2020**, *68*, 101767. [CrossRef]
51. Alqurashi, H.; Khurshid, Z.; Syed, A.U.Y.; Rashid Habib, S.; Rokaya, D.; Zafar, M.S. Polyetherketoneketone (PEKK) An Emerging Biomaterial for Oral Implants and Dental Prostheses. *J. Adv. Res.* **2021**, *28*, 87–95. [CrossRef]
52. Qin, L.; Yao, S.; Zhao, J.; Zhou, C.; Oates, T.W.; Weir, M.D.; Wu, J.; Xu, H.H.K. Review on Development and Dental Applications of Polyetheretherketone-Based Biomaterials and Restorations. *Materials* **2021**, *14*, 408. [CrossRef]
53. Alexakou, E.; Damanaki, M.; Zoidis, P.; Bakiri, E.; Mouzis, N.; Smidt, G.; Kourtis, S. PEEK High Performance Polymers: A Review of Properties and Clinical Applications in Prosthodontics and Restorative Dentistry. *Eur. J. Prosthodont. Restor. Dent.* **2019**, *27*. [CrossRef]
54. Bathala, L.; Majeti, V.; Rachuri, N.; Singh, N.; Gedela, S. The Role of Polyether Ether Ketone (Peek) in Dentistry—A Review. *J. Med. Life* **2019**, *12*, 5–9. [CrossRef]
55. Van de Vijfeijken, S.E.C.M.; Münker, T.J.A.G.; Spijker, R.; Karssemakers, L.H.E.; Vandertop, W.P.; Becking, A.G.; Ubbink, D.T. CranioSafe Group Autologous Bone Is Inferior to Alloplastic Cranioplasties: Safety of Autograft and Allograft Materials for Cranioplasties, a Systematic Review. *World Neurosurg.* **2018**, *117*, 443–452.e8. [CrossRef] [PubMed]
56. Binhammer, A.; Jakubowski, J.; Antonyshyn, O.; Binhammer, P. Comparative Cost-Effectiveness of Cranioplasty Implants. *Plast Surg* **2020**, *28*, 29–39. [CrossRef] [PubMed]
57. Kinsman, M.; Aljuboori, Z.; Ball, T.; Nauta, H.; Boakye, M. Rapid High-Fidelity Contour Shaping of Titanium Mesh Implants for Cranioplasty Defects Using Patient-Specific Molds Created with Low-Cost 3D Printing: A Case Series. *Surg Neurol. Int.* **2020**, *11*, 288. [CrossRef] [PubMed]
58. MedCAD | PEEK Cost and Price Comparison. Available online: https://medcad.com/peek-cost-and-price-comparison/ (accessed on 15 February 2022).
59. Thien, A.; King, N.K.K.; Ang, B.T.; Wang, E.; Ng, I. Comparison of Polyetheretherketone and Titanium Cranioplasty after Decompressive Craniectomy. *World Neurosurg.* **2015**, *83*, 176–180. [CrossRef] [PubMed]
60. Wyleżoł, M. Hybrid Modeling Methods of Cranial Implants. *Adv. Sci. Technol. Res. J.* **2018**, *12*, 35–47. [CrossRef]
61. Al-Ahmari, A.; Nasr, E.A.; Moiduddin, K.; Anwar, S.; Kindi, M.A.; Kamrani, A. A Comparative Study on the Customized Design of Mandibular Reconstruction Plates Using Finite Element Method. *Adv. Mech. Eng.* **2015**, *7*, 1687814015593890. [CrossRef]
62. High-Performance Materials. Available online: https://www.intamsys.com/high-performance-materials/ (accessed on 13 January 2022).
63. Marcián, P.; Narra, N.; Borák, L.; Chamrad, J.; Wolff, J. Biomechanical Performance of Cranial Implants with Different Thicknesses and Material Properties: A Finite Element Study. *Comput. Biol. Med.* **2019**, *109*, 43–52. [CrossRef]
64. Ridwan-Pramana, A.; Marcián, P.; Borák, L.; Narra, N.; Forouzanfar, T.; Wolff, J. Finite Element Analysis of 6 Large PMMA Skull Reconstructions: A Multi-Criteria Evaluation Approach. *PLoS ONE* **2017**, *12*, e0179325. [CrossRef]
65. Brandicourt, P.; Delanoé, F.; Roux, F.-E.; Jalbert, F.; Brauge, D.; Lauwers, F. Reconstruction of Cranial Vault Defect with Polyetheretherketone Implants. *World Neurosurg.* **2017**, *105*, 783–789. [CrossRef]
66. Kim, K.; Noh, H.; Park, K.; Jeon, H.W.; Lim, S. Characterization of Power Demand and Energy Consumption for Fused Filament Fabrication Using CFR-PEEK. *Rapid Prototyp. J.* **2022**, *ahead-of-print*. [CrossRef]
67. Jiang, C.-P.; Cheng, Y.-C.; Lin, H.-W.; Chang, Y.-L.; Pasang, T.; Lee, S.-Y. Optimization of FDM 3D Printing Parameters for High Strength PEEK Using the Taguchi Method and Experimental Validation. *Rapid Prototyp. J.* **2022**, *ahead-of-print*. [CrossRef]
68. Lalegani Dezaki, M.; Mohd Ariffin, M.K.A.; Hatami, S. An Overview of Fused Deposition Modelling (FDM): Research, Development and Process Optimisation. *Rapid Prototyp. J.* **2021**, *27*, 562–582. [CrossRef]
69. Mahendru, S.; Jain, R.; Aggarwal, A.; Aulakh, H.S.; Jain, A.; Khazanchi, R.K.; Sarin, D. CAD-CAM vs Conventional Technique for Mandibular Reconstruction with Free Fibula Flap: A Comparison of Outcomes. *Surg. Oncol.* **2020**, *34*, 284–291. [CrossRef] [PubMed]
70. Nayman, J.; Pearson, E.S., IX. On the Problem of the Most Efficient Tests of Statistical Hypotheses. *Phil. Trans. R. Soc. Lond. A* **1933**, *231*, 289–337. [CrossRef]
71. Lo Giudice, A.; Ronsivalle, V.; Grippaudo, C.; Lucchese, A.; Muraglie, S.; Lagravère, M.O.; Isola, G. One Step before 3D Printing—Evaluation of Imaging Software Accuracy for 3-Dimensional Analysis of the Mandible: A Comparative Study Using a Surface-to-Surface Matching Technique. *Materials* **2020**, *13*, 2798. [CrossRef]
72. Geomagics Control. X; 3D Systems. Available online: https://www.3dsystems.com/software#inspectionsoftware (accessed on 24 January 2022).
73. Hammad Mian, S.; Abdul Mannan, M.; Al-Ahmari, A. The Influence of Surface Topology on the Quality of the Point Cloud Data Acquired with Laser Line Scanning Probe. *Sens. Rev.* **2014**, *34*, 255–265. [CrossRef]
74. Verbruggen, S.W.; Vaughan, T.J.; McNamara, L.M. Strain Amplification in Bone Mechanobiology: A Computational Investigation of the in Vivo Mechanics of Osteocytes. *J. R Soc. Interface* **2012**, *9*, 2735–2744. [CrossRef]

75. Carter, D.R.; Fyhrie, D.P.; Whalen, R.T. Trabecular Bone Density and Loading History: Regulation of Connective Tissue Biology by Mechanical Energy. *J. Biomech* **1987**, *20*, 785–794. [CrossRef]
76. Mosley, J.R. Osteoporosis and Bone Functional Adaptation: Mechanobiological Regulation of Bone Architecture in Growing and Adult Bone, a Review. *J. Rehabil Res. Dev.* **2000**, *37*, 189–199.
77. Santos, P.O.; Carmo, G.P.; de Sousa, R.J.A.; Fernandes, F.A.O.; Ptak, M. Mechanical Strength Study of a Cranial Implant Using Computational Tools. *Appl. Sci.* **2022**, *12*, 878. [CrossRef]

Review

Processing and Quality Control of Masks: A Review

Sedigheh Farzaneh [1] and Mohammadali Shirinbayan [2,*]

1. P4Tech, 23 Rue du 8 Mai 1945, 94470 Paris, France; sedigheh.farzaneh@gmail.com
2. Arts et Metiers Institute of Technology, CNAM, PIMM, HESAM University, 75013 Paris, France
* Correspondence: mohammadali.shirinbayan@ensam.eu

Abstract: It is clear that viruses, especially COVID-19, can cause infection and injure the human body. These viruses can transfer in different ways, such as in air transfer, which face masks can prevent and reduce. Face masks can protect humans through their filtration function. They include different types and mechanisms of filtration whose performance depends on the texture of the fabric, the latter of which is strongly related to the manufacturing method. Thus, scientists should enrich the information on mask production and quality control by applying a wide variety of tests, such as leakage, dynamic respiratory resistance (DBR), etc. In addition, the primary manufacturing methods (meltblown, spunlaid, drylaid, wetlaid and airlaid) and new additive manufacturing (AM) methods (such as FDM) should be considered. These methods are covered in this study.

Keywords: face mask; medical devices; additive manufacturing; filtration performance

Citation: Farzaneh, S.; Shirinbayan, M. Processing and Quality Control of Masks: A Review. *Polymers* **2022**, *14*, 291. https://doi.org/10.3390/polym14020291

Academic Editor: Emin Bayraktar

Received: 22 October 2021
Accepted: 12 November 2021
Published: 11 January 2022

Publisher's Note: MDPI stays neutral with regard to jurisdictional claims in published maps and institutional affiliations.

Copyright: © 2022 by the authors. Licensee MDPI, Basel, Switzerland. This article is an open access article distributed under the terms and conditions of the Creative Commons Attribution (CC BY) license (https://creativecommons.org/licenses/by/4.0/).

1. Introduction

Over the past decades, having a healthy body has been a critical need in which different facilities and personal protection devices were developed. These devices can protect people against micro-organisms and biological aerosols, including bacteria, viruses and fungi, which are recognized as a part of causing diseases. In addition, these days, a new virus called COVID-19 has been detected that has caused many deaths worldwide [1] and has been associated with other biological effects [2,3]. Regarding this disease, more demands for adopting personal protection equipment (PPE) is required. There is a wide variety of transmission ways of micro-organisms, such as airborne and direct/indirect contact, which is classified based on particle diameters. For example, airborne transmission is defined for particle diameters of ≤ 5 μm. This form of transmission spreads without contact and raises demands for facial protection such as face masks. This transmission can happen either between healthcare workers and patients, or in different indoor areas [4].

For this fact, excellent protection by face masks in the atmosphere against particles and aerosols leads emphasis on research and development in processing and quality control of face masks [3]. For instance, the type of polymer for fabrication plays an important role both in the final performance of face masks and the environmental risks [5–8]. Besides basic, industrialized fabrication manners of face masks such as meltblown, spunlaid, drylaid, wetlaid and airlaid technologies, cutting-edge processes of additive manufacturing (AM) processes are applied to meet demands, which are discussed in the following subsections.

For quality control, different experiments have been performed on filtration, leakage, dynamic breathing resistance (DBR) performance, wearing comfort, etc. [9]. Each test has its related terms that should be understood and explained. For example, for the infiltration performance test, the mechanism of inertial impaction, interception, diffusion and electrostatic attraction have impacts, which are presented in this review paper.

2. Types of Applied Materials in Face Mask Production

Generally, in various investigations, it was explained that most starting materials for face mask fabrication include non-woven materials such as polypropylene, glass papers and woolen felt, which have been proven to have special characteristics such as high-temperature resistance in autoclaving while serving a stable structure and cost-effective final product [10,11]. Furthermore, disposable non-woven fabrics are another useful type of non-woven material that have gained attention due to the lower risk of contamination in comparison with other materials. In this regard, a comparison of some characteristics of this type with reusable materials was performed, as shown in Table 1. Reusable fabrics could also be sterilized for secondary applications [12].

Table 1. Comparison between disposable and reusable textiles characteristics [11].

Characteristic	Disposable Non-Woven	REUSABLE	
		Traditional Textile	Micro-Porous Textile
Mechanical behavior	1	2	3
Resistance to bacterial penetration	3	1	2
Resistance to liquid penetration	3	1	2
Flexibility	3	1	2

Remark: 1–3 represent poor to best criteria of properties.

However, generally speaking, usual applicable polymers in face mask production are polypropylene, polyethylene, polyesters, polyamides, polycarbonates, polyphenylene oxide and trifluorochloroethylene. Besides, some materials are applied together for better achievement of properties such as using polypropylene that is treated with dimethyl-dioctadecyl-ammonium bromide to improve bacterial attraction in order to import positive electrical charges [1].

Characteristic of Non-Woven Fabrics

Based on previous explanations, most non-woven materials are disposable and single-use; however, the second group needs sterilization before reuse. Nevertheless, there are some advantages and limitations to the application of non-woven materials. The main characteristic of these materials is the low cost of the final product. Furthermore, their permeability to air and non-adherence to wounds makes them an excellent dressing material [13,14]. Nevertheless, the term "single-use" is a limitation regarding their low resistance and poor drape ability in consideration of disposable non-woven materials [15]. At this stage, non-woven materials consist of various characteristics, which are listed below, and affecting the mask structure:

Fiber bonding. Non-woven materials are usually fabricated by the addition of an external chemical binder. Mechanical bonding has a negligible effect on the absorbency of fibers since inherent characteristics are not involved in this type of bonding. Yet, mechanical bonding causes two changes in the entanglement of fibers. First, the entanglement could limit the natural ability of the whole structure to swell. Second, the structure may prevent collapse in presence of external pressure. Considering these changes, mechanical bonding influences the capillary absorption of fluid [15,16].

Web assemblage. The manner of fiber arrangement to form a structure has a significant influence on the web properties, including packing, capillary orientation, pore size, capillary dimensions, etc. The absorbency of non-woven fibers is considered to be affected by their arrangement as well. Localized rearrangement of fibers also fulfills web formation and increases the wicking abilities of fabric [15,16].

Web finishing. In the nonwoven method, the fibers are assembled into the final structure and bonded by chemical or physical means. The absorbency of the nonwoven compound increases by chemical finishing since it modifies the wetting performance of a fiber surface and, as a result, affects the capillary behavior. Mechanical softening treatments can affect web properties and absorbency characteristics since fiber crimp could have an influence on packing efficiency and the resulting structure [13,14].

Fiber finishing. Fiber finishing is used to improve fiber's processing performance within the equipment utilized for the transformation of fibers into a web. Since the finishing is on the surface of the fibers, it can influence wetting and liquid wicking and can have a direct impact on absorbency. Other morphological features such as surface rugosity and core uniformity can, in some cases, affect absorbency. In addition, the performance requirements in the fabrication of non-woven materials involve an optimization of different properties: liquid interaction, fabric flexibility and air permeability, and tensile properties [16].

3. Classification of Face Masks

Here, different classifications of face masks based on the application, materials and methods of production are presented. However, Table 2 shows different types of face masks with respect to their categories and relative properties.

Table 2. Introduction to types of face masks [1,17–22].

Types	Pros and Cons	Appearance
Basic Cloth face masks	Easily fabrication, cost-effective and simplest type of face mask. The starting materials could be clothes sweatshirts, T-shirts, etc. However, not much applicability for aerosols with diameters of 20–1000 nm compared to the other types.	
Surgical face masks (SFMs)	This type serves the wearer for protection against fluid stream and bacteria capturing. It has three layers, with a role of filtering media, moisture absorbance, and water repelling. The effectiveness of this type is similar to the N95 respirator. However, they are not capable of reducing the emission of small-size droplets.	
N95 respirator	Known as electrets filters in the group of filtering facepiece respirators (FFR), with surgical and standard sorts, they filter particles with diameters of 0.3 μm with 95% efficiency. It has a ventilator fan and four layers of materials of non-woven polypropylene for outer/inner layers and modacrylic, non-woven polypropylene metlblown for middle ones. However, N95 respirators are not applicable for sufficient protection against aerosols with diameters of less than 300 nm.	
P100 respirator/gas mask	This is another type of filtering facepiece respirator (FFR), with a particle-filtering efficiency and penetration of 99.97% and 0.03%, respectively. In addition, this type is better than N95 respirators in terms of less leakage and keeping a standard form in changing temperature and humidity.	

Table 2. *Cont.*

Types	Pros and Cons	Appearance
Self-contained breathing apparatus (SCBA)	This type of mask is equipped with an air supply that is normally applied for firefighting protection that resists forms of airborne contamination. However, it limits the mobility of the user and restricts workplace moments.	
Full face respirator	This is made from rigid plastic materials with transparent parts for observation, which are fabricated for the aim of breathing trouble treatment. There are different types with respect to the size and shapes: air-purifying respirators (APR) and atmosphere-supplying respirators (ASR). Face supplies for holding the masks are made of adaptable elastomeric materials to well cover the face. Another element is straps that hold the mask body on the user head for the aim of leakage prevention. However, based on wearer behavior, these elements, especially the straps, can be broken.	
Full-length face shield	This kind of mask contains elastic headbands to cover the head and a transparent rigid polymeric (polycarbonate) full-length face shield. This could protect the user from liquid infected splashes in sneezing.	

Due to the COVID-19 consequences and application of the SFMs in different departments, categories of these types are discussed more. According to the ASTM F2100-11 standard, SFMs are generally categorized into three main groups: Level 1 (low) barrier, Level 2 (medium) barrier, and Level 3 (high) barrier. Level 1 has the lowest barrier of protection, while Level 3 has the highest barrier of protection. There are different criteria that have been implemented into the classification of SFMs:

Bacterial filtration efficiency (BFE): This criterion is designed for measuring bacterial filtration efficiency of SFMs using Staphylococcus aureus as the challenge organism. Staphylococcus aureus is based on its clinical relevance as a leading cause of nosocomial infections. A higher bacterial filtration efficiency percentage indicates a better protection level for the patient and healthcare professionals against transmission diseases from the source of the patient and healthcare professionals.

Breathing resistance: This is used to determine the resistance of airflow through the masks. The SFM is subjected to a controlled flow of air. A lower breathing resistance illustrates a better comfort level to the end-user. The following sections will provide more information about this test.

Quality evaluation: This controls the quality evaluation to avoid transmission diseases, and the critical requirements are performed before the marketing of SFMs. For example, one of the important parts of the quality evaluation is the investigation of toxicity and biocompatibility of the masks. Sipahi et al. [23] studied the biocompatibility of eight marketed masks with different brands through their cytotoxicity and inflammation-inducing capacity. They showed that widely used disposable medical masks induced a surprisingly high rate of cytotoxicity and inflammation. In addition, they showed that evaluation of inflammation with cytotoxicity can be used to study the biocompatibility of medical devices such as with surgical masks.

Based on the mentioned classifications, there are three main levels of protection for SFMs that are indicated in Table 3.

Table 3. ASTM F2100-11 levels of protection in SFMs.

Level of Protection	Characteristic of Each Level
Level 1 (Low barrier)	- Minimum BFE protection - Used for general procedures and respiratory etiquette - Designed to resist splash or spray at venous pressure
Level 2 (Moderate barrier)	- High BFE protection - More breathable than high barrier masks - Designed to resist a splash or spray at arterial pressure
Level 3 (High barrier)	- High BFE protection - Highest fluid resistance - Designed to resist a splash or spray during tasks such as orthopedic surgery

4. Primary Techniques of Processing

During the past several years, different technologies have been implemented in the fabrication of non-woven fabrics [24–26]. As explained, the manufacturing of these materials are divided into two main steps: preparation of fibers in the web and bonding of fiber in the web. Presumably, there is related technology to the formation of a web that will be explained in this section. Repartition of worldwide production according to the technologies is shown in Figure 1.

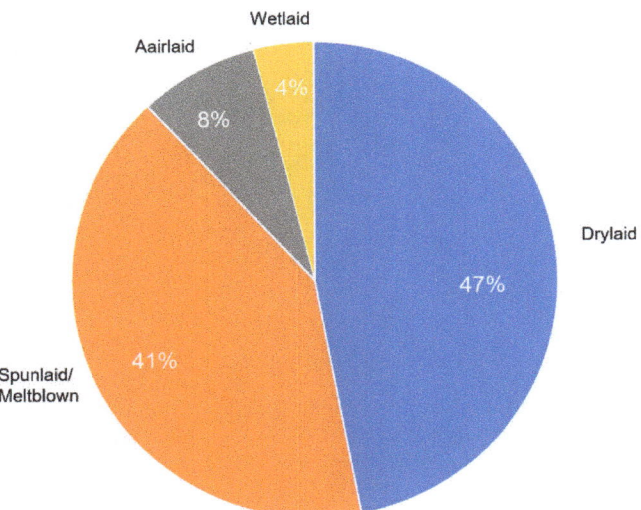

Figure 1. Repartition of worldwide production of nonwoven materials according to technologies [25].

4.1. Meltblown Process and Spunlaid Technology

The development of microfiber was first applied using a spray gun as a process to improve textile structures [27]. Following the expansion of microfibers, the technology was patented as a meltblown process [28,29]. In this process, numerous thermoplastic polymers such as polypropylene (PP), polystyrene, polyesters, polyurethane, nylon, polyethylene low and high density (LLDPE, LDPE, HDPE), and polycarbonate (PC) are used, of which the most popular polymer is polypropylene. Due to its low-melt viscosity, there is a possibility of passage through the micron-size holes. As an example, almost all meltblown webs are layered between two spunbond fabrics as shown in Figure 2 [30–32].

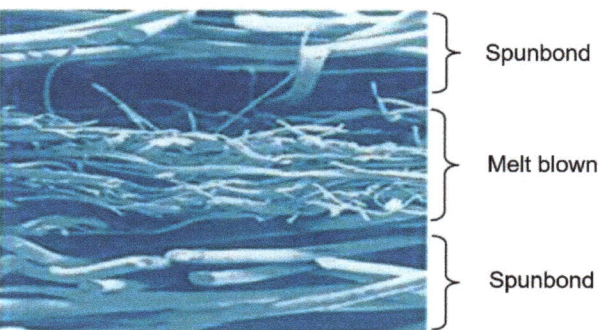

Figure 2. Typical microscopic image of a web representing the large fibers as spunbond and small fibers as meltblown [33].

4.2. Meltblown Process

This process was first presented in the early 1950s by the United States Naval Research Laboratories and was applied to thermoplastics to produce microfibers of less than ten microns diameter [34]. In this process, there are four different factors: die assembly, the extruder, metering pump, and winding. The polymer resin is heated to melting point by feeding into the extruder, it then passes through the metering pump, turning into a homogenized polymer that feeds into the die assembly. As soon as the formation of a self-bonded web is performed, the microfibers are collected on a drum (Figure 3).

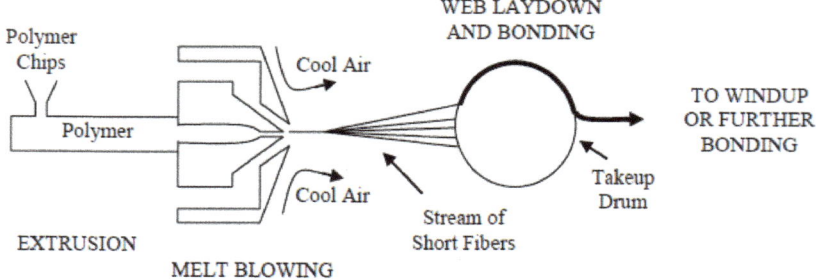

Figure 3. Schematic of the melt blowing process [35].

The definition and characteristics of applied polymers are not usually available, and hence researchers tried to publish the related works in order to perform the correlation of different parameters. However, it comprises a number of parameters, including machine, process, and materials. The interaction of these parameters is an important issue in the process, and the most important parameters are summarized in Table 4 [35].

Table 4. Definition of machine, process, and material parameters [31,34–37].

Machine Parameters	Process	Material
Air velocity	Polymer temperature	Polymer forms (granules, chips)
Air pressure	Air temperature	Polymer type
Air temperature	Die temperature	Polymer degradation
Die temperature	Die hole size	Polymer additives
Air flow rate	Die set-back	Melt viscosity
Melt flow index	Web collection type	-

Spunlaid technology. This technology, also called spunbond, is a machinery system adopted with polymer extrusion that manufactures fiber structures from molten filaments. These systems were presented commercially for the first time by DuPont and Rhone-Poulenc in the US and France in the mid-1960s, respectively [34]. This technique provides the chance for mass, cost-effective nonwoven products. There are different steps in this process that are illustrated in Figure 4.

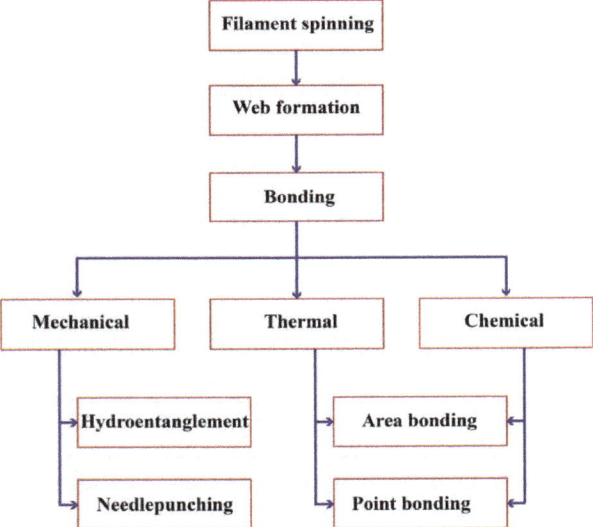

Figure 4. Sequences in spunlaid process [36].

As can be seen in Figure 5, the spunlaid process contains different parts and elements for production, including extruder, filter, metering pump, spinning block, quenching, drawing, web forming, bonding and winding.

4.3. Drylaid Technology

This technology was first designed for textile industries, of which the common applicable materials comprise different staple fibers such as polyester, polypropylene, and cotton. Normally, the chosen fibers are those capable of reaching the web properties. Drylaid webs processing generally consists of four steps: (i) staple fiber preparation, (ii) opening, cleaning, mixing, and blending, (iii) carding, and (iv) web laying [34]. Although it is necessary to cut the produced fibers into staple fibers, the fiber preparation process is affected by the manufacturing methods. As these procedures are successive, the opening, cleaning, and mixing should be without defects in the products in order not to apply negative effects on the final product. The carding step is then performed by a machine named "card". This step is conveyed by passing the entangled fibers between the closely spaced cloth surfaces [37].

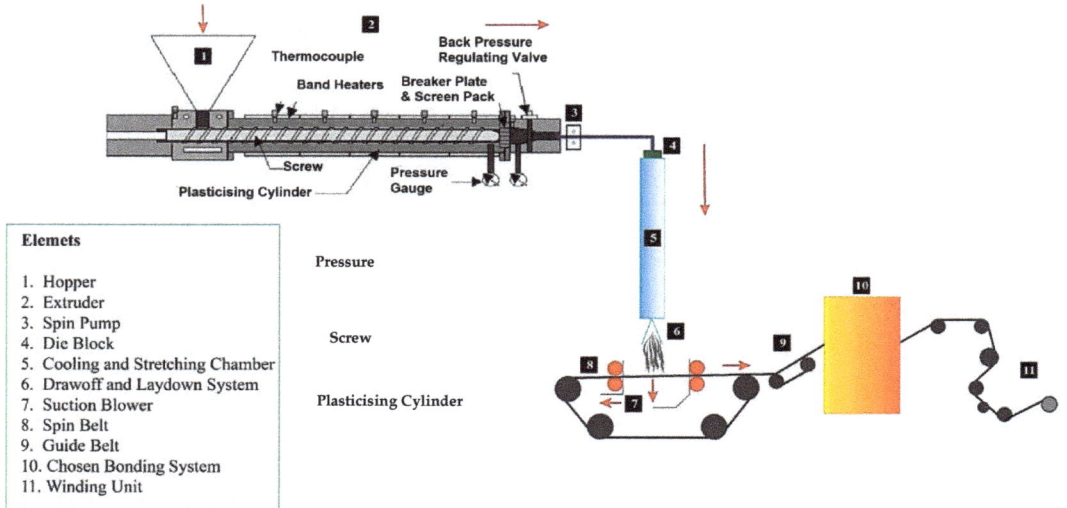

Figure 5. Schematic spunlaid process [36].

4.4. Wetlaid and Airlaid Technology

Wetlaid technologies. These technologies come directly from paper-making technologies that are designed to manipulate fibers suspended in fluid, defined as "wetlaid". They were primarily introduced in the 1930s by Dexter and was purchased by Ahlstrom in 2000, who is a leader in wetlaid products [34]. The green material comprises cellulosic fibers as wood pulp and a wide variety of other fibers. These fibers have a short length in the size range of 2 to 20 mm. The process consists of dispersion to be as homogeneous as possible, the blend of fibers in water to flow the fiber solution into a forming wire, and the extraction water through the forming wire to lay fibers into a web form. Regarding the size and fineness of the fibers, the webs will look extremely uniform and sometimes similar to paper.

Airlaid technologies. The arrival of this technique dates to the 1960s, with Karl Krøyer in Denmark, and was subsequently sold to the M&J Fibertech company in the 1980s. This technology applies the same type of raw materials as wetlaid and particularly short firers as wood fibers. This process comprises obtaining a homogenous suspension of fibers in the air and then filtering this suspension through a forming wire. Fibers held by the wire will form the web. As for the wetlaid, the webs will look exceptionally uniform [34].

5. Additive Manufacturing (AM) of Face Masks

Three-dimensional (3D) printing or additive manufacturing (AM) technologies (Figure 6) are known as the fourth industrial revolution in our scientific world, which was first presented in 1986 by Charles Hull through a manner of so-called stereolithography (SLA). This technology is expanding because of the wide variety of advantages such as minimum demands for postprocessing, less unusable wastes materials, and widely used applications, especially in polymers and face masks production [38,39]. AM technologies are replacing other technologies to become an accepted generic term for layer technology. Everyone is able to operate a 3D printing machine, even at home or inf an office, to print a 3D object [40]. Currently, AM machines play an important role in medical devices and biomaterial fabrication [41].

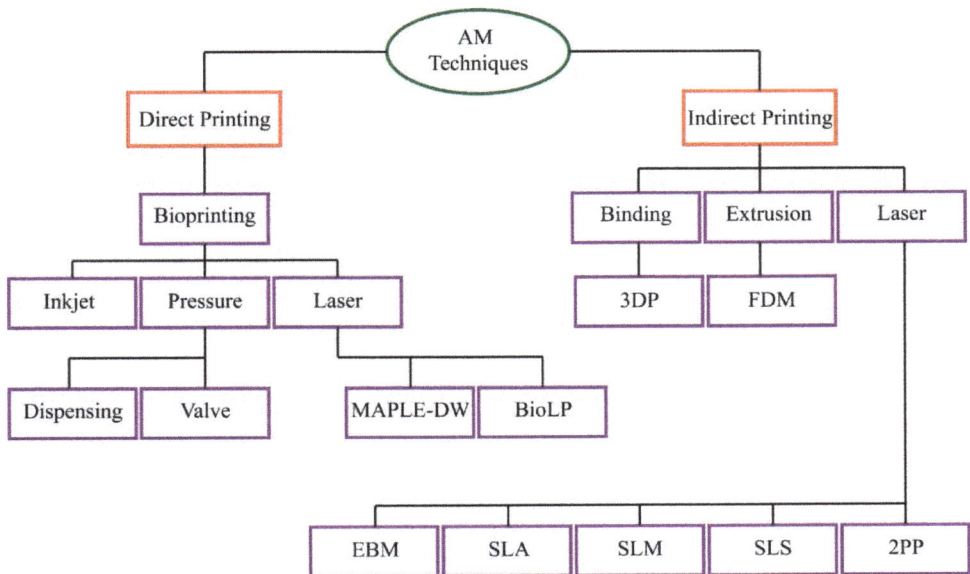

Figure 6. A review of additive manufacturing (AM) technologies and related subsessions [38].

In polymeric materials, the critical limitation is the contaminations against viruses or bacteria in processing [41,42]. Currently, polylactic acid (PLA) is a common polymer in AM technologies produced from renewable sources [43].

Apparently, there is a problem of sterilizing of the 3D-printed parts due to the porous structure in the range of 6–8 μm [44], which could be a weak point of using these structures in medical objects. However, applying the specific settings in the extrusion of layers to the antimicrobial materials can reduce dimensions to around 0.0002 μm, which is smaller than the size of viruses, such as in COVID-19 [44,45]. For instance, the 3D-printing and industrial production of PLA is presented in Figure 7.

Researchers are of the opinion that AM techniques could be used in the fabrication of medical devices to provide rapid production of final products such as ventilators, connectors, face masks, etc. [46]. As mentioned before, the importance of face masks for patients and health care workers is enticing because they categorize it as a critical medical device, especially against coronaviruses. There exists a limitation in full protection against viruses or bacteria due to the gap between the surfaces of the face masks and face (i.e., leakage) [47]. Despite the different efforts during the COVID-19 pandemic in the fabrication of medical devices [48–50], most researchers tried to propose the application of AM machines in the production of face masks in an efficient time. In Figure 8, general methodology workflow for face mask production is illustrated, which involves three main phases including: "Phase I", "Phase II" and "Phase III" that are digitizing, modeling and fabrication, respectively.

Swennen et al. [52] proposed a custom-made 3D printed face mask as a replacement against the lack of FFP2/3 SFMs. As shown in Figure 9, reusable polyamide 11 (PA11) was supplied for the 3D-printed SFM, and polypropylene (PP) non-woven meltblown particles were implemented for the filter membrane. They found that the 3D-printed SFM in combination with the FFP2/3 filter membranes could be an alternative; however, they require to show more validation of the proposed method.

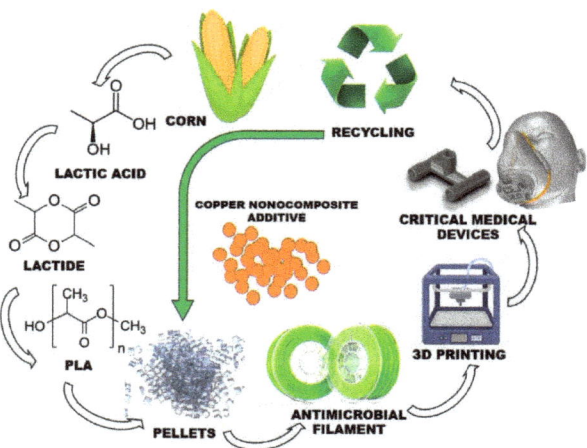

Figure 7. A representation of the manufacturing process for an antimicrobial polymer [44].

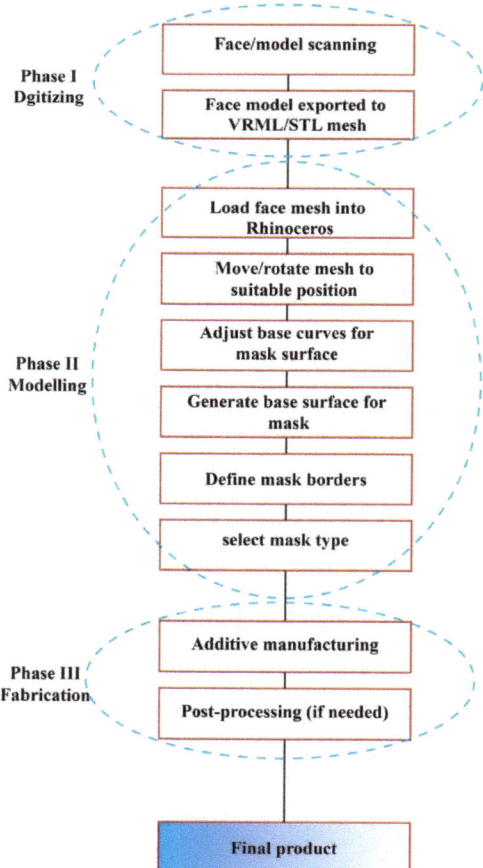

Figure 8. Methodology in additive manufacturing of a face mask [51].

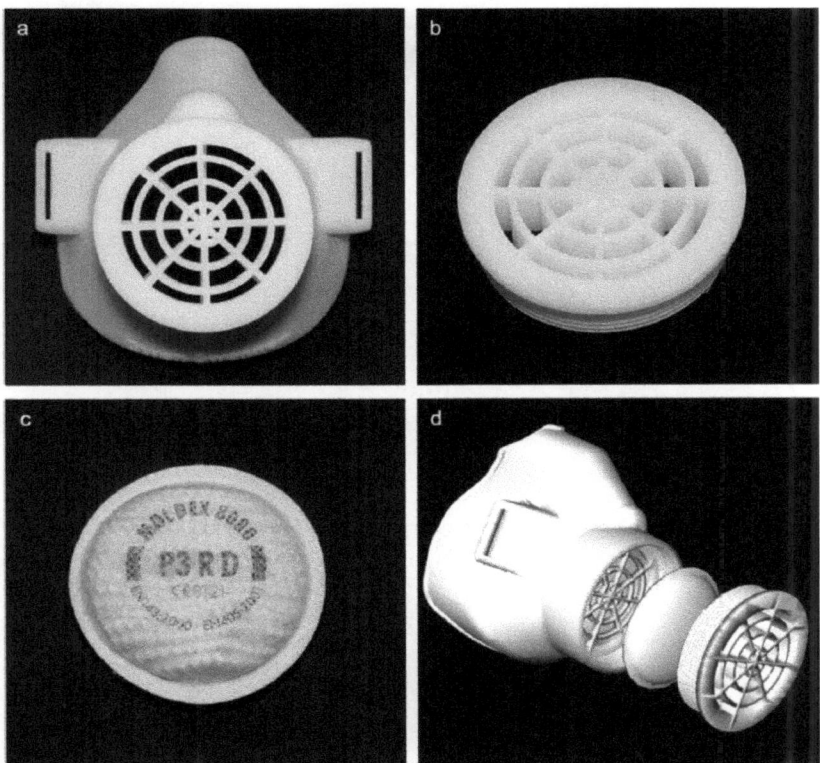

Figure 9. Typical representation of the 3D-printed face mask: (**a**) reusable 3D printed face mask, (**b**) filter membrane support, (**c**) polypropylene (PP) nonwoven meltblown particle filter, (**d**) 3D image of the prototype [52].

Consequently, Provenzano et al. [53] worked on the fabrication of reusable 3D-printed N95 face masks in different conditions by using many 3D-printing machines with PLA and ABS. Based on the outcomes, they found that PLA has better quality in comparison with ABS (Figure 10).

Finally, AM technology in the fabrication of face masks requires more developments in order to obtain high-quality products. However, recent efforts are appreciable to be applied in the medical industry for the appropriate applications.

Figure 10. Typical image of fabricated masks using PLA and ABS [53].

6. Standards in Quality Controls of Face Masks

Currently, due to this unrecognized virus (COVID-19), companies have produced a vast variety of masks; furthermore, minding the standards are essential. For this, the required information about the most important standards should be covered. The standard of "EN 14683:2019+AC:2019" is attributed to medical face masks (i.e., requirements and test methods), with the scope of construction, design, performance requirements and test methods for medical face masks that aim to decline the transition of ineffective agents from staff to patients. Table 5 presents the needed terms of SFMs for acceptable performance.

Table 5. Demands for the excepted performance of medical face masks (EN 14683:2019+AC:2019).

Evaluation	Type I [a]	Type II [a]	Type IIR [b]
Bacterial filtration efficiency (BEF), %	≥95	≥98	≥98
Differential pressure (Pa·cm^{-2})	<40	<40	<60
Splash resistance pressure (kPa)	NR *	NR	≥16.0
Microbial cleanliness (cfu. g^{-1})	≤30	≤30	≤30

* not required. [a] Classified as bacterial filtration efficiency, which is "type I" and should only be used for patients and other persons to limit infections spreads. [b] This type is divided with respect to splash resistance "R".

7. Filtration Performance (FP) Tests

The first step to perform quality control of a face mask is a filtration test, which plays an important role in mask quality evaluation. Different researchers consumed time for investigating this area, which introduced the mechanism of filtration in masks and respirators. In this protocol, four mechanisms work together: inertial impaction, interception, diffusion and electrostatic attraction. They are presented in Figure 11 [1,54].

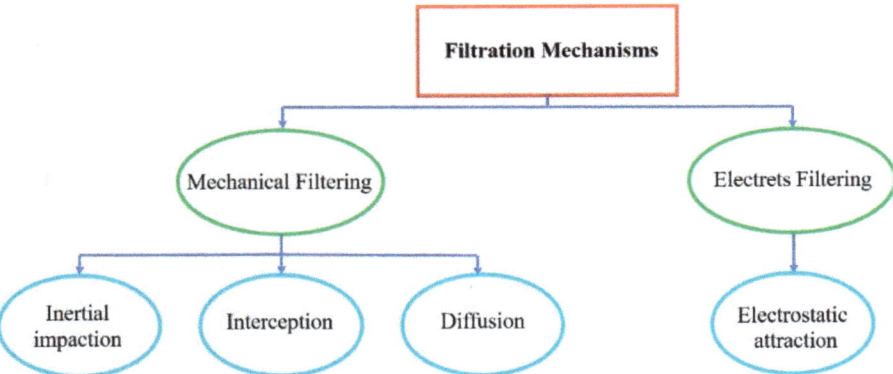

Figure 11. Different mechanisms in the filtration process [1].

The activation of each mechanism depends on particle size, face velocity and density in the airflow atmosphere. Figure 12 shows the relation of particle size and mechanisms of activation.

The mechanism of inertial impaction occurs when the size of the particle is more than 1 µm, which causes an increase of inertia in each particle, altering the direction of the particle in the atmosphere. The interception mechanism takes place when the particle size lowers to around 0.6 µm, which is not dependent on the face velocity of the particle, and no deviation is observed during the progress in comparison with former mechanisms. Besides, the most productive mechanism in the filtering of the particles is diffusion progress that accounts for particle sizes of less than 0.2 µm and in low velocity, based on the Brownian motion of particles. This motion increases the probability of particle accidents with fibers, and reduced velocity broadens the holding time of particles that consequently improves the probability of particle accidents and efficiency of filtering. Finally, the last mechanism is an electrostatic attraction that occurs by charging either the media or the particles, which is in addition to the mechanical mechanisms employed in NIOSH (National Institute for Occupational Safety and Health) accepted filters. In this mechanism, velocity has a negative impact on efficiency [55].

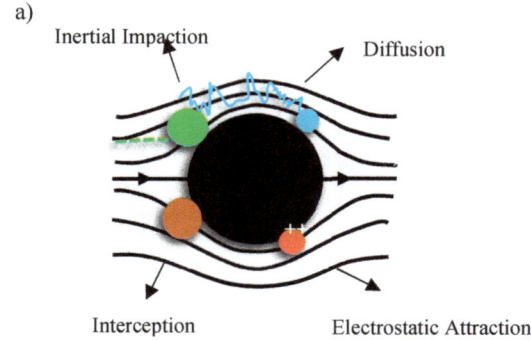

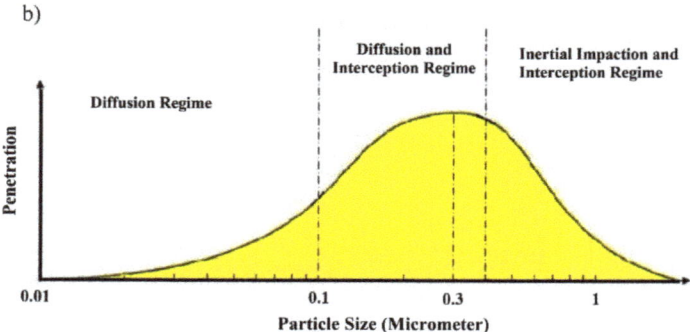

Figure 12. (**a**) Different collision of particles based on the four main filtering mechanisms. (**b**) The relationship between particle size distribution and type of filtering mechanisms [54,55].

Filtration efficiency is defined as the capability and capacity of reserving viruses and particles in the atmosphere [6] and is related to different factors such as thermal rebound, face velocity, airflow rate, humidity and particle charge states, which are briefly described in Table 6 [56].

Table 6. Test parameters and details of related roles in filtration efficiency [56,57].

Factors	Remarks
Thermal Rebound	Definition: Based on critical velocity and kinetic energy, which depends on particle diameter, yield pressure, particle density, etc. Effects: Negatively affect filtration efficiency in nanoscale particles, which depends on high temperature behavior of nanoparticles that is difficult to define the exact critical diameter of the boundary condition.
Face Velocity	Definition: Has an impact on diffusion, interception and electrostatic attraction of the fibrous filtration, which contributes to particle shape and velocity range. Effects: Generally, in high velocities (e.g., 20 cm.s^{-1}), it causes an outweighing interception mechanism to become a diffusion mechanism, which reduces the filtration efficiency.
Airflow Rate	Definition: Used for filtration efficiency evaluation of respiratory and fibrous filtration. Effects: This factor directly increases the penetration of the particles by increasing airflow rate. The suggestion for the test is 85 and 350 l.min^{-1} for similarity with real situation.
Relative Humidity (RH)	Definition: In large scale particles, elevation in capillary force, which consequently improves the adherence of particles to the fibers in charged filters, takes a part with ions and electrons.

Table 6. *Cont.*

Factors	Remarks
	Effects: Depending on the filtration mechanism, it has negative and positive impacts on the filtration process, which, in mechanical and electrets filtration, shows an increase and decrease in the process, respectively. Generally, it was reported that the type of effect is completely related to the fabrication of the masks and filters.
Particle Charge States	Definition: This considers charged/uncharged particles with mechanical and electrets filtration in the view of coulomb and image force interaction with mask medium and particles. *Effects:* The best performance of filtration was observed in incidence of neutralized particles to the electrets filtration.

Besides these kinds of evaluation, Pacitto et al. [58] researched exposure evaluation of nine different face masks based on price (1–44 Euros) in the reduction of exposure to particle mass concentration (PM2.5), particle number concentration (PNC), lung deposition surface area (LDSA) and black carbon concentration (BC), with breathing rates of 32, 42 and 52 l.min^{-1}. The test set-up is illustrated in Figure 13. Dummy heads were used as adult human heads in special dimensions, and they were occupied with different masks and different additional equipment such as airflow splitters, pumps, dust track, etc. The dummy heads were placed outdoors at a height of 1.60 m, and the mouth of each head employed an anti-electrostatic inlet tube and splitter separating airflow in 4 channels. It was reported that the effectiveness is directly related to the PM2.5 concentration.

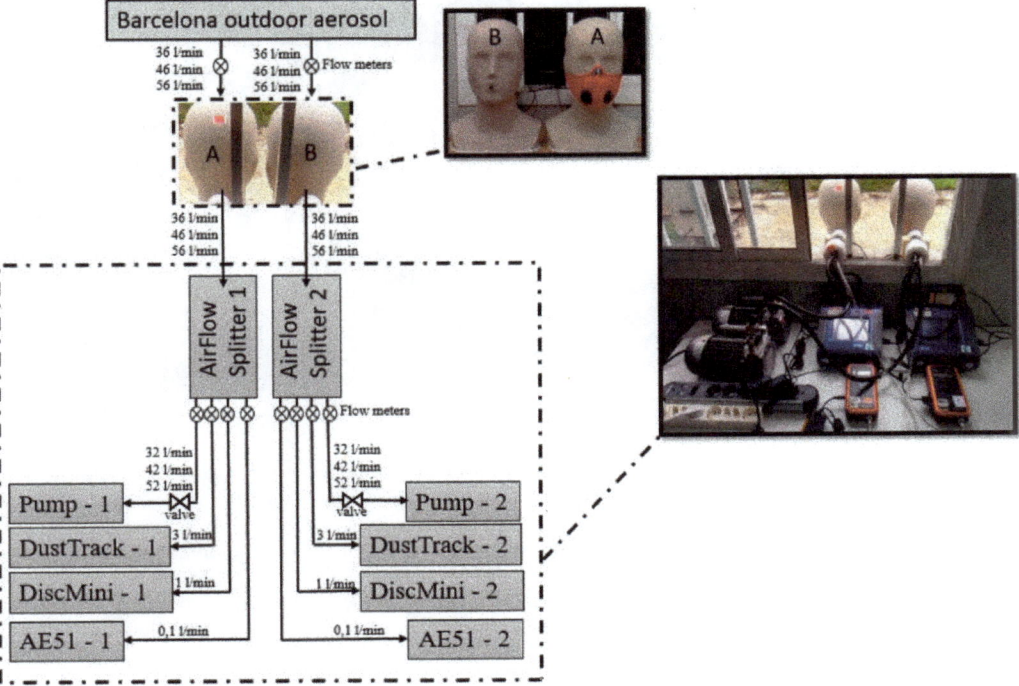

Figure 13. Set-up illustrations for measuring performance [58].

8. Leakage Test

Besides using a filtration test, the leakage problem is another contributing factor that should be considered [59]. When the leakage happened, filter penetration disappeared, causing the consideration of this term to be studied as a part of the quality controlling of masks. For instance, Guha et al. [60] conducted a study to understand the contribution of leakage of aerosols through the gaps in SFMs and surgical respirators. They searched into the leakage of charge-neutralized, polydisperse, dried sodium-chloride aerosols in different personal protective equipment (PPE), with altering breathing rates, aerosol particle sizes and gap sizes. The ration of aerosols concentration between the input and output of SFMs is defined as intrinsic penetration without gaps, or total inward leakage (TIL), with consideration of gaps based on percentage. As mentioned, the protection is related to the intrinsic penetration and amount of leakage in the site. Generally, the summation of the two terms gives total inward leakage (TIL), and the penetration was separated from TIL from studying the effect of particle size on leakage. Thus, the leakage is noted as:

$$\text{Leakage (size)\%} = \text{TIL(size)} - \text{penetration(size)}$$

For the experimental part, an artificial hole on the mask was created to perform the leakage test (Figure 14).

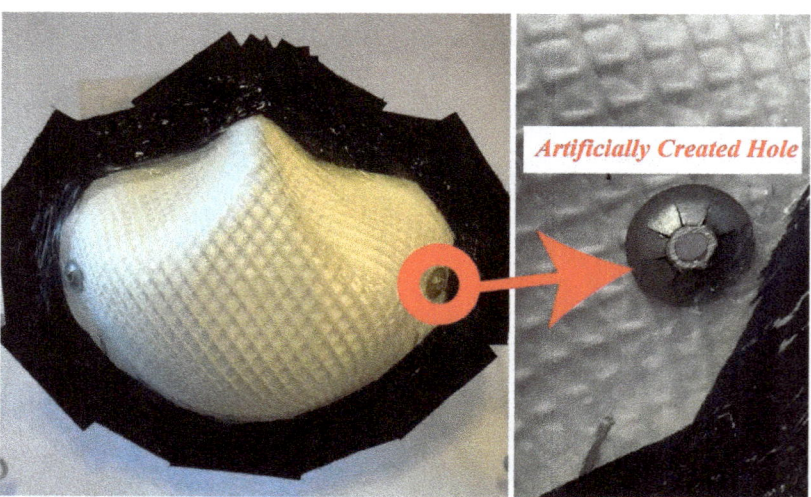

Figure 14. The holes are created to evaluate leakage performance of the mask [60].

Finally, they announced that aerosol leakage is not related to size, especially above 100 nm in used masks. In addition, more TLI normally does not attribute to higher risk and is considered in parallel with the breathing flow rate [60].

For instance, Rengasamy et al. [61] researched the evaluation of filter penetration and face seal leakage to TIL with submicron-size bioaerosols (NaCl). In this study, different artificially created holes were placed into two N95 FFR models, using SFM models applied to a manikin that breathed minute volumes of 8 and 40 L. Figure 15 illustrates the set up of this research that two modes were investigated: (a) no artificial leaks and (b) with some artificial leaks induced through the needle. In addition, for better understanding of the research, the breathing simulator serves various changing terms, such as tidal volume and breathing rate.

Finally, the results showed that N95 FFRs outweigh the two SFMs in terms of filtration efficiency and good fitting characteristics.

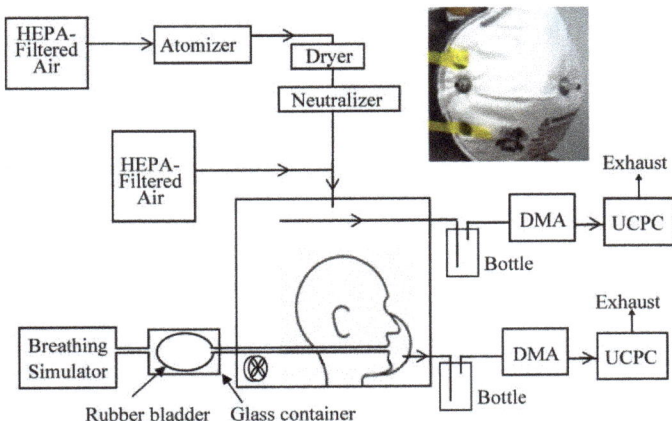

Figure 15. Experimental set-up for testing and evaluation of submicron-sized bioaerosols leakage [61].

9. Dynamic Breathing Resistance (DBR) Test

Yao et al. [62] designed an experimental set-up including breathing simulator, mass flow controller, virtual instrument system, microelectronic system and head model (Figure 16). In addition, the role of each part is presented in Table 7.

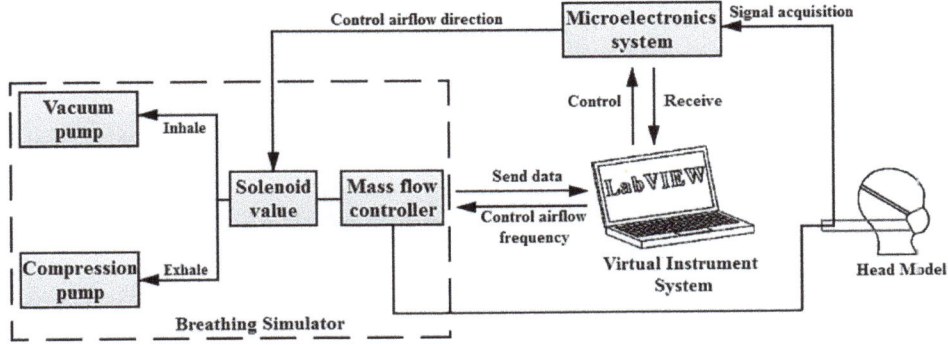

Figure 16. A designed experimental set up for measuring dynamic breathing resistance (DBR) [55].

Table 7. The key elements of DBR machine are illustrated in Figure 16 [62].

Components	Role
Vacuum pump	simulates inhalation process
Compression pump	simulates exhalation process
Mass flow controller	monitors airflow rate with respect to certain breathing frequency
Virtual instrument	controls microelectronics system, mass flow controller and obtains the dynamic altering of airflow rate from mass flow controller and breathing resistance signals from microelectronics system
Microelectronic system	manages solenoid valve for changing the direction of air flow for the aim of exhalation and inhalation simulation
Pressure sensors	records dynamic changes of breathing resistance with regard to time

Moreover, six indices were proposed to evaluate the dynamic performance of face masks in the breathing process, which is presented in Table 8.

Table 8. The six indices presented for DBR measurement [62].

Indices	Unit	Diagram	Remarks
Maximum exhalation resistance (MER)	Pa		MER is defined as: $MER = BRE(t)|_{max}$ in which BRE(t) shows breathing resistance with respect to time for exhalation process.
Maximum inhalation resistance (MIR)	Pa		MIR is defined as: $MIR = |BRE(t)||_{max}$ in which BRI(t) shows breathing resistance with respect to time for inhalation process.
Average change rate of exhalation resistance (ACE)	Pa·S^{-1}		The slope of the exhalation resistance curve that is center 60% with regard to breathing resistance: $ACE = \frac{BRE_b - BRE_a}{t_b - t_a}$ a and b refer to limits of the center 60% of the exhalation resistance curve according to BRE_a and BRE_b, respectively.
Average change rate of inhalation resistance (ACI)	Pa·S^{-1}		It is defined as center 60% of the slope of the inhalation resistance curve: $ACI = \frac{BRI_j - BRI_i}{t_j - t_i}$ i and j refer to limits of the center 60% of the inhalation resistance curve according to BRI_i and BRI_j, respectively.
Maximum change rate of exhalation resistance (MCE)	Pa·S^{-1}		It is defined as the maximum slope of the exhalation resistance curve: $MCE = SBRE(t)|_{max}$ SBRE(t) is the slope of exhalation resistance with regard to time.
Maximum change rate of inhalation resistance (MCI)	Pa·S^{-1}		It is defined as the maximum slope of the inhalation resistance curve: $MCI = SBRI(t)|_{max}$ SBRI(t) is the slope of inhalation resistance with regard to time.

Based on this research frame, twelve types of facemasks with various varieties, such as shape, respiratory valve, and basic materials, have been tested in which the results show that there are noteworthy differences between the indices in each type of applied mask. It was proven that the maximum breathing resistance of the dynamic measurement in comparison with the breathing resistance of the static measurement revealed a linear relationship. In addition, DBR provides an altered rate of breathing resistance [62].

10. Conclusions

The main purpose of this review was to present different techniques in quality control and processing of face masks. These days, due to the COVID-19 consequences, public attention is drawn to face mask application for reducing death and infection. For this, productive information about face masks in terms of starting materials, primary and advance processing, mechanisms of filtration and related required application tests were considered.

Face masks made of different polymers such as polypropylene, glass papers, woolen felt, polyethylene, polyesters, polyamides, polycarbonates, and polyphenylene oxide have the own properties, and they need more detailed evaluation.

The families of face masks include basic cloth face masks, surgical face masks (SFMs), N95 respirator, P100 respirator/gas mask, self-contained breathing apparatus (SCBA), full face respirator and full-length face shield. Each type of face mask has special advantages concerning application. For instance, basic cloth face masks are easily fabricated materials (e.g., can be produced from a T-shirt, etc.) at low cost but lack efficient infiltration. Surgical face masks (SFMs) and N95 respirators show almost similar efficiency infiltration. For a P100 respirator, it was reported that the efficiency of filtering is 99.97% and presents less leakage, which is better than SFMs and N95.

For face mask fabrication, there are different methods such as airlaid, wetlaid, spunlaid/meltblown and drylaid, and each one shows specific properties. A new generation of fabrication methods called additive manufacturing (AM) is also applied for face mask production, which is expanding. However, AM techniques need more development to obtain high-quality products in terms of mechanical and physical properties.

After face mask production, quality control is the final step before marketing. Generally, the tests of filtration performance (FP), leakage, and static/dynamic breathing resistance (DBR) are passed to inspect the efficiency of face masks. Different set ups for validation of face masks were presented and reviewed.

In future studies, it is recommended to study the recycling of used face masks and mechanical properties of AM machined ones for enriching more information and improving the quality of face masks. Also created steam during the respiration cycle can provide the environment with high humidity which leads to the accelerated mechanism of penetration and faster spread of microorganisms to the inner parts of the mask. Regarding, production of masks to deal with this phenomenon, it seems necessary, especially in masks such as Surgical face masks (SFMs), Basic Cloth face masks and N95 respirators.

Author Contributions: Writing—original draft preparation, S.F. and M.S. All authors have read and agreed to the published version of the manuscript.

Funding: This research received no external funding.

Institutional Review Board Statement: Not applicable.

Informed Consent Statement: Not applicable.

Data Availability Statement: The data presented in this study are available on request from the corresponding author.

Conflicts of Interest: The authors declare no conflict of interest.

References

1. Tcharkhtchi, A.; Abbasnezhad, N.; Seydani, M.Z.; Zirak, N.; Farzaneh, S.; Shirinbayan, M. An overview of filtration efficiency through the masks: Mechanisms of the aerosols penetration. *Bioact. Mater.* **2021**, *6*, 106–122. [CrossRef] [PubMed]
2. Montano, L.; Donato, F.; Bianco, P.M.; Lettieri, G.; Guglielmino, A.; Motta, O.; Bonapace, I.M.; Piscopo, M. Semen quality as a potential susceptibility indicator to SARS-CoV-2 insults in polluted areas. *Environ. Sci. Pollut. Res.* **2021**, *28*, 37031–37040. [CrossRef] [PubMed]
3. Montano, L.; Donato, F.; Bianco, P.; Lettieri, G.; Guglielmino, A.; Motta, O.; Bonapace, I.; Piscopo, M. Air pollution and COVID-19: A possible dangerous synergy for male fertility. *Int. J. Environ. Res. Public Health* **2021**, *18*, 6846. [CrossRef] [PubMed]
4. Bunyan, D.; Ritchie, L.; Jenkins, D.; Coia, J.E. Respiratory and facial protection: A critical review of recent literature. *J. Hosp. Infect.* **2013**, *85*, 165–169. [CrossRef] [PubMed]
5. Aragaw, T.A. Surgical face masks as a potential source for microplastic pollution in the COVID-19 scenario. *Mar. Pollut. Bull.* **2020**, *159*, 111517. [CrossRef]
6. Akalin, M.; Usta, I.; Kocak, D.; Ozen, M.S. Investigation of the filtration properties of medical masks. *Med. Healthc. Text.* **2010**, 93–97. [CrossRef]
7. Mahdavi, A. *Efficiency Measurement of N95 Filtering Facepiece Respirators against Ultrafine Particles under Cyclic and Constant Flows*; Concordia University: Montreal, QC, Canada, 2013.
8. Smith, J.D.; MacDougall, C.C.; Johnstone, J.; Copes, R.A.; Schwartz, B.; Garber, G.E. Effectiveness of N95 respirators versus surgical masks in protecting health care workers from acute respiratory infection: A systematic review and meta-analysis. *Cmaj Can. Med. Assoc.* **2016**, *188*, 567–574. [CrossRef]
9. Yuasa, H.; Kumita, M.; Honda, T.; Kimura, K.; Nozaki, K.; Emi, H.; Otani, Y. Breathing simulator of workers for respirator performance test. *Ind. Health* **2014**. [CrossRef]
10. O'Dowd, K.; Nair, K.M.; Forouzandeh, P.; Mathew, S.; Grant, J.; Moran, R.; Bartlett, J.; Bird, J.; Pillai, S.C. Face masks and respirators in the fight against the COVID-19 pandemic: A review of current materials, advances and future perspectives. *Materials* **2020**, *13*, 3363. [CrossRef]
11. Chellamani, K.P.; Veerasubramanian, D.; Balaji, R.S.V. Surgical face masks: Manufacturing methods and classification. *J. Acad. Ind. Res.* **2013**, *2*, 320–324.
12. Kocak, D.; Akalin, M.; Usta, I.; Merdan, N. New approach to produce absorbent pads for new end uses. *Med. Text. Biomater. Healthc.* **2006**, 320–326. [CrossRef]
13. Hall, D.M.; Adanur, S.; Broughton, RM., Jr. Natural and man made fibers. In *Wellingt Sears Handb Ind Text*, 1st ed.; CRC Press New Holland: Boca Rat, FL, USA, 1995; pp. 37–52.
14. Lou, C.-W.; Lin, C.-W.; Chen, Y.-S.; Yao, C.-H.; Lin, Z.-S.; Chao, C.-Y.; Lin, J.-H. Properties evaluation of tencel/cotton nonwoven fabric coated with chitosan for wound dressing. *Text. Res. J.* **2008**, *78*, 248–253. [CrossRef]
15. Dias, T. *Electronic Textiles: Smart Fabrics and Wearable Technology*; Woodhead Publishing: Sawston, UK, 2015.
16. Karthik, T.; Rathinamoorthy, R. *Nonwovens: Process, Structure, Properties and Applications*; WPI Publishing: Worcester, MA, USA, 2017.
17. Rengasamy, S.; Eimer, B.; Shaffer, R.E. Simple respiratory protection—Evaluation of the filtration performance of cloth masks and common fabric materials against 20–1000 nm size particles. *Ann. Occup. Hyg.* **2010**, *54*, 789–798. [PubMed]
18. Milton, D.K.; Fabian, M.P.; Cowling, B.J.; Grantham, M.L.; McDevitt, J.J. Influenza virus aerosols in human exhaled breath: Particle size, culturability, and effect of surgical masks. *PLoS Pathog.* **2013**, *9*, e1003205. [CrossRef]
19. Bałazy, A.; Toivola, M.; Adhikari, A.; Sivasubramani, S.K.; Reponen, T.; Grinshpun, S.A. Do N95 respirators provide 95% protection level against airborne viruses, and how adequate are surgical masks? *Am. J. Infect. Control* **2006**, *34*, 51–57. [CrossRef] [PubMed]
20. Dreger, R.W.; Jones, R.L.; Petersen, S.R. Effects of the self-contained breathing apparatus and fire protective clothing on maximal oxygen uptake. *Ergonomics* **2006**, *49*, 911–920. [CrossRef] [PubMed]
21. Atangana, E.; Atangana, A. Facemasks simple but powerful weapons to protect against COVID-19 spread: Can they have sides effects? *Results Phys.* **2020**, *19*, 103425. [CrossRef]
22. Ji, D.; Fan, L.; Li, X.; Ramakrishna, S. Addressing the worldwide shortages of face masks. *BMC Mater.* **2020**, *2*, 1–11. [CrossRef]
23. Sipahl, H.; Bayram, F.E.O.; Palabiyik, S.S.; Bayram, D.; Aydin, A. Investigation of the biocompatibility of surgical masks. *Pteridines. Sciendo* **2018**, *29*, 80–86.
24. Suikkanen, T. Analytics for the Modified Kraft Pulps. Master's Thesis, LUT University, Lappeenranta, Finland, 2015.
25. Montefusco, A.F. The use of Nonwovens in air filtration. *Filtr. Sep.* **2005**, *42*, 30–31. [CrossRef]
26. Fu, J.; Yamaguchi, M.; Muroga, S.; Tanaka, T.; Okamura, C.; Obi, L.; Kato, K. Development of meltblown non-woven fabric type non-magnetic noise suppressor. In Proceedings of the 2016 Asia-Pacific Symposium on Electromagnetic Compatibility (APEMC), Shenzhen, China, 17–21 May 2016; pp. 656–658.
27. Malkan, S.R. An overview of spunbonding and meltblowing technologies. *Tappi J.* **1995**, *78*, 185–190.
28. Turbak, A.F. *Nonwovens: Theory, Process, Performance, and Testing*; TAPPI PRESS: Atlanta, Georgia, 1993.
29. McCulloch, J.G. The history of the development of melt blowing technology. *Int. Nonwovens J.* **1999**, 1558925099OS-800123. [CrossRef]
30. Jirsák, O.; Wadsworth, L.C. *Nonwoven Textiles*; Carolina Academic Press: Durham, NC, USA, 1999.
31. Červík, R. *Netkané Filtrační Textilie pro Čištění Kapalin*; Univerzita Tomáše Bati ve Zlíně: Zlin, Czech, 2020.

32. Dutton, K.C. Overview and analysis of the meltblown process and parameters. *J. Text. Apparel. Technol. Manag.* **2009**, *6*.
33. Drabek, J.; Zatloukal, M. Meltblown technology for production of polymeric microfibers/nanofibers: A review. *Phys. Fluids* **2019**, *31*, 91301. [CrossRef]
34. Wilson, A. The formation of dry, wet, spunlaid and other types of nonwovens. *Appl. Nonwovens Tech. Text.* **2010**, 3–17.
35. Lichstein, B.M. *The Nonwovens Handbook*; INDA, Association of the Nonwoven Fabrics Industry: New York, NY, USA, 1988.
36. Midha, V.K.; Dakuri, A. Spun bonding technology and fabric properties: A review. *J. Text. Eng. Fash. Technol.* **2017**, *1*, 1–9. [CrossRef]
37. Patel, B.M.; Bhrambhatt, D. Nonwoven technology. *Text. Technol.* **2008**, 1–54.
38. Rezaie, H.R.; Rizi, H.B.; Khamseh, M.M.R.; Öchsner, A. 3D-Printing Technologies for Dental Material Processing. *Rev. Dent. Mater.* **2020**, 201–210.
39. Li, L.; Sun, Q.; Bellehumeur, C.; Gu, P. *Solid Freeform Fabrication Symposium*; The University of Texas at Austin: Austin, Texas, 2001; pp. 400–407.
40. Gray, R.W.; Baird, D.G.; Bøhn, J.H. Effects of processing conditions on short TLCP fiber reinforced FDM parts. *Rapid Prototyp. J.* **1998**, *4*. [CrossRef]
41. Rezaie, H.R.; Rizi, H.B.; Khamseh, M.M.R.; Öchsner, A. *A Review on Dental Materials*; Springer: Berlin/Heidelberg, Germany, 2020.
42. González-Henríquez, C.M.; Sarabia-Vallejos, M.A.; Rodríguez Hernandez, J. Antimicrobial polymers for additive manufacturing. *Int. J. Mol. Sci.* **2019**, *20*, 1210. [CrossRef] [PubMed]
43. Zuniga, J.M. 3D printed antibacterial prostheses. *Appl. Sci.* **2018**, *8*, 1651. [CrossRef]
44. Zuniga, J.M.; Cortes, A. The role of additive manufacturing and antimicrobial polymers in the COVID-19 pandemic. *Expert Rev. Med. Devices* **2020**, *17*, 477–481. [CrossRef] [PubMed]
45. Barnard, D.L.; Kumaki, Y. Recent developments in anti-severe acute respiratory syndrome coronavirus chemotherapy. *Future Virol.* **2011**, *6*, 615–631. [CrossRef] [PubMed]
46. Branson, R.D. A single ventilator for multiple simulated patients to meet disaster surge. *Acad. Emerg. Med.* **2006**, *13*, 1352–1353. [CrossRef] [PubMed]
47. Borkow, G.; Zhou, S.S.; Page, T.; Gabbay, J. A novel anti-influenza copper oxide containing respiratory face mask. *PLoS ONE* **2010**, *5*, e11295. [CrossRef]
48. Khurana, S.; Singh, P.; Sinha, T.P.; Bhoi, S.; Mathur, P. Low-cost production of handrubs and face shields in developing countries fighting the COVID19 pandemic. *Am. J. Infect. Control* **2020**, *48*, 726–727. [CrossRef] [PubMed]
49. Cavallo, L.; Marcianò, A.; Cicciù, M.; Oteri, G. 3D printing beyond dentistry during COVID 19 epidemic: A technical note for producing connectors to breathing devices. *Prosthesis* **2020**, *2*, 46–52. [CrossRef]
50. Wesemann, C.; Pieralli, S.; Fretwurst, T.; Nold, J.; Nelson, K.; Schmelzeisen, R.; Hellwig, E.; Spies, B.C. 3-D printed protective equipment during COVID-19 pandemic. *Materials* **2020**, *13*, 1997. [CrossRef]
51. Cazon, A.; Aizpurua, J.; Paterson, A.; Bibb, R.; Campbell, R.I. Customised design and manufacture of protective face masks combining a practitioner-friendly modelling approach and low-cost devices for digitising and additive manufacturing: This paper analyses the viability of replacing conventional practice with AM me. *Virtual Phys. Prototyp.* **2014**, *9*, 251–261. [CrossRef]
52. Swennen, G.R.J.; Pottel, L.; Haers, P.E. Custom-made 3D-printed face masks in case of pandemic crisis situations with a lack of commercially available FFP2/3 masks. *Int. J. Oral Maxillofac. Surg.* **2020**, *49*, 673–677. [CrossRef]
53. Provenzano, D.; Rao, Y.J.; Mitic, K.; Obaid, S.N.; Pierce, D.; Huckenpahler, J; Berger, J.; Goyal, S.; Loew, M.H. *Rapid Prototyping of Reusable 3D-Printed N95 Equivalent Respirators at the George Washington University*; MDPI AG: Basel, Switzerland, 2020.
54. Hinds, W.C. *Aerosol Technology: Properties, Behavior, and Measurement of Airborne Particles*; John Wiley & Sons: Hoboken, NJ, USA, 1999.
55. Haghighat, F.; Bahloul, A.; Lara, J.; Mostofi, R.; Mahdavi, A. Development of a procedure to measure the effectiveness of N95 respirator filters against nanoparticles. *Stud. Res. Pro. R-754* **2012**.
56. Mostofi, R.; Wang, B.; Haghighat, F.; Bahloul, A.; Jaime, L. Performance of mechanical filters and respirators for capturing nanoparticles—Limitations and future direction. Ind Health. *Natl. Inst. Occup. Saf. Health* **2010**, *48*, 296–304.
57. Mahdavi, A.; Haghighat, F.; Bahloul, A.; Brochot, C.; Ostiguy, C. Particle loading time and humidity effects on the efficiency of an N95 filtering facepiece respirator model under constant and inhalation cyclic flows. *Ann. Occup. Hyg.* **2015**, *59*, 629–640.
58. Pacitto, A.; Amato, F.; Salmatonidis, A.; Moreno, T.; Alastuey, A.; Reche, C.; Buonanno, G.; Benito, C.; Querol, X. Effectiveness of commercial face masks to reduce personal PM exposure. *Sci. Total Environ.* **2019**, *650*, 1582–1590. [CrossRef]
59. Ingle, M.A.; Talmale, G.R. Respiratory mask selection and leakage detection system based on canny edge detection operator. *Procedia Comput. Sci.* **2016**, *78*, 323–329. [CrossRef]
60. Guha, S.; McCaffrey, B.; Hariharan, P.; Myers, M.R. Quantification of leakage of sub-micron aerosols through surgical masks and facemasks for pediatric use. *J. Occup. Environ. Hyg.* **2017**, *14*, 214–223. [CrossRef] [PubMed]
61. Rengasamy, S.; Eimer, B.C.; Szalajda, J. A quantitative assessment of the total inward leakage of NaCl aerosol representing submicron-size bioaerosol through N95 filtering facepiece respirators and surgical masks. *J. Occup. Environ. Hyg.* **2014**, *11*, 388–396. [CrossRef] [PubMed]
62. Yao, B.; Wang, Y.; Ye, X.; Zhang, F.; Peng, Y. Impact of structural features on dynamic breathing resistance of healthcare face mask. *Sci. Total Environ.* **2019**, *689*, 743–753. [CrossRef]

Review

Toward Polymeric and Polymer Composites Impeller Fabrication

Nader Zirak [1,2,*], Mohammadali Shirinbayan [1,2], Michael Deligant [2] and Abbas Tcharkhtchi [1]

[1] Arts et Metiers Institute of Technology, CNRS, CNAM, PIMM, HESAM University, F-75013 Paris, France; mohammadali.shirinbayan@ensam.eu (M.S.); abbas.tcharkhtchi@ensam.eu (A.T.)
[2] Arts et Métiers Institute of Technology, CNAM, LIFSE, HESAM University, F-75013 Paris, France; michael.deligant@ensam.eu
* Correspondence: nader.zirak@ensam.eu

Abstract: Impellers are referred to as a core component of turbomachinery. The use of impellers in various applications is considered an integral part of the industry. So, increased performance and the optimization of impellers have been the center of attention of a lot of studies. In this regard, studies have been focused on the improvement of the efficiency of rotary machines through aerodynamic optimization, using high-performance materials and suitable manufacturing processes. As such, the use of polymers and polymer composites due to their lower weight when compared to metals has been the focus of studies. On the other hand, methods of the manufacturing process for polymer and polymer composite impellers such as conventional impeller manufacturing, injection molding and additive manufacturing can offer higher economic efficiency than similar metal parts. In this study, polymeric and polymer composites impellers are discussed and conclusions are drawn according to the manufacturing methods. Studies have shown promising results for the replacement of polymers and polymer composites instead of metals with respect to a suitable temperature range. In general, polymers showed a good ability to fabricate the impellers, however in more difficult working conditions considering the need for a substance with higher physical and mechanical properties necessitates the use of composite polymers. However, in some applications, the use of these materials needs further research and development.

Keywords: impeller; polymers; polymer composites; manufacturing process; additive manufacturing; conventional manufacturing; injection moulding; performance

1. Introduction

Impellers are referred to as a key component of turbomachinery [1]. By rapidly rotating the impeller can force the working fluid by converting the velocity of the fluid to pressure [2]. Considering the use of impellers in different rotary machinery systems, they have played a vital role in various applications such as aerospace [3], automotive [4] and medical [5] applications. The improvement of a system's efficiency by impellers has attracted attention in a lot of studies [6]. To this end, in general, the studies based on geometry optimization [7], the use of high-performance materials [8] and suitable manufacturing processes [1] have tried to improve the system. Weight loss of the impeller, along with the optimization of the impeller with a proper manufacturing process, can lead to achieving an ideal impeller. So that the use of lighter materials with high strength and an ability to withstand forces during working is considered as an effective step to increasing efficiency [9].

Metal [10], polymeric [11] and composite [12] materials are the main categories of materials that have been used for the fabrication of rotors and impellers. In general, the weight, high cost of raw materials, the fact that common methods of fabrication have led to increasingly high manufacturing costs, and the high maintenance cost of metallic rotors, have all proved to be main disadvantages of metals [10]. All these problems have led to polymers and composites being at the center of attention with regard to studies. Micro Organic Rankine Cycle (mORC), Heating, Ventilation, and Air Conditioning (HVAC) and refrigeration systems are among the systems which have the potential to replace their rotary components with polymer or polymer composites. For example, in the case of micro Organic Rankine cycle, which is one of the important systems for handling fossil fuel sources (flue gases and waste heat) and renewable energies, the replacement of metal rotors with polymer and composite rotors has been mentioned as an important factor in dealing with the limited use of these turbines due to their uneconomical cost [11].

Using polymers and composites due to low weight, good chemical resistance, and good strength have been a good choice for use instead of metals in the manufacture of rotors and impellers [13]. Also, the use of these materials in manufacturing can involve methods with lower costs compared to traditional methods for producing metal parts such as forging and casting. In general, thermoplastic polymers and thermoset polymers were among the polymers that have been used to fabricate the rotors. In thermoplastic polymers, acrylonitrile butadiene styrene (ABS) [11], polylactic acid (PLA) [14], polyether ether ketone (PEEK) [15] are the examples that have been used for fabricating the impellers. On the other hand, composites have been used in situations where a material with a higher strength is required [12]. Among the composites, PEEK-GF30 has been one of the composites that have been used for this purpose.

In this review, polymers and polymer composites that were considered for the fabrication of impellers in different applications were studied. Due to the importance of manufacturing methods and their effect on the final product, different manufacturing processes such as Additive manufacturing, Injection molding and Conventional impeller manufacturing were explained. Regarding, the importance of computational fluid dynamic for simulation and the interaction structural fluid, with respect to the main stresses, were discussed. Finally, performance tests for the evaluation of fabricated impellers were mentioned.

2. Materials Used for Fabrication of Impellers

Given the vital role of materials used in rotary components fabrication, with respect to the production of the different components and working conditions, the selection and manufacture of materials have been among the most important thing [4]. In this regard, great progress has been made in the materials used for the fabrication of impellers. As mentioned, metals, polymers, and composites have been among the main group of materials used to fabricate impellers. In general, in the metals class, stainless steel, titanium, aluminum and nickel alloys are among the metal-based alloys widely used to produce rotors, impellers and fans [16]. In high-temperature applications, such as a combustion chamber or turbine inlet, which are known as "hot zones", Nickel alloys have been used. On the other hand, titanium alloys have been used in zones with lower operating temperatures, known as "cold zones", such as compressor inlets and turbine outlets [3].

Considering the reasons mentioned, the focus of this study will be on polymers and polymer composites. Good resistance to impact loads [17], fatigue [18], erosion [19], and a high ratio between mechanical resistance and material density [9] are among the properties that are exemplary when using polymers and polymer composites in the production of impellers, and, as such, they have been a decisive choice.

2.1. Polymeric Impellers

In different kinds of systems, in order to increase the efficiency of compressors and pumps and economic efficiency, polymers have been introduced for the fabrication of rotors [4,11,20,21]. Polymers defined as macromolecules consist of large numbers of smaller molecules, or repeating units, called monomers, which are formed chemically bonded. Polymer molecules can have a degree of order, relative orientation and a kind of monomer that can vary within the same polymer molecules [22]. Low price, ease of manufacture, resistance to water and versatility have been among the advantages of polymers, and these factors have led to their application in industry [23]. Polymers can exist in different forms of powders, granolas, filaments and resins, which are selected depending on the fabrication process. In general, polymers used for the fabrication of impellers can be divided in two categories of thermoplastic and thermoset polymers.

2.1.1. Impellers Based on Thermoplastics

Thermoplastics have been used to fabricate impellers in many studies. Generally, considering the ability of this group to be soften and melt when heated, two-state fabrication based on heat-softening or liquid state is preferable [24]. Injection molding [25] and 3-D printing-based processes such as fused deposition modeling (FDM) [26] or selective laser sintering (SLS) [27] are the methods by which thermoplastics are used for parts fabrication. Thermoplastics are divided in two groups of amorphous and semi-crystalline. Powders, granola and filament are different forms of thermoplastics polymers. Recyclability, good ductility and impact resistance compared to thermosets are the advantages of this class of polymer. In general, thermoplastic parts show a modulus lower than 5 GPa, which depends on the chemical composition and fabrication method as it can be changeable [24]. Figure 1 shows the different types of thermoplastic polymers with respect to ultra-performance, engineering-grade and general-purpose categories that represent the different classes of polymer materials.

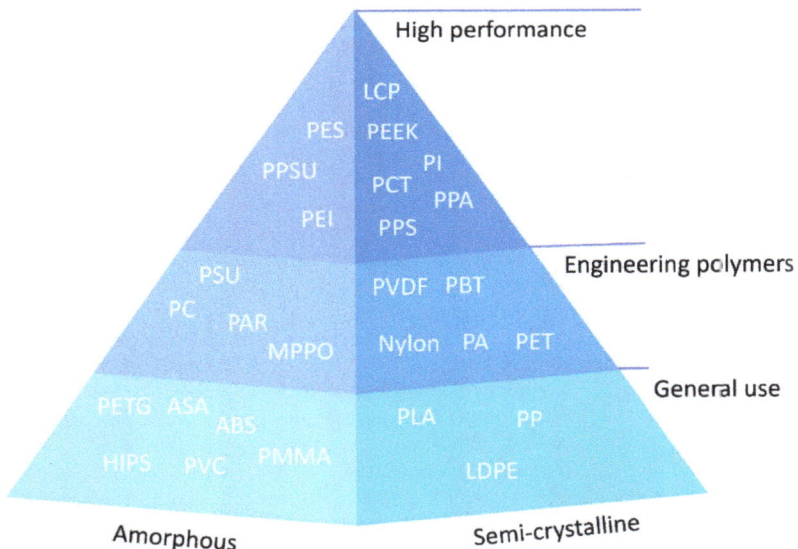

Figure 1. Different types of thermoplastics polymer.

Among the different types of thermoplastic polymers, ABS [11], PLA [14], Polyethylene terephthalate (PETG) [28], PEEK [15] and Polyphenylene sulfide (PPS) [29] are the examples that have been used for the fabrication of rotating components. Table 1 shows the physical and mechanical properties of these polymers. Modulus, cost, degradability, and water absorbability were the important parameters that impacted upon the candidate selection, and this concept will be discussed in Section 4.

Table 1. Physical and mechanical properties of polymers used in the fabrication of impellers.

	PLA	ABS	PPS	PETG	PEEK
Glass transition temperature (°C)	56–63	102–115	75–85	49–75	142.85
Melting temperature (°C)	125–178	-	285	-	342.85
Modulus (GPa)	1.03–4.0	1.8–2.39	3.9–4.1	0.9–1.6	3.6
Tensile strength (MPa)	51.7–80.9	42.5–44.8	79	44.12–57	107
Ref.	[30]	[31]	[32]	[33,34]	[35]

ABS is an amorphous thermoplastic polymer that has been applied to the fabrication of rotary components in micro Organic Rankin Cycle (mORC) [11,36], pumps [13,37–39] and the rotor blades of drones [40,41]. Hernandez-Carrillo et al. [11] studied the use of ABS impeller in the mORC. This study was performed by increasing the efficiency of the mORC by reducing the conventional weight of the impeller. Working conditions of the impeller, such as inlet temperature and pressure were 55 °C, 4Bara and the outlet temperature and pressure were 44.9 °C, 2.5Bare, respectively. Also, the rotational speed was 36,000 rpm. Considering the working condition and factor of safety (FoS), which represents the ratio of yield strength per the maximum equivalent stress, the ABS impeller provided the expected operating condition. Reducing the cost of the fabrication of the impeller by using the ABS, which can cause the mass production of mORC, was one of the important advantages of applying this polymer. Also, according to their simulation results, with respect to the working fluid of Penta-fluoro-propane (R45fa), the isentropic efficiency of the impeller was estimated to be 76–86%. However, the limitated operation of ABS under 89.9 °C was mentioned as one of the narrow operation capabilities. Pavlovic et al. [37] investigated the mechanical properties of ABS in the fabrication of impeller pumps and showed that ABS can be considered as a good candidate for the fabrication of impellers. Polak [38] studied the ABS impeller for a radial centrifugal pump by verifying the hydraulic parameters. The results showed an increase in efficiency in rotational speed of 2950 rpm. The surface smoothness of the ABS impeller was mentioned as an effective parameter in increasing efficiency.

PLA is a semi-crystalline thermoplastic polymer that is derived from renewable resources, such as corn starch or sugarcane. Biodegradability and composability of the PLA are among the properties of this polymer [42]. Economic cost, environment-friendly biocompatibility and suitable physicomechanical properties of this polymer have made this a suitable choice when compared to other polymers. PLA has been used for the fabrication of impellers for pumps [13,43,44], compressors [45] and marine [14] applications. For the fabrication of impellers in pumps and marine applications, PLA has been used. However, considering the accessibility of two polymers of PLA and ABS, these two kinds of thermoplastic have been compared in many studies as an impeller of the pumps. In general, considering the high level of brittleness of PLA when compared to ABS [37], it can be said that application by more stress ABS has been preferred. Birosz et al. [45] studied the PLA wheel for compressors given the importance of creep and orientation properties of the material, which are essential to designing the impeller of the compressor during rotation. Regarding tensile strength, creep and bending properties, these were considered when analyzing the PLA. Creep performance results showed that PLA creep behavior was most similar to the weakly cross-linked elastomer so that at low loads, the creep curve was held to a constant limit. According to their results, PLA was introduced as a material worth choosing when seeking long-term service.

PETG is an amorphous thermoplastic [46]. PETG impeller has been used in pump [47,48] and mORC [36] applications. Good water-resistance and biodegradability of this polymer [49] are reasons for choosing this polymer in the manufacture of pump blades. Odetti et al. [48] investigated the PTEG impeller in the application of a Pump-Jet Module (PJM). Considering the rotational speed of 1200 rpm that led to a thrust of 14 N, a PTEG impeller showed suitable properties during the working for this application.

Polyether-ether-ketone (PEEK) is a high-performance semi-crystalline thermoplastic polymer [35]. Excellent mechanical and thermal properties, as well as good chemical resistance, are among the bold advantages of this polymer [50]. PEEK impeller has a high position in pump and compressor applications for different industries such as automotive [51], aerospace [52] and medical [5]. In the case of medical applications, using the PEEK impeller in centrifugal pump due to the improved durability and strength it offers has attracted a lot of studies' attention [5,53,54]. Similarly, in the case of heart failure, using the PEEK impeller in the centrifugal pump of a HeartWare Ventricular Assist Device (HVAD) due to the improved durability and strength it offers has attracted a lot of studies' attention. In the HVAD, the rotational speed of the PEEK impeller is in the range of 1800–4000 rpm and generates flows up to 10 l/mL [5]. Also using the PEEK impeller with the aim of reduction in wear, reduced noise levels and more consistent running properties by replacing stainless steel for regenerative pumps was another application of this polymer [51]. In general thermomechanical properties, due to the thermal stress during increased temperature, is one of the important factors that can lead to the limitation of using the polymers as impellers in various applications.

Zywica et al. [15] studied the use of plastics with the aim of using them as an impeller for the ORC system. In this study, PPS and PEEK were considered as thermoplastic polymers. The rotational speed for the impeller was 120,000 rpm. The simulation results based on heat resistance, chemical resistance, strength properties, and thermal expansions showed that PEEK polymer can be considered as a good material for the fabrication of impellers.

2.1.2. Impellers Based on Thermosets

Thermosetting polymer can be defined as a soft solid or viscous state prepolymer that can be changed to the infusible, insoluble polymer network (thermoset) by curing. Curing of the prepolymer can be performed based on heating or suitable radiation. During the curing, cross linking the materials leads to them setting and they can no longer flow [55]. The main components of thermosets consist of monomers, co-monomers (hardeners), catalysts and initiators. Also for improving the mechanical properties and reducing the costs, some fillers such as calcium carbonate, sawdust, recycled powdered thermosets, etc., can be used in the formulation of thermosets [56]. In addition, the use of the short fiber to improve the mechanical properties is one of the important ways to increase the mechanical properties [57], which will be discussed in the composite section. Thermosets are divided into epoxy resins, phenolic resins, amine–formaldehyde, polyurethanes, silicones, cyanates, vinyl esters, dicyclopentadiene and other metathesis thermosets. Depending on the different formulations, different physical and mechanical properties can be achieved from the thermosets. For example, glass transition temperature can vary in the range of 20 to 200 °C [58]. Modulus can be achieved for light-cured resin [59] and continuous carbon fiber reinforced thermosetting composites [60] in the range of 0.18 and 161.4 GPa, respectively. Thermoset manufacturing processing is divided into categories: additive manufacturing techniques; solid thermoset processing; and liquid thermoset processing [61].

Matveev et al. [62] studied the thermoset impellers fabricated by the SLA method. High chemical resistance, practically inert to liquid hydrocarbons (gasoline, kerosene, petroleum and synthetic oils) and hot streams (up to 100 °C) of water and air fabrication were some requirements of their study. By their study, thermoset impellers fabricated by 3D printing were considered as parts that fully meet the requirements of the experimental samples for gas-dynamic studies. In the case of the turbocharger, Andrearczyk et al. [9] investigated the wheel printed by a thermoset. The size of rotor was 42.5 mm in diameter (Figure 2a) and the maximum rotational speed was selected at 100,000 rpm. According to the simulation results, the maximum stress on the impeller for a rotational speed of 90,000 rpm was 27 MPa, whereas the yield stress of the resin printed was 54 MPa. We should mention that the results obtained from the compressor wheels, which were fabricated by polymer and aluminum, showed that at 90,000 rpm the polymeric wheel can operate like an aluminum wheel (Figure 2b).

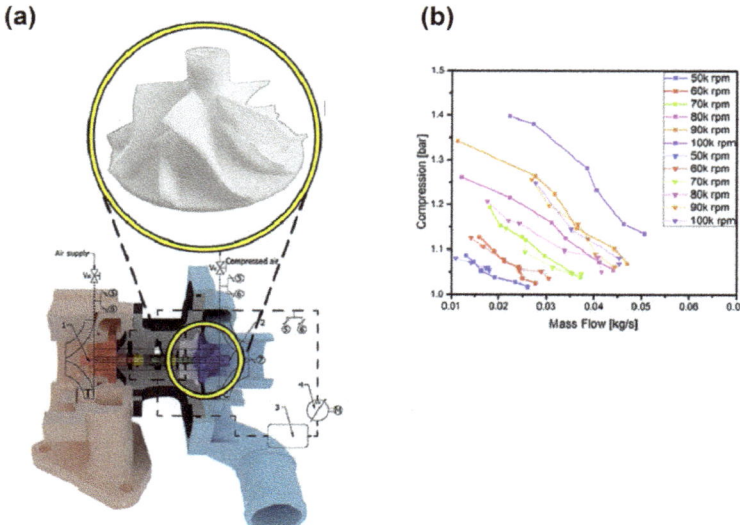

Figure 2. (**a**) Cross section of turbocharger with compressor wheel fabricated by MJP and (**b**) experimental results of polymeric (dotted lines) and aluminum (solid lines) compressor wheels (reprinted with permission from [9]).

2.2. Polymer Composites Impellers

The key role of composite materials, considering their weight reduction effect without sacrificing robustness, has been shown in the modern industry [63]. In this section, only the polymer matrix composite will be discussed. In the polymer matrix composite, the mechanical properties of the materials will be improved by using the fiber as reinforcement in the matrix of the polymer. Considering the type of matrix and reinforcement, the composites can have different categories. In general, reinforcement fiber can be divided into inorganic, glass and carbon fibers [64], and composite reinforced with either glass fibers (GF) or carbon fibers (CF) have been included in more than 90% of the studies [65]. Thermoplastic [66] or thermoset [67] polymers have been used as polymeric matrices for fabrication of impellers.

2.2.1. Carbon Fiber as Reinforcement in Fabrication of Impellers

Carbon fiber polymer-matrix composites have been introduced as one of the efficient classes of material, in the place of metals. Depending on the type of fiber condition, be it short or continuous, these types of composites can be classified. PPS, PEEK, PI and PEI are the thermoplastics and epoxy is the thermoset, which has been widely used as a matrix for these composites [68]. PEEK [52,69] and epoxy [69,70], carbon-fiber-reinforced, are the composites that have been most used in the fabrication of impellers.

PEEK reinforced with carbon fibers has been one of the exemplary composites used in the fabrication of rotary components of pumps and compressors [71,72]. PEEK composites reinforced with polyacrylonitrile short carbon fibers, 30% in weight, which is called CF30 PEEK is a famous commercial type of this composite [73]. Garcia-Gonzalez et al. [74] investigated the energy absorbed to analyze the mechanical impact behavior of short carbon fiber reinforced PEEK composites and unfilled PEEK. Tensile elastic modulus of GF30 PEEK in transversal, longitudinal conditions and unfilled fiber PEEK were 12.6, 24 and 3.6 GPa, respectively. According to their results, reinforced composites showed a brittle failure. The direction of fibers and degree of crystallization played a key role in the mechanical properties. The homogenization of elastic material and anisotropic damage for failure prediction has been proposed in their study. Investigating the vapor-grown carbon nanofibers for use as reinforcement for PEEK showed that by increasing the nanofiber, the modulus of composites increases, which refers to the effect of fiber on the crystallization of PEEK. Also, the effect of carbon nanofiber as a lubricant, which was associated with significant decreases in the wear rate of the composite, was shown [75]. Yang et al. [76] studied the effect of the surface modification of carbon fiber on the mechanical properties of CF PEEK composites. Their method was introduced as one of the main solutions to enhancing the interface and led to reaching an interfacial shear strength of 83.13 MPa.

The investigation of the PEEK carbon-reinforced impeller in the case of a microturbine-generator introduced this composite as a suitable material instead of an aluminum impeller [11,36]. We should mention that the mechanical properties of the impeller at the rotational speed of 32,040 and 40,500 rpm were appropriate. However, considering the importance of the chemical reaction of working fluid with impeller in ORC and refrigeration systems, the final approval of the substance was considered dependent on more studies [36]. Martynyuk et al. [12] investigated using the polymer reinforced with carbon fiber for the fabrication of a centrifugal compressor wheel. The maximum working temperature was 287 °C, and the outlet pressure was 7 bar. The calculations showed that carbon fibers UMT 49S and phthalonitrile binder PN-3M, which have been used as a reinforcing part, can be used to fabricate the wheel of a centrifugal compressor. Also, the use of composite resulted in a 45% reduction in rotor weight compared to the similar aluminum specimen.

Using the carbon fiber in the matrix of thermosets (especially epoxy) to improve the properties, has been used in order to fabricate impellers in pump and compressor applications [67,68,70]. In general, resisting moisture and other environmental influences, offers lower shrinkage and better mechanical properties are among the points that lead to the selection of epoxy resins as a polymeric matrix [77]. Shah et al. [78] studied the thermomechanical characterization of different types of epoxy resin of HinpoxyC, HinpoxyVB, ARL135 and ARL136 epoxy resin systems reinforced by HCU200/A45 carbon fiber. Tensile strength of the reinforced epoxy resin systems of HinpoxyC, HinpoxyVB, ARL135 and ARL136 were achieved 745, 752, 698 and 830 MPa, respectively. In general, the final properties can be variable depending on the type of matrix, carbon reinforcement, and fabrication process. For example, Ming et al. [79] studied the different parameters in 3D printed continuous carbon fiber reinforced thermosetting epoxy, such as printing speed, printing space, printing thickness, curing pressure and curing temperature with the aim of optimizing the parameters. According to their results, optimized conditions with 58 wt.% fiber led to them achieving the maximum flexural strength and modulus of 952.89 MPa and 74.05 GPa, respectively. Furthermore, Pérez-Pacheco et al. [80] studied the effect of moisture absorption on damage accumulation in carbon fiber–epoxy composites laminates with respect to the two different superficial carbon fiber treatments. In their studies, the interphase microstructure has been mentioned as a critical aspect of the moisture diffusion mechanism. Considering the number of hydrogen bonds between the water and epoxy resin network, and the two different activations of energy, subsequently, different phenomenon such as swelling or degradation can happen. Also, the sensibility of matrix failure mechanisms caused by hydrolysis has been discussed.

Uhlig et al. [81] studied the highly stressed bladed rotor fabricated by epoxy resin and reinforced with 60% carbon fiber. According to their results, the explosion rational frequency of the rotor was in the range of 1080–1100 Hz. Also, stress exposure factor for the fiber fracture at the explosion frequency range was about 0.8, so that this factor for the interfiber fracture should not have exceeded the failure limit of 1. The composite rotor has been considered as a suitable candidate to improve the efficiency compared to aluminum alloy rotors. Liu et al. [82] used the carbon fiber reinforced shape memory epoxy composites to fabricate the wind blades. The stiffness under good shape memory fixation at room temperature and switching temperature (Tsw) reached 37 and 4.4 GPa, respectively. According to their results, a sustainable continuous stable mechanical state has been observed. Also, variable wind speed in the range of 9–10 m s^{-1} was provided.

2.2.2. Glass Fiber as Reinforcement in Fabrication of Impellers

Composites of the polymeric matrix reinforced with glass fibers have been used for the fabrication of impellers [83]. In general glass fiber is an inorganic non-metallic material. Heat resistance, high tensile strength, and excellent chemical stability are among the properties of these fibers [64]. The composition of the glass fiber consists of SiO_2, Al_2O_3, TiO_2, B_2O_3, CaO, MgO, Na_2O, K_2O and Fe_2O_3. Differences in the composition of fibers lead to the appearance of various properties, which puts them in different categories. Different Young's modulus can be achieved in the range of 51.7 to 86.9 GPa [82,84].

Fan et al. [85] investigated the diffusion of water in glass fiber reinforced polymer composites, with respect to room temperature and 50 °C. According to the micromechanics model, moisture diffusion in the GFRP in deionized water for matrix at room temperature and 50 °C were 1.41×10^{-7} and 4.57×10^{-7} mm^2/s, respectively. Also, there was good agreement between the model and the experimental data. The diffusivity of fluid in the GFRP composites in the fiber was smaller than that of the polymeric resin, however it was not negligible. In such a way, the lateral fiber diffusivity was determining the factor that would control the thickness diffusivity of the GFRP plates. Nayak and Ray [86] investigated the residual mechanical properties of nano-Al_2O_3 filled glass fiber reinforced polymer composites (nano-GFRP) in the hydrothermal environment. According to the results, the addition of 0.1 wt.% of nano-Al_2O_3 into the GFRP composite has reduced the moisture diffusion coefficient to 10%. However, this addition has led to the improvement in the residual flexural and interlaminar shear strength by 16 and 17%, respectively, compared to the GFRP. They showed that nano-GFRP has created an opportunity to use this composite in a hydrothermal environment. Improvement in the wear properties of composites reinforced with glass fiber is another important factor that attracted the attention of a lot of studies [87,88]. Öztürk et al. [89] analyzed the erosive wear behavior of the epoxy resin and the glass-fiber-reinforcement was evaluated with respect to the different parameters, such as various impingement angles (from 20° to 90°), velocity in the range of 70–200 mm/s, exposure time and erodent size. According to their results, the best erosion resistance has been achieved for composite filled with 16 wt.% silica fume.

Umaras et al. [29] studied the impeller fabricated by reinforced PPS with 40% fiberglass in the application of an automotive water pump. The rotational speed was 4500 rpm, and the working fluid was water and ethylene-glycol in the temperature range of 80–100 °C. Maximum radial and tangential stresses due to the press fit on the impeller have been calculated 100 and 83.5 MPa, respectively. Considering the physicomechanical properties of PPS reinforced with glass and maximum stress on the impeller this material has been considered a suitable selection in this situation. In the case of a micro-turbine-generator, PEEK-GF30 radial turbine impeller has been considered as one of the suitable candidates. Isaias et al. [36] investigated the fabrication of a micro-turbine-generator for an Organic Rankine Cycle (ORC) with the aim of replacing polymers with metals, and they considered the high technical and economic potential of polymers. For this purpose, the rotor was produced by the FDM process by using polymer and composite materials and the diameter of impeller was 45 mm. The results showed the ability to rotate the rotor at a rotational speed of 32,040 rpm and a peak rotational speed of 40,500 rpm. Also, due to the importance of the final surface obtained for the rotors and impellers on the final efficiency [90], it was shown that an acceptable final surface has been obtained by this method for the fabricated rotor. In another study, Organic Rankine Cycle microturbines fabricated by aluminum, ABS and PEEK-G30 were compared together. The diameter and rotational speed of the impeller were 49 mm and 36,000 rpm and the R245fa fluid was used as a working fluid. According to their study, in addition to the fact that the PEEK-GF30 and ABS showed they were suitable for mass production processes, the economic benefits, properties such as chemical resistance and lower inertia, with the latter characteristic helping to minimize imbalance, shaft fatigue, and damage of the casing in case of failure, were all among the other advantages they found. In addition, results of a simulation showed that PEEK-GF30 and ABS can be good candidates in these operating conditions and are good alternatives to aluminum in this application [11].

In general, required properties for working conditions, economic efficiency, and manufacturing methods are among the parameters that can impact the selection of polymers or polymer composites to fabricate the impellers. Also, recyclability of the materials, and any ecological problems the present, are also important parameters to consider when choosing the materials. So, thermoplastics can show more compatibility to this end, compared to the thermosets and polymer composites.

3. Manufacturing Process

The important effect of the manufacturing method can be clearly be linked to the energy consumed during the process and impeller performance [1,3]. The ability to be mass produced, attainability of suitable mechanical properties, a good surface, high precision for complex geometries, and economic efficiency are among the criteria to be considered when selecting the fabrication process of impellers. Additive manufacturing, milling and injection molding have been among the methods used to fabricate the polymeric and composite impellers.

3.1. Conventional Impeller Manufacturing

Conventional impeller manufacturing has consisted of a process in which the impeller is fabricated through the machining processes in a subtractive way. In this way, by removing additional layers, the desired shape with different accuracy will be manufactured depending on the selected machining and parameters process. Turning, milling, drilling and grinding are among the conventional manufacturing which can be applied for machining polymeric and composite materials [91].

Mainly, parameters that impact the final polymer or composite machined product can be divided into three categories: machine and environmental variables; tool design and machining conditions; and composition of the substances. The machine and environmental parameters such as slide straightness, temperature stability and vibration are the general parameters that controlled the dimension on a large scale. Surface roughness and delamination factor have been the effect by tool design and machining conditions such as rake angle, tip radius, depth of cut and cutting speed. Another important parameter referred to is the composition, which depends on the different physical and chemical properties as machinability will be variant. Also when comparing the polymers and composites, polymers are more homogeneous and have been accompanied by better machining capabilities. Delamination, cracking, fiber pull-out, and burning are among the defects which can happen during composites machining.

Among the different machining process, milling has been severally applied to the fabrication of polymeric and composite impellers [11,36,92,93]. The most common milling machining can be divided into the peripheral milling or profiling and end milling. In this method, extra material will be removed by rotating a cutterhead with control based on computer numerical control (CNC) which is called CNC milling [94].

Hernandez-Carrillo et al. [36] investigated the PEEK-GF30 impeller fabricated by five-axis CNC milling. According to their study, fabricated impeller showed good surface and acceptable mechanical properties against the centrifugal force load. Mentzos et al. [92] investigated the polymeric impeller fabricated by the CNC milling process in a pump application. Effect of process parameters such as cutting speed, feed rate and depth of cut has been considered in the final roughness surface of the impeller. Their results showed that by reducing the tool step-over and feed rate a smoother surface was obtained.

3.2. Injection Molded Impellers

Injection molding consists of four-steps of the cyclic process that includes the phases of filling, packing, cooling, and ejection to fabric the parts. Granola, or powder under pressure, and temperature will be molted and used to fill the mold. Depending on several parameters, such as raw materials, mold design, and process-specific parameters the final quality of the parts can be different [95]. Cost reduction and production in a short time have been among the parameters which attracted the attention of studies to fabricate the injected impellers [96,97]. In general, the cost of fabricated parts can be estimated by the parameters such as mold base, number of cavities, and injection mold. As to increasing the performance of the hydraulic pump from the injected impeller, adjusting the distance of the front shroud and rear shroud, namely the impeller outlet width, has been the most economical way to increase efficiency through the injected impeller [29,96,98].

Process parameters of injection molding such as molding temperature, melt temperature, injection pressure, and injection time have been introduced as important parameters of the properties and cost of fabricated parts [99]. The optimization of this parameter for the fabrication of the injected impellers attracted the attention of a lot of studies.

Rosli et al. [100] have studied the optimization of process parameters of injected blower impeller fans with respect to the melting temperature, molding temperature, injection time and injection pressure processing parameters. Polypropylene has been used to fabrication the impellers. Given that in the response surface methodology the results optimum of mold temperature, melt temperature, injection time and injection pressure have been 110 °C, 210 °C, 0.8 s and 212.81 MPa., Shen et al. [101] has investigated the mold cooling design optimization for fabrication of the injected impeller. The melting temperature of the polymer was 230 °C. According to their study, the maximum ejection temperature of the impeller had reduced from 53 °C to 33 °C with a conformal cooling channel. Also, their mold had provided a higher cooling efficiency and a more uniform cooling. Topology optimization to reduce the mass of the mold showed the total mass reduction was about 20%.

3.3. Additive Manufacturing (AM) in Fabrication of Impellers

Additive manufacturing (AM) is a manufacturing process based on the fabrication of parts by joining the materials, directly from the 3d model. Since 1980 some sources have fabricated part of what was produced directly by a suitably formatted data file in a layer-by-layer fashion, and this can therefore be considered as the start date of AM processing. Over the past years, due to the significant increase in mechanical properties obtained by the production of AM, this method has been introduced as a desirable and reliable method for the production of parts. AM technology can be classified to the liquid polymer, discrete particle, molten material and solid shield systems. The use of AM to rotor and impeller fabrications is no exception and has attracted the attention of many studies. Selective laser melting (SLM) [102,103], electron beam melting (EBM) [104,105], stereolithography (SLA), fusion deposit modelling (FDM) and MultiJet printing technique (MJP) are the AM methods which have been used for rotor and impeller fabrications.

One of the most common and low-cost methods among the AM process is FDM. In this method the melting of the thermoplastic filament from a nozzle at a certain speed means parts can be prepared [106]. In several studies FDM has been used to produce the compressor and pump impellers. Quail et al. investigated the pump impeller by FDM method. Caille et al. [107] studied the pump impeller fabricated by FDM method. The impeller was 75 mm in diameter and 1.3 mm in thickness. The impeller was fabricated for use in a pump by a 3-kW induction motor and rotational speed of 3000 rpm. According to their results the impeller produced by this method was ensured to show qualities that included strength, performance and speed of manufacture. In a study by Fernandez et al. [13] FDM-impeller fabricated for pump purpose has been analyzed. Their study showed that, the impeller fabricated by FDM had a similar performance to the original impeller of the rotodynamic hydraulic pump. Also, considering the important effect of impeller roughness on the performance of compressors and pumps, the inherent roughness of the external impeller surfaces had no limitation in the results of the head-flow curve of the pump. In another study by Priyanka and Varaprasada Rao [108] they concluded that an impeller produced by FDM can be introduced as a deserving method, which can replace traditional manufacturing techniques in the industries. However, the difference of mechanical properties in various directions (anisotropy) has been one of the problems that have always existed in parts made by the FDM process [109]. Badarinath et al. [110] investigated the development and characterizes the performance of a robotic FDM system instead of FDM machines based on the three-axis cartesian system. They showed that an impeller fabricated by this method had regions of infill and perimeter overlap in the base and was free from voids and over-deposition. The results also indicated that a uniform deposition at regions of directional changes have parts with complex geometry. This method can be used for the fabrication of the parts with more complex geometries. Weiß et al. [111] studied the potential of additive manufacturing for the fabrication of different components of the ORC system by introducing a micro-turbine-generator-construction-kit (MTG-c-kit) in a customized turbogenerator (Figure 3a). As such, the air turbine was fabricated at 120 mm in diameter. Selective Laser Sintering (SLS) and FDM were the additive manufacturing methods used for the fabrication of this turbine through the Nylon and ABS, respectively. In this study, achieving a good surface quality of fabricated rotor was mentioned as an important parameter that can impact performance. The surface quality of the fabricated rotor through the SLS and FDM is shown in Figure 3b. According to their results for the fabricated wheel with Nylon by SLS, the first performance of the system was around 10%, which with a sufficient sealing of the surroundings of the stator nozzles increased to 20%. This maximum efficiency was achieved at the design pressure ratio and rotational speed of 1.4 bar and 6000 rpm, respectively. Also considering the better surface quality of the FDM compared to the SLS, the performance in the pressure ratio of 1.6 was similar. We should be mention that the notches on the surface of the FDM rotor were perpendicular to the airflow.

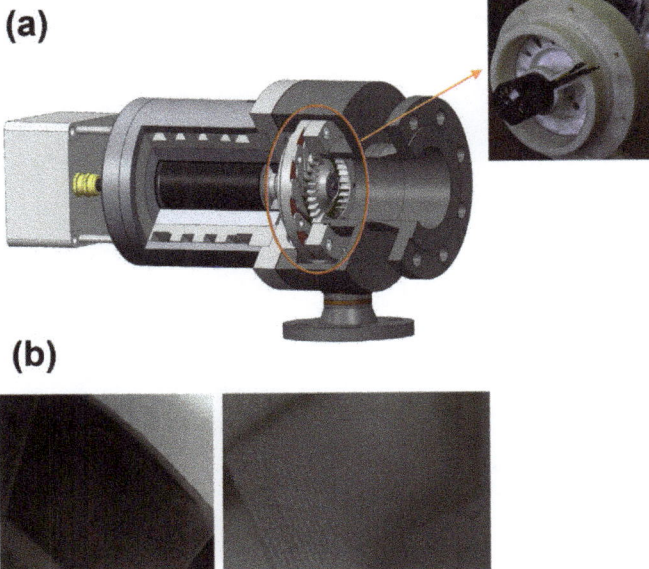

Figure 3. (a) Schematic of developed radial quasi impulse cantilever design and (b) left: fabricated rotor by SLS right: fabricated stator by FDM+SLS (reprinted with permission from [111]).

Stereolithography (SLA) is a technique where the polymerization of a photocurable liquid monomer in a spatially selective manner occurs using an ultraviolet light (or a laser). The 3D structure is achieved by alternating between the thin liquid films and spatially controlled photopolymerization steps [112]. Generally, rotors or impellers fabricated by this method are used in two goals: directly used as a rotor or impeller, or used in casting methods [113]. In an industrial case, Przybylski et al. [114] investigated the SLA-impeller fabricated for pump. The pump rotor, which was in the medium range size, analyzed the pump at the nominal load and at 3000 rpm rotational speed. According to their study, the rotor produced by this method had acceptable results compared to rotors that were commonly used.

In the study by Isaias et al. [11], the AM method was applied for the fabrication of an impeller for the Organic Rankine Cycle (ORC) radial microturbine. In this study, a thermoplastic impeller (ABS) with a diameter of 49 mm was used for rotation at a speed of 36,000 rpm. The results of their study showed that the rotor made by this method could compete with the aluminum sample. In addition, the latter characteristic helped to minimize imbalance, shaft fatigue, and damage of the casing in case of failure and were among the advantages achieved by this method.

Inkjet printing (IJP) can be defined as a technology for printing by depositing tiny droplets onto a substrate without dependence on the high-speed operation of mechanical printing elements. Polyjet and multijet printing are famous processes based on this technique [115]. In this method photocurable resin by a piezo printed process can be layer by layer fabricated. The high precision of this method to fabricate the complex geometries is one of the highlight features of this technique. Studies have introduced this technique as a high-performance method for making rotors and impellers [116]. Andrearczyk and Żywica [117] fabricated the compressor wheel and turbine wheel by MJP with the purpose of using them in the turbocharger. According to their study, the pressure range of the compressor was in the range of 0 to 1.6 MPa. Also, considering the tensile strength of this photopolymer, which mentioned around 65 MPa, and the working temperature of 88 °C, this process by this type of photopolymer was an assured method to fabricate the mentioned parts of the turbocharger. A study by Artur Andrearczyk et al. [9] using the MJP investigated the range of applications of this method concerning design, testing, and optimization of the elements of fluid-flow machines. In this study, the polymeric fabricated rotor was in the range of 42.5 mm, which was applied in the compressor inlet of the turbocharger (turbochargers are the fluid-flow machines with one of the highest nominal rotational speeds). The temperature range was 50 to 150 °C for the inlet air of this machine. The rotational speed up to 100,000 rpm was analyzed in this experiment. We should mention that the results obtained from the compressor wheels, which were fabricated by polymer and aluminum, showed that at 90,000 rpm the polymeric wheel can operate like the aluminum wheel. While the tensile strength and glass transition of the standard experiment of this polymer, which has been used in this study, differed from the aluminum wheel. Khalil et al. [118] studied the effect of different blade heights of rotors on the performance of a micro-scale axial turbine. Different parts of this system, with respect to the exploded assembly drawing, are shown in Figure 4a. In this study, different parts of this system such as reducer, stator, disc, rotor, rotor case, and closing were fabricated by a Polyjet printing technique and using a resin (RGD525) material. The three different blade height sizes of the rotors were 4, 6, and 8 mm (Figure 4b). The experiment was performed in environment temperature and at pressure ratios in the range of 1.2 to 1.75, and rotational speeds of 4000 to 16,000 rpm. According to their results, fabricated rotors with blade heights of 4, 6, and 8 mm were able to produce power up to 630.75, 694.1, and 796.89 W, respectively, at an expansion ratio around 1.75 and through the rotational speed of around 16,000 rpm.

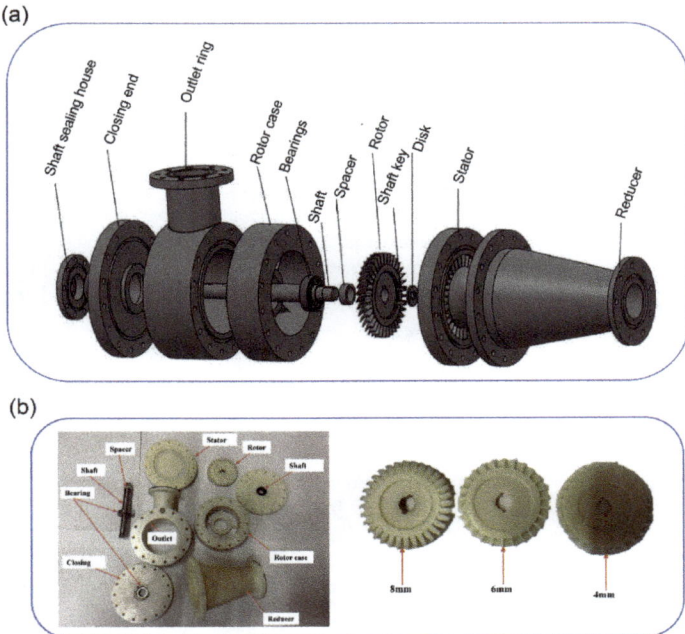

Figure 4. (**a**) Exploded schematic of micro-scale axial turbine and (**b**) left: fabricated all turbine parts and right: fabricated rotors with different blades heights (reprinted with permission from [118]).

4. Structural Stress Analysis

Numerical analyses for parts designed with consideration of working conditions have been one of the most critical steps in the investigation of materials [119]. Regarding impellers there are no exceptions to this, and numerical evaluations have always been one of the vital steps in the investigation of materials for this component. In other words, numerical analyses can be considered as a tool to balance performance and reliability during the development and design of the products [120].

In this section, numerically analyzing the centrifugal impellers and rotors of compressors will be discussed. As mentioned, properties such as modulus of elasticity, thermal expansion, fracture toughness, fatigue strength, thermal conductivity, specific heat capacity, corrosion resistance, and thermal stability have been among the parameters considered in the selection of materials for the fabrication of impellers and rotors of compressors [121]. We should mention that centrifugal stress due to rotational forces, bending stress due to fluid pressure and change of momentum, and thermal stresses due to thermomechanical load are the properties that were considered as effective factors in the simulations. So, by examining the materials under the mentioned forces, the authority of the material will be assayed in simulation [122].

Structural analysis is the most common application of finite element analysis which allows for the investigation of different types of loads, including stress, strain, deformation, and so on. The linear structural static equations are as follow [123]:

$$\frac{\partial \sigma_x}{\partial x} + \frac{\partial \tau_{xy}}{\partial y} + \frac{\partial \tau_{xz}}{\partial z} + F_{bx} = 0 \qquad (1)$$

$$\frac{\partial \tau_{yx}}{\partial x} + \frac{\partial \sigma_y}{\partial y} + \frac{\partial \tau_{yz}}{\partial z} + F_{by} = 0 \qquad (2)$$

$$\frac{\partial \tau_{zx}}{\partial x} + \frac{\partial \tau_{zy}}{\partial y} + \frac{\partial \sigma_z}{\partial z} + F_{bz} = 0 \qquad (3)$$

where σ represents the normal stress, τ shows the shear stress, Fbx, Fby and Fbz are the body forces per unit volume acting along the directions x, y, and z, respectively.

Centrifugal force per area of the blade appears as centrifugal stresses, which can be written generally as follows:

$$F_C = mr\omega^2 \qquad (4)$$

where F_C represents the centrifugal force, ω is a rotational speed, r and m show the radius and mass of the considered section, respectively.

In the thermal condition, control equations for linear elastic and isotropic three-dimensional solid materials to considering the thermal load are as follows [124]:

$$\varepsilon = D^{-1} \cdot \sigma + \alpha \cdot \Delta T \qquad (5)$$

$$\Delta T = T - T_{Ref} \qquad (6)$$

where ε, σ, T_{Ref}, D and α are total strain vector, stress vector, referenced temperature, material elastic stiffness matrix and matrix of thermal expansion coefficient, respectively.

Fluid-structure interaction (FSI) is the approach that can be used for structural examination of the impeller. In this method, the effect of the fluid dynamics on the structural mechanics of the impeller based on computational fluid dynamics (CFD) and structural finite element analysis (FEA) can be analyzed [125]. Solving the Reynolds-averaged Navier–Stokes (RANS) equations has been one of the most used methods in CFD [126]. The equations are as follows [124]:

$$\nabla(\rho V) = 0 \qquad (7)$$

$$\nabla(\rho V V) = \rho f - \nabla p + \nabla T \qquad (8)$$

$$\nabla(\rho h V) = \rho f \cdot V + \nabla T \cdot V - \nabla q \qquad (9)$$

where V, f, p, T, h and q are velocity vector of the fluid, body force vector per unit mass, pressure, viscous stress tensor, volumetric enthalpy and heat flux vector, respectively.

Andrearczyk et al. [9] investigated the plastic wheel in turbocharger application. In this experiment, the rotor has been fabricated by the MultiJet 3D printing method with a diameter of 42.5 mm. For calculation strength analyses, rotational speed and temperature were sets of 100,000 rpm and 55 °C, respectively. According to their results, the maximum deformation and the maximum stress on the impeller geometry were 192 μm and 27 MPa, respectively. Figure 5 shows the stress and deformation results. Also, the performance of the compressor was simulated based on solving RANS by ANSYS to achieve the compressor performance map. Their results showed the streamlines in a relative reference frame, in which the rotational speed and mass flow rate were 200,000 rpm and 0.09 kg/s, respectively.

Kar et al. [18] studied a polymeric impeller in centrifugal pump applications. The impeller had a diameter of around 5.08 cm and has been fabricated by polyetherimide. Structural stress analysis was performed by finite element analysis (FEA) under maximum centrifugal conditions with respect to the rotational speed of 72,000 rpm and gravitational load. Their results showed that the maximum stress on the impeller was 5.45 MPa.

Isaias et al. [11] investigated a different polymer, composite and metal impeller together with the aim of evaluating the feasibility of developing a simplified turboexpander. Calculations under different loadings were examined in this study. Full load conditions, a rotor blocked, which is full flow and being supplied with the rotor stopped, and 27% overspeed due to the consideration of international standards were the conditions which were considered. The rotational speed was 36,000 rpm and the diameter of impeller was 49 mm. Their results showed that the obtained stress showed high sensitivity to rotational force and pressure loading. The equivalent stress on the PEEK impeller in different conditions of loading has been performed and according to their results the maximum stress on the blade was 10 MPa. Also, the value of the factor of safety for aluminum, PEEK reinforced with

30% glass fiber and ABS were 19.92, 22.25 and 13.32, respectively, with an analysis of 27% over-speed. These values were greater than the minimum requirement. Considering their simulation, additional stress caused by temperature was analyzed. The simulated efficiency was 86%, 0%, and 84% for the full load, rotor blocked and 27% over-speed situations, respectively.

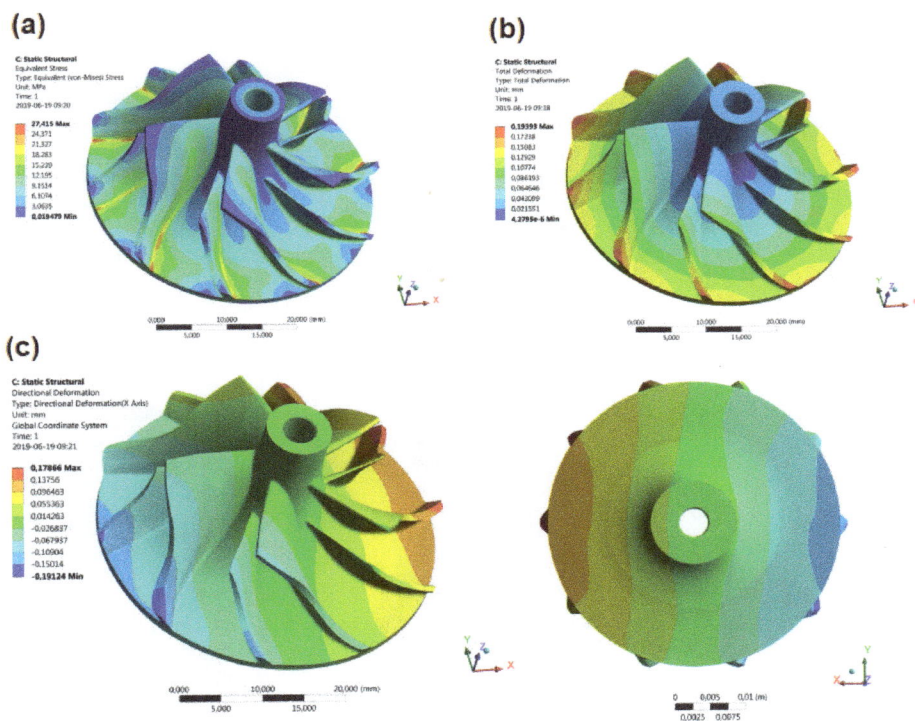

Figure 5. (a) stress distribution, (b) deformation distribution (c) deformation in the X as isometric bottom view on the rotor (reprinted with permission from [9]).

5. Performance Evaluation

An accurate performance evaluation of the impeller in the pump, compressor or turbine is essential due to it confirming the ability of the machine to respond to working conditions, as well as the correct energy consumption. In this regard, performance investigation is possible through the analysis of work done on working fluid. In general, inlet temperature, inlet pressure, discharge temperature, discharge pressure, rotational speed, differential pressure across flow meter (or pitot traverse), temperature and pressure at the flow meter are among the measurements to determine the machine performance [127,128]. In addition, investigation of vibration has been considered to confirm the dynamic performance. Furthermore, the tribological behavior of the impeller under working conditions is another parameter that should be considered. Depending on the application, such as the pump or compressor, the environment of the test bench would be different [129]. We should mention that performance tests will be performed to achieve the compressor map, which represents the corrected flow versus the pressure rise at various aerodynamic speeds [130].

5.1. Pressure Measurements

Measuring the pressure is essential to investigating the performance of the system. So, during the experiment input and output pressure are measured by a pressure gauge. Depending on the kind of pressure being measured such as static, or dynamic pressure, the installation place of the gauge can be different. For example, output pressure can be measured by a pressure gauge installed along the discharge path.

Mojaddam and Torshizi [131] studied the impeller hub and shroud of a radial flow compressor by implementing different meridional contours on the same impeller characteristics. Evaluation of performance was performed by considering the pressure ration from a compressor inlet to a diffuser outlet. According to the comparison of compressor performances for both cases, in circular and elliptical hub and shroud curves, pressure ratio and the isentropic efficiency in different mass flow rates at a fixed rotational speed have been shown. The difference in pressure ratio was minimal at low rotational speed, so that the maximum difference was 1.4% at the highest mass flow rate. For their design rotational speed differences were considerable at 3% and at the maximum mass flow rate at approximately 10%. Also at high rotational speed, the pressure ratio was the same. In conclusion, pressure ratio and total-to-total isentropic efficiency for both impellers along with inlet section and vane-less diffuser were selected to evaluate the newly suggested curves. Their results showed that the elliptical curves have acceptable performance in comparison with circular curves.

Li et al. [132] investigated the impact of the blade angle of the plastic impeller on the performance of the centrifugal pump. They analyzed the pressure fluctuation at the outlet of the impeller. According to their results increasing the outlet angle and inlet angle played a key role in machining the optimal performance. The best results have been achieved for an outlet angle of 35° plastic impeller, in which efficiency and head were 81.02% and 35.80 m, respectively. In general, the optimal performance of the printed impeller has been shown according to the simulation and experimental results.

5.2. Mass Flow Measurements

To analyze the system performance and achieve the compressor and pump maps, measurement of the mass flow is essential. This parameter can be measured by a flow meter. Mass flow rate refers to the product of the working fluid density, the cross-sectional area and the flow velocity [133]:

$$\dot{m} = \rho \, A \, V \tag{10}$$

In the case of compressors, the corrected mass flow rate can be calculated by the following equation [134]:

$$\dot{m}^* = \dot{m} \left[\frac{P_{ref}}{P_{in,0}} \right] \sqrt{\frac{T_{in,0}}{T_{ref}}} \tag{11}$$

where $T_{in,0}$ and $P_{in,0}$ are the total temperature and total pressure at the compressor inlet, respectively, and T_{ref} and P_{ref} are 288.15 K and 1 atm, respectively.

Sun et al. [135] evaluated the influence of humidity on the performance of a centrifugal compressor. Their results showed that pressure ratio and peak isentropic efficiency have been decreased by increasing the humidity. Also, the variation in performance was analyzed by measuring the mass flow in different rotational speeds in humid and dry air conditions. The mechanism of influence on performance was analyzed by measuring the mass flow in dry and humid conditions. According to their results, at the same rotational speed, the mass flow of humid air was smaller than that of dry air. Figure 6 shows the differences between saturated humid air at 100% design rotational speed and its corresponding dry air.

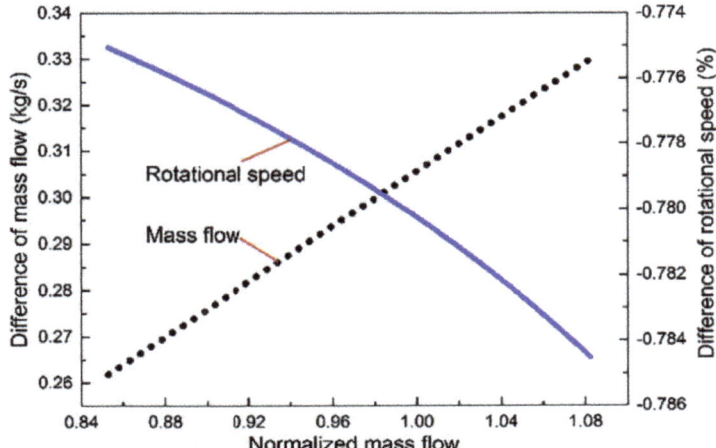

Figure 6. Differences between saturated humid and corresponding dry air (reprinted with permission from [135]).

5.3. Vibration Analysis of Impeller

The importance of analyzing the vibration of impellers has been mentioned in a lot of studies [136]. Increasing cyclic stress and fatigue failure, collision of the rotor with stationary parts, seized bearings, vibrating force transmission to stationary parts, and induced vibration of peripheral units have been among the problems due to rotors vibration. The natural frequency from the impeller or rotor vibration by the 3-D finite element method has been analyzed by considering the difference between the inertial coordinate system fixed to the stationary side and the rotational coordinate system fixed with the rotor. Due to the symmetry of the impellers with respect to the center of rotation, the analysis of rotating structures by 3-D finite elements is considered cyclic symmetry [137].

Vibration measurements can be performed by a vibration analyzer [138] or digital image correlation (DIC) by cameras [139]. In this way, amplitude in the different range of frequency or displacement magnitude against the time in a certain frequency is achievable. Neri et al. [140] described a measurement system to investigate the impeller damping (Figure 7a) by measuring the excitation force during the test with respect to the response and the load amplitude. The vibrational measurements in the high-frequency range of 2568 Hz, 6239 Hz and 6357 Hz showed a harmonic response and frequency values in combination with the low response amplitudes, in the range of 10 μm. Their results in the high frequency of 6239 Hz in the cylindrical reference frame coordinates showed a smooth map for all three directions (Figure 7b).

Mousmoulis et al. [141] studied the vibration of pump impeller considering the important effect of the cavitation in the steady and dynamic operation of a pump. In this regard, vibration measurements for the inception of cavitation have been performed in the frequency range of 5–10 kHz. Given that the impellers with different geometries have been considered, the lower incidence angle and the use of splitter blades showed the milder noise and vibration characteristics through the entire Thoma number range tested. Also, they mentioned that increases vibration band power at part load conditions can be due to the increasing turbulence intensity and the backflow cavitation mechanism.

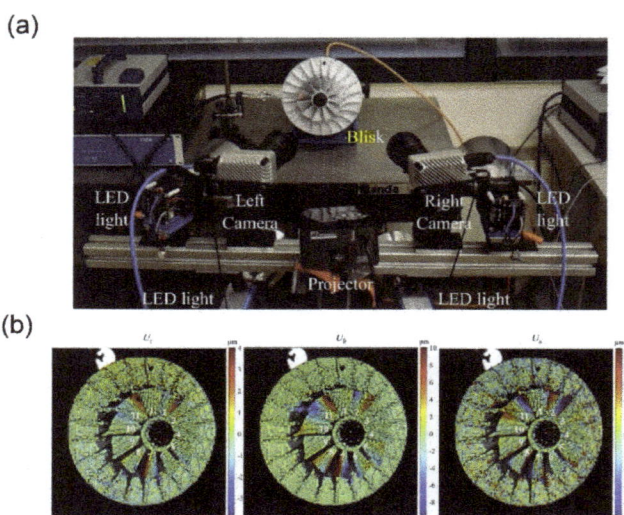

Figure 7. (a) Optical setup of the stereo-digital image correlation measurement and (b) Displacement maps for the 6239 Hz excitation frequency (reprinted with permission from [140]).

5.4. Tribology Behavior Analysis

Tribology can be defined as a science of surfaces in contact with each other and consists of friction, lubricant and wear [142]. Considering the interaction between working fluid and impeller, investigation of wear resistance and friction can play an important role in the efficiency of the machine. So, analyzing the tribology behavior, which is the science of interacting surfaces in relative motion, has a high position [143]. For example, different friction on the impeller caused by various viscosity of fluids impacts the performance of the system. On the other hand, the viscosity of fluid has an effect on the friction losses and can change the performance, such that increasing the fluid viscosity causes a reduction in performance [144,145]. So, the viscosity of the fluid can change the characteristic curve of the systems. Considering the characteristic curves, viscosity correction factors can be obtained and viscosity correction factor can be defined through the $\eta/\eta water$, where η and $\eta water$ are efficiency of viscous fluid and water, respectively [145].

The wear of the impeller during working is another important parameter that should be considered in analyzing the performance of systems. Wear is defined as the removal of material from a solid surface caused by friction or impact. Jiang [19] studied the wearing properties of the fabricated impellers of PLA, ABS and VeroGray by a 3D printing method. The wear test was performed in different concentrations of erodent material for 110 h. During the experiment after every 5 h, the weight loss of the impellers to calculate wear rate was measured. According to their primary results, the VeroGray impeller fabricated by Polyjet 3D printer highly reduced the experimental time and cost and was chosen as the impeller for analysis of the wear test. VeroGray impeller at a rotational speed of 1200 rpm and in presence of 5% concentrations of erodent during the 110 h showed a mass loss percentage around 8.08%.

Upadhyay et al. [14] investigated the tribotechnological and mechanical properties of PLA propeller blades in a marine application. They studied the friction and wear behavior with its sliding wear mechanism, due to the importance of tribotechnological, and studied soft and hard interfaces with degradation properties. Figure 8a shows the PLA-propeller blades and (b) represents the result of coefficient of friction (COF) of PLA sample for smooth and rough surfaces at a relative humidity of 40%. The average COF of PLA sample at the top and the bottom surface was 0.158 and 0.56, respectively, with respect to the immersion time, which was 30 days in seawater. The wear microstructure at the top and bottom surface of the PLA sample was shown in Figure 8c. Sliding at the polymer sample's top surface provides a smooth transition to the ball material due to low surface roughness. According to their results, they justified the immediate suitability of 3D printed PLA parts for practical marine application by designated tests of sliding and degradation.

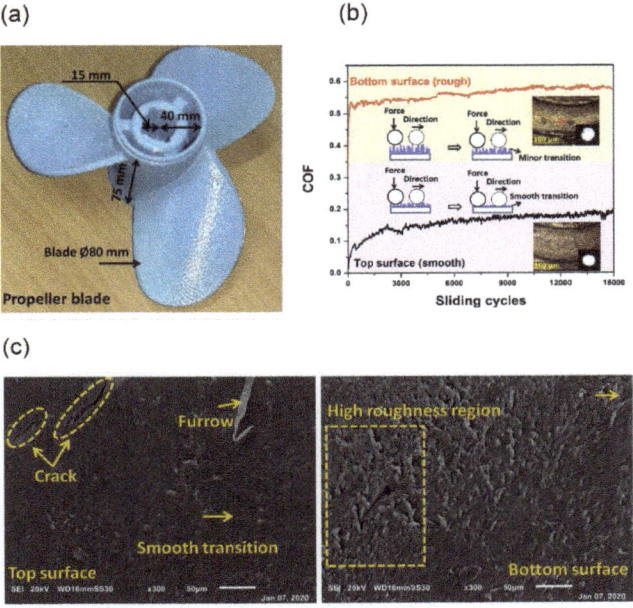

Figure 8. (**a**) 3D printed propeller blade, (**b**) Friction performance of PLA sample at the top and the bottom surface and (**c**) wear microstructure of PLA samples (reprinted with permission from [14]).

6. Conclusions

Studies have shown an increase in efficiency and effectiveness through the use of polymers. Among thermoplastic polymers and composites, PEEK and PPS with and without reinforcements were introduced as suitable options for the fabrication of impellers. On the other hand, in the case of turbochargers, the use of resins to make the wheels was recognized as a suitable option. Conventional impeller manufacturing, injection molding and additive manufacturing were the common methods in producing the impellers. In general, the softness of the surface obtained by various methods had an acceptable level for use in the field of compressors and pumps. However, the time-consuming nature of the 3D-printing method, for example, compared to the injection method will require further improvements. However, among the 3D printing methods, the SLA and MultiJet printing methods can provide a more appropriate surface than other methods due to the production process. High anisotropy in FFF samples is also a challenge in using this method in the construction of impellers, which requires more study. In injection molding and conventional machining methods, they can be examined in order to fabricate the geometries with less complexity compared to additive manufacturing. Fluid-structure interaction, as well as performance evaluations to analyze the impeller, are key steps in the final evaluation of impellers. In conclusion, the use of polymers and polymer composites were promising options as alternatives to metals in some applications. Limitations of plastic impellers have also been identified, including the potentially shorter lifespan of blades (compared with its metal counterpart) when moisture or impurities are present in the working fluid and need more research.

Author Contributions: Conceptualization, N.Z., M.S., M.D. and A.T.; validation, N.Z., M.S., M.D. and A.T.; formal analysis, N.Z., M.S., M.D. and A.T.; writing—original draft preparation, N.Z., M.S., M.D. and A.T.; writing—review and editing, N.Z., M.S., M.D. and A.T.; visualization, N.Z., M.S., M.D. and A.T.; supervision, M.S., M.D. and A.T.; project administration, M.S., M.D. and A.T.; funding acquisition, M.S., M.D. and A.T. All authors have read and agreed to the published version of the manuscript.

Funding: The authors thankfully acknowledge financial support provided by ADEME (ECOLOCAR project).

Institutional Review Board Statement: Not applicable.

Informed Consent Statement: Not applicable.

Data Availability Statement: Not applicable.

Conflicts of Interest: The authors declare no conflict of interest.

References

1. Peng, S.; Li, T.; Wang, X.; Dong, M.; Liu, Z.; Shi, J.; Zhang, H. Toward a sustainable impeller production: Environmental impact comparison of different impeller manufacturing methods. *J. Ind. Ecol.* **2017**, *21*, S216–S229. [CrossRef]
2. Boyce, M.P. *Centrifugal Compressors: A Basic Guide*; PennWell Corporation: Nashville, TN, USA, 2003.
3. González-Barrio, H.; Calleja-Ochoa, A.; Lamikiz, A.; de Lacalle, L.N.L. Manufacturing Processes of Integral Blade Rotors for Turbomachinery, Processes and New Approaches. *Appl. Sci.* **2020**, *10*, 3063. [CrossRef]
4. Andrearczyk, A.; Konieczny, B.; Sokołowski, J. Additively Manufactured Parts Made of a Polymer Material Used for the Experimental Verification of a Component of a High-Speed Machine with an Optimised Geometry—Preliminary Research. *Polymers* **2021**, *13*, 137. [CrossRef] [PubMed]
5. Agarwal, S.; High, K.M. Newer-generation ventricular assist devices. *Best Pract. Res. Clin. Anaesthesiol.* **2012**, *26*, 117–130. [CrossRef] [PubMed]
6. Bozorgasareh, H.; Khalesi, J.; Jafari, M.; Gazori, H.O. Performance improvement of mixed-flow centrifugal pumps with new impeller shrouds: Numerical and experimental investigations. *Renew. Energy* **2021**, *163*, 635–648. [CrossRef]
7. Novaković, T.; Ogris, M.; Prezelj, J. Validating impeller geometry optimization for sound quality based on psychoacoustics metrics. *Appl. Acoust.* **2020**, *157*, 107013. [CrossRef]
8. Dhere, M.M.C.; Badadhe, A.M.; Patil, A.S.; Bankar, M.B.; Tarange, A.K. Design, Analysis & Material Optimization of Submersible Pump Impeller by FEA & Experimentation. *Int. J. Sci. Technol. Eng.* **2018**, *5*, 46–55.
9. Andrearczyk, A.; Bagiński, P.; Klonowicz, P. Numerical and experimental investigations of a turbocharger with a compressor wheel made of additively manufactured plastic. *Int. J. Mech. Sci.* **2020**, *178*, 105613. [CrossRef]
10. Sun, J.; Chen, S.; Qu, Y.; Li, J. Review on stress corrosion and corrosion fatigue failure of centrifugal compressor impeller. *Chin. J. Mech. Eng.* **2015**, *28*, 217–225. [CrossRef]
11. Hernandez-Carrillo, I.; Wood, C.J.; Liu, H. Advanced materials for the impeller in an ORC radial microturbine. *Energy Procedia* **2017**, *129*, 1047–1054. [CrossRef]
12. Martynyuk, L.A.; Afanasiev, D.V.; Bykov, L.V.; Ezhov, A.D.; Mezintsev, M.A. The study of the applicability of polymer composite materials for the manufacture of the impeller of a centrifugal compressor. *IOP Conf. Ser. Mater. Sci. Eng.* **2021**, *1060*, 12026. [CrossRef]
13. Fernández, S.; Jiménez, M.; Porras, J.; Romero, L.; Espinosa, M.M.; Dominguez, M. Additive manufacturing and performance of functional hydraulic pump impellers in fused deposition modeling technology. *J. Mech. Des.* **2016**, *138*, 24501. [CrossRef]
14. Upadhyay, R.K.; Mishra, A.K.; Kumar, A. Mechanical degradation of 3D printed PLA in simulated marine environment. *Surf. Interfaces* **2020**, *21*, 100778. [CrossRef]
15. Zywica, G.; Kaczmarczyk, T.Z.; Ihnatowicz, E.; Baginski, P.; Andrearczyk, A. Application OF a heat resistant plastic IN a high-speed microturbine designed for the domestic ORC system. *Int. Semin. ORC Power Syst.* **2019**, 1–8.
16. Azevedo, T.F.; Cardoso, R.C.; da Silva, P.R.T.; Silva, A.S.; Griza, S. Analysis of turbo impeller rotor failure. *Eng. Fail. Anal.* **2016**, *63*, 12–20. [CrossRef]
17. Maier, R.; VINTILĂ, S.; Mihalache, R.; Vilag, V.; Sima, M.; Dragan, V. Decreasing the Mass of Turbomachinery Subansamblies Using Advanced Polymer Composites. *Mater. Plast.* **2019**, *56*, 687–692. [CrossRef]
18. Kar, N.K.; Hu, Y.; Kar, N.J.; Kar, R.J. Failure analysis of a polymer centrifugal impeller. *Case Stud. Eng. Fail. Anal.* **2015**, *4*, 1–7. [CrossRef]
19. Jiang, C. Investigating Impeller Wear and Its Effect on Pump Performance Using Soft Materials. 2019. Available online: https://doi.org/10.7939/r3-cdza-bx11 (accessed on 20 December 2020).
20. Gebäck, T.; Heintz, A. A Pore Scale Model for Osmotic Flow: Homogenization and Lattice Boltzmann Simulations. *Transp. Porous Media* **2019**, *126*, 161–176. [CrossRef]

21. Škorpík, J. 15. Shapes of Parts and Materials of Turbomachines. Available online: https://www.transformacni-technologie.cz/en_15_shapes-of-parts-and-materials-of-turbomachines_part-1.pdf (accessed on 20 December 2020).
22. Gad, S.E. Polymers. In *Encycl Toxicol*, 3rd ed.; Wexler, P., Ed.; Academic Press: Oxford, UK, 2014; pp. 1045–1050. Available online: https://www.sciencedirect.com/science/article/pii/B978012386454300912X (accessed on 20 December 2014).
23. Jafferson, J.M.; Chatterjee, D. A review on polymeric materials in additive manufacturing. *Mater. Today Proc.* **2021**, *46*, 1349–1365. [CrossRef]
24. Mallick, P.K. Thermoplastics and thermoplastic–matrix composites for lightweight automotive structures. In *Materials, Design and Manufacturing for Lightweight Vehicles*; Woodhead Publishing: Sawston, UK, 2021; pp. 187–228.
25. Ogorodnyk, O.; Martinsen, K. Monitoring and control for thermoplastics injection molding a review. *Procedia Cirp.* **2018**, *67*, 380–385. [CrossRef]
26. Garzon-Hernandez, S.; Arias, A.; Garcia-Gonzalez, D. A continuum constitutive model for FDM 3D printed thermoplastics. *Compos. Part B Eng.* **2020**, *201*, 108373. [CrossRef]
27. Drummer, D.; Rietzel, D.; Kühnlein, F. Development of a characterization approach for the sintering behavior of new thermoplastics for selective laser sintering. *Phys. Procedia* **2010**, *5*, 533–542. [CrossRef]
28. Ponticelli, G.S.; Tagliaferri, F.; Venettacci, S.; Horn, M.; Giannini, O.; Guarino, S. Re-Engineering of an Impeller for Submersible Electric Pump to Be Produced by Selective Laser Melting. *Appl. Sci.* **2021**, *11*, 7375. [CrossRef]
29. Umaras, E.; Barari, A.; Tsuzuki, M.S.G. Tolerance analysis based on Monte Carlo simulation: A case of an automotive water pump design optimization. *J. Intell. Manuf.* **2020**, *32*, 1883–1897. [CrossRef]
30. Garlotta, D. A literature review of poly (lactic acid). *J. Polym. Environ.* **2001**, *9*, 63–84. [CrossRef]
31. Olivera, S.; Muralidhara, H.B.; Venkatesh, K.; Gopalakrishna, K.; Vivek, C.S. Plating on acrylonitrile–butadiene–styrene (ABS) plastic: A review. *J. Mater. Sci.* **2016**, *51*, 3657–3674. [CrossRef]
32. Sastri, V.R. 8-High-Temperature Engineering Thermoplastics: Polysulfones, Polyimides, Polysulfides, Polyketones, Liquid Crystalline Polymers, and Fluoropolymers. In *Plast Med Devices*, 2nd ed.; Sastri, V.R., Ed.; William Andrew Publishing: Oxford, UK, 2014; pp. 173–213.
33. Dolzyk, G.; Jung, S. Tensile and fatigue analysis of 3D-printed polyethylene terephthalate glycol. *J. Fail. Anal. Prev.* **2019**, *19*, 511–518. [CrossRef]
34. Latko-Durałek, P.; Dydek, K.; Boczkowska, A. Thermal, rheological and mechanical properties of PETG/RPETG blends. *J. Polym. Environ.* **2019**, *27*, 2600–2606. [CrossRef]
35. Garcia-Gonzalez, D.; Rusinek, A.; Jankowiak, T.; Arias, A. Mechanical impact behavior of polyether–ether–ketone (PEEK). *Compos. Struct.* **2015**, *124*, 88–99. [CrossRef]
36. Hernandez-Carrillo, I.; Wood, C.; Liu, H. Development of a 1000 W organic Rankine cycle micro-turbine-generator using polymeric structural materials and its performance test with compressed air. *Energy Convers. Manag.* **2019**, *190*, 105–120. [CrossRef]
37. Pavlović, A.; Šljivić, M.; Kraisnik, M.; Ilić, J.; Anić, J. Polymers in additive manufacturing: The case of a water pump impeller. *FME Trans.* **2017**, *45*, 354–359. [CrossRef]
38. Polák, M. Behaviour of 3D printed impellers in performance tests of hydrodynamic pump. In Proceedings of the 7th International Conference on Trends in Agricultural Engineering, Prague, Czech Republic, 17–20 September 2019; pp. 17–20.
39. Premkumar, T.M.; Pushpak, V.; Krishna, K.V.; Reddy, D.G.; Kumar, N.S.; Hariram, V.; Seralathan, S.; Nakandhrakumar, R. Design and fusion deposit modelling of radial flow centrifugal pump. *Mater. Today Proc.* **2020**, *33*, 3497–3503. [CrossRef]
40. Jang, J.; Cho, K.; Yang, G.-H. Design and experimental study of dragonfly-inspired flexible blade to improve safety of drones. *IEEE Robot. Autom. Lett.* **2019**, *4*, 4200–4207. [CrossRef]
41. Li, T.; Wen, B.; Tian, Y.; Li, Z.; Wang, S. Numerical simulation and experimental analysis of small drone rotor blade polarimetry based on RCS and micro-Doppler signature. *IEEE Antennas Wirel. Propag. Lett.* **2018**, *18*, 187–191. [CrossRef]
42. Zirak, N.; Shirinbayan, M.; Farzaneh, S.; Tcharkhtchi, A. Effect of molecular weight on crystallization behavior of Poly Lactic Acid (PLA) under isotherm and non-isotherm conditions. *Polym. Adv. Technol.* **2021**. [CrossRef]
43. Kyzyrov, U.; Turgali, D. *Performance Enhancement of a Centrifugal Pump by Impeller Retrofitting*; Nazarbayev University School of Engineering and Digital Sciences: 2019. Available online: https://nur.nu.edu.kz/bitstream/handle/123456789/4476/Performance%20Enhancement%20of%20a%20Centrifugal%20Pump%20by%20Impeller%20Retrofitting.pdf;jsessionid=3A3B30B32BD593170B53A63F8FFC63C9?sequence=5 (accessed on 20 December 2020).
44. Kopparapu, R.; Mathew, S.; Siciliano, E.; Stasick, G.; Dias, M. Designing a Centrifugal Pump System for High Altitude Water Crises. 2017.
45. Birosz, M.T.; Andó, M.; Jeganmohan, S. Finite Element Method modeling of Additive Manufactured Compressor Wheel. *J. Inst. Eng. Ser. D* **2021**, *102*, 79–85. [CrossRef]
46. Yu, Z.; Lei, J.; Ou, Y.; Yang, G. Toughening of polyethylene terephthalate/amorphous copolyester blends with a maleated thermoplastic elastomer. *J. Appl. Polym. Sci.* **2003**, *89*, 797–805. [CrossRef]
47. Machalski, A.; Skrzypacz, J.; Szulc, P.; Błoński, D. Experimental and numerical research on influence of winglets arrangement on vortex pump performance. *J. Phys. Conf. Ser.* **2021**, *1741*, 12019. [CrossRef]
48. Odetti, A.; Altosole, M.; Bruzzone, G.; Caccia, M.; Viviani, M. Design and construction of a modular pump-jet thruster for autonomous surface vehicle operations in extremely shallow water. *J. Mar. Sci. Eng.* **2019**, *7*, 222. [CrossRef]

49. Huynh, A.V.; Stein, P.; Buhr, E.D. 3D-printed assistive pipetting system for gel electrophoresis for technicians with low acuity vision. *Biotechniques* **2020**, *70*, 49–53. [CrossRef] [PubMed]
50. Ling, X.; Jing, X.; Zhang, C.; Chen, S. Polyether Ether Ketone (PEEK) Properties and Its Application Status. *IOP Conf. Ser. Earth Environ. Sci.* **2020**, *453*, 12080. [CrossRef]
51. Platt, D.K. *Engineering and High Performance Plastics Market Report: A Rapra Market Report*; Smithers Rapra Publishing: Shrewsbury, UK, 2003.
52. Berry, D. Use of Victrex® PEEK™ Thermoplastic to Drive New Designs, Processing Flexibility, and Cost Reduction in Aerospace Components. *SAE Trans.* **2002**, *111*, 426–431.
53. Pedersen, K.; Bengtsson, A.F.; Edlund, J.S.; Eriksson, L.C. Sulphate-controlled diversity of subterranean microbial communities over depth in deep groundwater with opposing gradients of sulphate and methane. *Geomicrobiol. J.* **2014**, *31*, 617–631. [CrossRef]
54. Pedersen, K. Analysis of copper corrosion in compacted bentonite clay as a function of clay density and growth conditions for sulfate-reducing bacteria. *J. Appl. Microbiol.* **2010**, *108*, 1094–1104. [CrossRef] [PubMed]
55. Gotro, J.; Prime, R.B. Thermosets. *Encycl. Polym. Sci. Technol.* **2002**, 1–75.
56. Pascault, J.-P.; Williams, R.J.J. Overview of thermosets: Structure, properties and processing for advanced applications. *Thermosets* **2012**, 3–27.
57. Li, J.H.; Huang, X.D.; Durandet, Y.; Ruan, D. A review of the mechanical properties of additively manufactured fiber reinforced composites. *IOP Conf. Ser. Mater. Sci. Eng.* **2021**, *1067*, 12105. [CrossRef]
58. Mullins, M.J.; Liu, D.; Sue, H.-J. *Mechanical properties of thermosets*; Thermosets: Woodhead, UK, 2018; pp. 35–68.
59. Sano, Y.; Matsuzaki, R.; Ueda, M.; Todoroki, A.; Hirano, Y. 3D Printing of Discontinuous and Continuous Fibre Composites Using Stereolithography. *Addit. Manuf.* **2018**, *24*, 521–527. Available online: https://www.sciencedirect.com/science/article/pii/S2214860418303282 (accessed on 20 December 2018). [CrossRef]
60. Hao, W.; Liu, Y.; Zhou, H.; Chen, H.; Fang, D. Preparation and Characterization of 3D Printed Continuous Carbon Fiber Reinforced Thermosetting Composites. *Polym. Test.* **2018**, *65*, 29–34. Available online: https://www.sciencedirect.com/science/article/pii/S0142941817314149 (accessed on 10 December 2018). [CrossRef]
61. Biron, M. *Thermosets and Composites: Material Selection, Applications, Manufacturing and Cost Analysis*; Elsevier: Amsterdam, The Netherlands, 2013.
62. Matveev, V.N.; Shabliy, L.S.; Krivcov, A.V. Application of stereolithography prototypes for gas dynamic tests and visualization. *J. Phys. Conf. Ser.* **2017**, *803*, 12097. [CrossRef]
63. Paul, R.C.; Ramachandran, B.; Sushma, G.; Harshavardhan, K.H.A.; Rohith, I. An empirical research on areca fiber polymer composite for automotive components in modern industry. *Mater. Today Proc.* **2020**, *33*, 4493–4497. [CrossRef]
64. Wang, R.-M.; Zheng, S.-R.; Zheng, Y.G. *Polymer Matrix Composites and Technology*; Elsevier: Amsterdam, The Netherlands, 2011.
65. Kaundal, R. Role of process variables on solid particle erosion of polymer composites: A critical review. *Silicon* **2017**, *9*, 223–238. [CrossRef]
66. Mallick, V. Thermoplastic Composite Based Processing Technologies for High Performance Turbomachinery Components. *Compos. Part A Appl. Sci. Manuf.* **2001**, *32*, 1167–1173. Available online: https://www.sciencedirect.com/science/article/pii/S1359835X01000641 (accessed on 12 December 2001). [CrossRef]
67. Li, Q.; Piechna, J.; Mueller, N. Static, Dynamic and Failure Behavior of a Novel Axial Composite Impeller for Water Chiller. *ASME Int. Mech. Eng. Congr. Expo.* **2010**, *44298*, 81–87.
68. Chung, D. *Carbon Fiber Composites*; Elsevier: Amsterdam, The Netherlands, 2012.
69. Rashedi, A.; Sridhar, I.; Tseng, K.J. Multi-objective material selection for wind turbine blade and tower: Ashby's approach. *Mater. Des.* **2012**, *37*, 521–532. [CrossRef]
70. Patil, M.; Müller, N. Structural Analysis of Continuous Fiber Wound Composite Impellers of a Multistage High-Speed Counter Rotating Axial Compressor for Compressing Water Vapor (R-718) as Refrigerant Using Finite Element Analysis. *Mater. Des.* **2013**, *50*, 683–693. Available online: https://www.sciencedirect.com/science/article/pii/S0261306913002045 (accessed on 21 December 2013). [CrossRef]
71. Goerke, D.; Le Denmat, A.-L.; Schmidt, T.; Kocian, F.; Nicke, E. Aerodynamic and mechanical optimization of CF/PEEK blades of a counter rotating fan. In *Turbo Expo: Power for Land, Sea, and Air*; American Society of Mechanical Engineers: New York, NY, USA, 2012; pp. 21–33.
72. Wu, D.; Liu, Y.; Li, D.; Zhao, X.; Li, C. Effect of Materials on the Noise of a Water Hydraulic Pump Used in Submersible. *Ocean Eng.* **2017**, *131*, 107–113. Available online: https://www.sciencedirect.com/science/article/pii/S0029801816306060 (accessed on 20 December 2017). [CrossRef]
73. Henriques, B.; Fabris, D.; Mesquita-Guimarães, J.; Sousa, A.C.; Hammes, N.; Souza, J.C.; Silva, F.S.; Fredel, M.C. Influence of laser structuring of PEEK, PEEK-GF30 and PEEK-CF30 surfaces on the shear bond strength to a resin cement. *J. Mech. Behav. Biomed. Mater.* **2018**, *84*, 225–234. [CrossRef] [PubMed]
74. Garcia-Gonzalez, D.; Rodriguez-Millan, M.; Rusinek, A.; Arias, A. Investigation of mechanical impact behavior of short carbon-fiber-reinforced PEEK composites. *Compos. Struct.* **2015**, *133*, 1116–1126. [CrossRef]
75. Kumar, D.; Rajmohan, T.; Venkatachalapathi, S. Wear behavior of PEEK matrix composites: A review. *Mater. Today Proc.* **2018**, *5*, 14583–14589. [CrossRef]

76. Yang, Y.; Wang, T.; Wang, S.; Cong, X.; Zhang, S.; Zhang, M.; Luan, J.; Wang, G. Strong Interface Construction of Carbon Fiber–reinforced PEEK Composites: An Efficient Method for Modifying Carbon Fiber with Crystalline PEEK. *Macromol. Rapid Commun.* **2020**, *41*, 2000001. [CrossRef] [PubMed]
77. Suresha, B.; Ramesh, B.N.; Subbaya, K.M.; Chandramohan, G. Mechanical and three-body abrasive wear behavior of carbon-epoxy composite with and without graphite filler. *J. Compos. Mater.* **2010**, *44*, 2509–2519. [CrossRef]
78. Shah, D.B.; Patel, K.M.; Joshi, S.J.; Modi, B.A.; Patel, A.I.; Pariyal, V. Thermo-mechanical characterization of carbon fiber composites with different epoxy resin systems. *Thermochim. Acta* **2019**, *676*, 39–46. [CrossRef]
79. Ming, Y.; Zhang, S.; Han, W.; Wang, B.; Duan, Y.; Xiao, H. Investigation on process parameters of 3D printed continuous carbon fiber-reinforced thermosetting epoxy composites. *Addit. Manuf.* **2020**, *33*, 101184. [CrossRef]
80. Pérez-Pacheco, E.; Cauich-Cupul, J.I.; Valadez-González, A.; Herrera-Franco, P.J. Effect of moisture absorption on the mechanical behavior of carbon fiber/epoxy matrix composites. *J. Mater. Sci.* **2013**, *48*, 1873–1882. [CrossRef]
81. Uhlig, K.; Spickenheuer, A.; Bittrich, L.; Heinrich, G. Development of a highly stressed bladed rotor made of a cfrp using the tailored fiber placement technology. *Mech. Compos. Mater.* **2013**, *49*, 201–210. [CrossRef]
82. Liu, Y.; Guo, Y.; Zhao, J.; Chen, X.; Zhang, H.; Hu, G.; Yu, X.; Zhang, Z. Carbon fiber reinforced shape memory epoxy composites with superior mechanical performances. *Compos. Sci. Technol.* **2019**, *177*, 49–56. Available online: https://www.sciencedirect.com/science/article/pii/S0266353818325223 (accessed on 25 February 2019). [CrossRef]
83. Higbee, R.W.; Giacomelli, J.J.; Wyczalkowski, W.R. Advanced impeller design: Anti-ragging impeller, ARI2. *Chem. Eng. Res. Des.* **2013**, *91*, 2190–2197. [CrossRef]
84. Morampudi, P.; Namala, K.K.; Gajjela, Y.K.; Barath, M.; Prudhvi, G. Review on glass fiber reinforced polymer composites. *Mater. Today Proc.* **2021**, *43*, 314–319. Available online: https://www.sciencedirect.com/science/article/pii/S2214785320393445 (accessed on 16 January 2021). [CrossRef]
85. Fan, Y.; Gomez, A.; Ferraro, S.; Pinto, B.; Muliana, A.; La Saponara, V. Diffusion of water in glass fiber reinforced polymer composites at different temperatures. *J. Compos. Mater.* **2019**, *53*, 1097–1110. [CrossRef]
86. Nayak, R.K.; Ray, B.C. Water absorption, residual mechanical and thermal properties of hydrothermally conditioned nano-Al_2O_3 enhanced glass fiber reinforced polymer composites. *Polym. Bull.* **2017**, *74*, 4175–4194. [CrossRef]
87. Pun, A.K.; Singh, A.K. Thermo-mechanical and erosion wear peculiarity of hybrid composites filled with micro and nano silicon dioxide fillers–a comparative Study. *Silicon* **2019**, *11*, 1885–1901. [CrossRef]
88. Mohan, N.; Mahesha, C.R.; Rajaprakash, B.M. Erosive wear behaviour of WC filled glass epoxy composites. *Procedia Eng.* **2013**, *68*, 694–702. [CrossRef]
89. Öztürk, B.; Gedikli, H.; Kılıçarslan, Y.S. Erosive wear characteristics of E-glass fiber reinforced silica fume and zinc oxide-filled epoxy resin composites. *Polym. Compos.* **2020**, *41*, 326–337. [CrossRef]
90. Gülich, J.F. Effect of Reynolds number and surface roughness on the efficiency of centrifugal pumps. *J. Fluids Eng.* **2003**, *125*, 670–679. [CrossRef]
91. Caggiano, A. Machining of fibre reinforced plastic composite materials. *Materials* **2018**, *11*, 442. [CrossRef]
92. Mentzos, M.D.; Markopoulos, A.P.; Galanis, N.I.; Margaris, D.P.; Manolakos, D.E. Design, Numerical Analysis and Manufacture of Radial Pump Impellers with Various Blade Geometries. *Int. Rev. Mech. Eng. (IREME)* **2015**, *9*, 104. [CrossRef]
93. Quail, F.J.; Scanlon, T.; Strickland, M. Development of a regenerative pump impeller using rapid manufacturing techniques. *Rapid Prototyp. J.* **2010**, *16*, 337–344. [CrossRef]
94. Sheikh-Ahmad, J.Y. *Machining of Polymer Composites*; Springer: Berlin/Heidelberg, Germany, 2009.
95. Khosravani, M.R.; Nasiri, S. Injection molding manufacturing process: Review of case-based reasoning applications. *J. Intell. Manuf.* **2020**, *31*, 847–864. [CrossRef]
96. Shi, W.; Zhou, L.; Lu, W.; Pei, B.; Lang, T. Numerical prediction and performance experiment in a deep-well centrifugal pump with different impeller outlet width. *Chin. J. Mech. Eng.* **2013**, *26*, 46–52. [CrossRef]
97. Kim, J.-M.; Chai, S.-H.; Yoon, M.-H.; Hong, J.-P. Plastic injection molded rotor of concentrated flux-type ferrite magnet motor for dual-clutch transmission. *IEEE Trans. Magn.* **2015**, *51*, 1–4. [CrossRef]
98. Bari, K. Experimental and simulation performance for fan extraction system. *Glob. J. Res. Eng.* **2017**, *17*, 5–26.
99. Fernandes, C.; Pontes, A.J.; Viana, J.C.; Gaspar-Cunha, A. Modeling and Optimization of the Injection-Molding Process: A Review. *Adv. Polym. Technol.* **2018**, *37*, 429–449. [CrossRef]
100. Rosli, M.U.; Termizi, S.N.A.A.; Khor, C.Y.; Nawi, M.A.M.; Omar, A.A.; Ishak, M.I. Optimisation of Process Parameters in Plastic Injection Moulding Simulation for Blower Impeller's Fan Using Response Surface Methodology. In *Intelligent Manufacturing and Mechatronics*; Springer: Singapore, 2021; pp. 309–318.
101. Shen, Z.; Zheng, J.; Hu, D.; Meng, W.; Jiao, Z. Intelligent Mold Tooling Design with Plastic Injection, CFD and Structural Simulation. Available online: https://www.3ds.com/fileadmin/PRODUCTS-SERVICES/SIMULIA/Resources-center/PDF/2018-SAoE-Intelligent_Mold_Tooling_Design_with_Plastic_Injection__CFD_and_Structural_Simulation.pdf (accessed on 20 December 2021).
102. Arifin, M.; Wahono, B.; Junianto, E.; Pasek, A.D. Process manufacture rotor radial turbo-expander for small scale organic Rankine cycles using selective laser melting machine. *Energy Procedia* **2015**, *68*, 305–310. [CrossRef]
103. Jia, D.; Li, F.; Yuan, Z. 3D-printing process design of lattice compressor impeller based on residual stress and deformation. *Sci. Rep.* **2020**, *10*, 1–11. [CrossRef] [PubMed]

104. Tang, H.P.; Wang, Q.B.; Yang, G.Y.; Gu, J.; Liu, N.; Jia, L.; Qian, M. A honeycomb-structured Ti-6Al-4V oil–gas separation rotor additively manufactured by selective electron beam melting for aero-engine applications. *JOM* **2016**, *68*, 799–805. [CrossRef]
105. Tan, X.; Kok, Y.; Tor, S.B.; Chua, C.K. Application of electron beam melting (EBM) in additive manufacturing of an impeller. In Proceedings of the 1st International Conference on Progress in Additive Manufacturing (Pro-AM 2014), Singapore, 26–28 May 2014. pp. 327–332.
106. Solomon, I.J.; Sevvel, P.; Gunasekaran, J. A review on the various processing parameters in FDM. *Mater. Today Proc.* **2021**, *37*, 509–514. [CrossRef]
107. Quail, F.; Stickland, M.; Scanlon, T. *Rapid Manufacturing Technique Used in the Development of a Regenerative Pump Impeller*; International Association of Engineers: Hong Kong, China, 2009; pp. 1730–1736.
108. Priyanka, G.; Rao, M.V. Design and Additive Manufacturing of Pump Impeller using 3D Printing Technology. *Int. J. Sci. Res. Sci. Eng. Technol.* **2018**, *4*, 2394–4099.
109. Gao, X.; Qi, S.; Kuang, X.; Su, Y.; Li, J.; Wang, D. Fused filament fabrication of polymer materials: A review of interlayer bond. *Addit. Manuf.* **2020**, *37*, 101658. [CrossRef]
110. Badarinath, R.; Prabhu, V. Integration and evaluation of robotic fused filament fabrication system. *Addit. Manuf.* **2021**, *41*, 101951. [CrossRef]
111. Weiß, A.P.; Novotný, V.; Popp, T.; Streit, P.; Špale, J.; Zinn, G.; Kolovratník, M. Customized ORC micro turbo-expanders-From 1D design to modular construction kit and prospects of additive manufacturing. *Energy* **2020**, *209*, 118407. [CrossRef]
112. Dikshit, V.; Goh, G.D.; Nagalingam, A.P.; Goh, G.L.; Yeong, W.Y. Recent progress in 3D printing of fiber-reinforced composite and nanocomposites. *Fiber-Reinf. Nanocomposites Fundam Appl.* **2020**, *1*, 371–394.
113. Budzik, G. Properties of made by different methods of RP impeller foundry patterns. *Arch. Foundry Eng.* **2007**, *7*, 83–86.
114. Przybylski, W.; Dzionk, S. Impeller pump development using rapid prototyping methods. *Adv. Manuf. Sci. Technol.* **2011**, *35*, 15–23.
115. Manthiram, A.; Zhao, X.; Li, W. Developments in membranes, catalysts and membrane electrode assemblies for direct methanol fuel cells (DMFCs). In *Functional Materials for Sustainable Energy Applications*; Woodhead Publishing: Sawston, UK, 2012; pp. 312–369.
116. Mieloszyk, M.; Andrearczyk, A.; Majewska, K.; Jurek, M.; Ostachowicz, W. Polymeric structure with embedded fiber Bragg grating sensor manufactured using multi-jet printing method. *Measurement* **2020**, *166*, 108229. [CrossRef]
117. Andrearczyk, A.; Żywica, G. A concept of a test stand for the investigation of a 3D printed turbochargers and selected fluid-flow machinery. *Trans. Inst. Fluid-Flow Mach.* **2016**, *133*, 3–11.
118. Khalil, K.M.; Mahmoud, S.; Al-Dadah, R.K. Experimental and numerical investigation of blade height effects on micro-scale axial turbines performance using compressed air open cycle. *Energy* **2020**, *211*, 118660. [CrossRef]
119. Mourtzis, D.; Doukas, M.; Bernidaki, D. Simulation in manufacturing: Review and challenges. *Procedia Cirp.* **2014**, *25*, 213–229. [CrossRef]
120. Delhelay, D.S. Nonlinear Finite Element Analysis of the Coupled Thermomechanical Behaviour of Turbine DISC Assemblies. Ph.D. Thesis, University of Toronto, Toronto, ON, Canada, 2001.
121. Dowson, P.; Bauer, D.; Laney, S. Selection of materials and material related processes for centrifugal compressors and steam turbines in the oil and petrochemical industry. In Proceedings of the 37th Turbomachinery Symposium, Houston, TX, USA, 8–11 September 2008; Texas A&M University, Turbomachinery Laboratories: College Station, TX, USA, 2008.
122. Hannun, R.M.; Radhi, H.I.; Essi, N.A. The types of mechanical and thermal stresses on the first stage rotor blade of a turbine. *Rev. Innovaciencia* **2019**, *7*, 1–11. [CrossRef]
123. Essi, N.A.; Hunnun, R.M.; Radhi, H.I. Prediction of mechanical stresses of single rotor blade of low pressure of Nasiriya power plant steam turbine. *Univ Thi-Qar J. Eng. Sci.* **2019**, *10*, 71–78. [CrossRef]
124. Zheng, X.; Ding, C. Effect of temperature and pressure on stress of impeller in axial-centrifugal combined compressor. *Adv. Mech. Eng.* **2016**, *8*, 1687814016653547. [CrossRef]
125. Benra, F.-K.; Dohmen, H.J.; Pei, J.; Schuster, S.; Wan, B. A comparison of one-way and two-way coupling methods for numerical analysis of fluid-structure interactions. *J. Appl. Math.* **2011**, 853560. [CrossRef]
126. *Turbulent Flows*; IOP Publishing: Bristol, UK, 2001.
127. Siggeirsson, E.M.V.; Gunnarsson, S. Conceptual Design Tool for Radial Turbines. Master's Thesis, Chalmers Univ. Technol Repro-Service Gothenburg, Gothenburg, Sweden, 2015.
128. Lüdtke, K.H. *Process Centrifugal Compressors: Basics, Function, Operation, Design, Application*; Springer Science & Business Media: Berlin/Heidelberg, Germany, 2004.
129. Gresh, T. *Compressor Performance: Aerodynamics for the User*; Butterworth-Heinemann: Oxford, UK, 2018.
130. Thin, K.C.; Khaing, M.M.; Aye, K.M. Design and performance analysis of centrifugal pump. *World Acad. Sci. Eng. Technol.* **2008**, *46*, 422–429.
131. Mojaddam, M.; Torshizi, S.A.M. Design and optimization of meridional profiles for the impeller of centrifugal compressors. *J. Mech. Sci. Technol.* **2017**, *31*, 4853–4861. [CrossRef]
132. Li, J.; Tang, L.; Zhang, Y. The influence of blade angle on the performance of plastic centrifugal pump. *Adv. Mater. Sci. Eng.* **2020**, 7205717. [CrossRef]

133. DiPippo, R. (Ed.) Chapter 8—Binary Cycle Power Plants. In *Geotherm Power Plants*, 4th ed.; Butterworth-Heinemann: Oxford, UK, 2016; pp. 193–239. Available online: https://www.sciencedirect.com/science/article/pii/B9780081008799000082 (accessed on 20 January 2016).
134. Sharma, S.; Broatch, A.; García-Tíscar, J.; Allport, J.M.; Nickson, A.K. Acoustic characterisation of a small high-speed centrifugal compressor with casing treatment: An experimental study. *Aerosp. Sci. Technol.* **2019**, *95*, 105518. [CrossRef]
135. Sun, J.; Zuo, Z.; Liang, Q.; Zhou, X.; Guo, W.; Chen, H. Theoretical and experimental study on effects of Humidity on Centrifugal compressor performance. *Appl. Therm. Eng.* **2020**, *174*, 115300. [CrossRef]
136. Quyet, N.T. Evaluation of spring-mass-damping models of mistuned turbomachine impellers to analyse vibration and fatigue life. *B oo oaoo oo a* **2020**, *24*, 756–767.
137. Matsushita, O.; Tanaka, M.; Kanki, H.; Kobayashi, M.; Keogh, P. *Vibrations of Rotating Machinery*; Springer: Berlin/Heidelberg, Germany, 2017.
138. More, K.C.; Dongre, S.; Deshmukh, G.P. Experimental and numerical analysis of vibrations in impeller of centrifugal blower. *SN Appl. Sci.* **2020**, *2*, 1–14. [CrossRef]
139. Barone, S.; Neri, P.; Paoli, A.; Razionale, A.V. Low-frame-rate single camera system for 3D full-field high-frequency vibration measurements. *Mech. Syst. Signal Process.* **2019**, *123*, 143–152. [CrossRef]
140. Neri, P.; Paoli, A.; Santus, C. Stereo-DIC Measurements of a Vibrating Bladed Disk: In-Depth Analysis of Full-Field Deformed Shapes. *Appl. Sci.* **2021**, *11*, 5430. [CrossRef]
141. Mousmoulis, G.; Karlsen-Davies, N.; Aggidis, G.; Anagnostopoulos, I.; Papantonis, D. Experimental analysis of cavitation in a centrifugal pump using acoustic emission, vibration measurements and flow visualization. *Eur. J. Mech.* **2019**, *75*, 300–311. [CrossRef]
142. McKeen, L.W. *Fatigue and Tribological Properties of Plastics and Elastomers*; William Andrew: Norwich, NY, USA, 2016.
143. Kumar, S.; Singh, K.K. Tribological behaviour of fibre-reinforced thermoset polymer composites: A review. *Proc. Inst. Mech. Eng. Part L J. Mater. Des. Appl.* **2020**, *234*, 1439–1449. [CrossRef]
144. Abazariyan, S.; Rafee, R.; Derakhshan, S. Experimental study of viscosity effects on a pump as turbine performance. *Renew. Energy* **2018**, *127*, 539–547. [CrossRef]
145. Shojaeefard, M.H.; Tahani, M.; Ehghaghi, M.B.; Fallahian, M.A.; Beglari, M. Numerical study of the effects of some geometric characteristics of a centrifugal pump impeller that pumps a viscous fluid. *Comput. Fluids* **2012**, *60*, 61–70. [CrossRef]

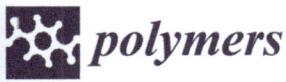

Article

Mathematical Modelling of Temperature Distribution in Selected Parts of FFF Printer during 3D Printing Process

Tomáš Tichý [1,*], Ondřej Šefl [2], Petr Veselý [1], Karel Dušek [1] and David Bušek [1]

1. Department of Electrotechnology, Czech Technical University in Prague, Technická 2, 166 27 Praha, Czech Republic; veselp13@fel.cvut.cz (P.V.); dusekk1@fel.cvut.cz (K.D.); busekd1@fel.cvut.cz (D.B.)
2. Department of Electrical Power Engineering, Czech Technical University in Prague, Technická 2, 166 27 Praha, Czech Republic; seflondr@fel.cvut.cz
* Correspondence: tichyto4@fel.cvut.cz

Abstract: This work presented an FEM (finite element method) mathematical model that describes the temperature distribution in different parts of a 3D printer based on additive manufacturing process using filament extrusion during its operation. Variation in properties also originate from inconsistent choices of process parameters employed by individual manufacturers. Therefore, a mathematical model that calculates temperature changes in the filament (and the resulting print) during an FFF (fused filament fabrication) process was deemed useful, as it can estimate otherwise immeasurable properties (such as the internal temperature of the filament during the printing). Two variants of the model (both static and dynamic) were presented in this work. They can provide the user with the material's thermal history during the print. Such knowledge may be used in further analyses of the resulting prints. Thanks to the dynamic model, the cooling of the material on the printing bed can be traced for various printing speeds. Both variants simulate the printing of a PLA (Polylactic acid) filament with the nozzle temperature of 220 °C, bed temperature of 60 °C, and printing speed of 5, 10, and 15 m/s, respectively.

Keywords: FFF; fused filament fabrication; additive manufacturing; temperature distribution modelling; dynamic and static model; thermal history

1. Introduction

3D printing (also known as Additive Manufacturing (AM)) has become widespread among industry; small series production and prototyping greatly benefits from its simplicity and especially versatility. AM can also be used to produce location-specific material compositions (heterogenous parts), also in terms of their structure [1], but the technique that has become widespread among general public recently are 3D printers based on polymer material extrusion. They are often referred to as Fused Filament Fabrication (FFF) technology and are popular since they usually have simple controls and are affordable. Increasingly, more and more companies around the world are becoming specialized in 3D printers or 3D printing materials production. The principle of the FFF printer is based on the extrusion of melted thermoplastic material and its deposition on the printing platform (bed) layer by layer. The material is in the form of a string (filament) with a standard diameter of 1.75 mm or 3 mm. Further description is based on the design of the 3D printer MK3S from Prusa Research company (Czech Republic), which represents one of the most common types of FFF printers. They are based on the RepRap concept, which is characterized by that part of the printer components is printed by another printer. It reduces the overall cost of the printer [2].

The string is pulled by drive gear through the printing head. It consists of a heatsink that dissipates heat and ensures that the filament is not melted prematurely, and a hot-end part with an extrusion nozzle. The hot-end is separated from the heatsink by a heat-break, which is a thin metal tube. It also helps to keep the filament unmelted before the hot-end and thereby ensures smooth movement of the filament through the extrusion head [2].

Common materials used in filament extrusion printing are PET-G (polyethylene terephthalate glycol-modified), ASA (acrylonitrile styrene acrylate), or PLA (polylactic acid). In this paper, we focused on PLA, which is probably the most widespread material for 3D printing due to its good printability and biodegradability. Mechanical properties of structures printed from PLA are comparable to other mentioned materials, and it also has potential for use in electrical engineering [3].

To achieve good printing quality, proper setting of the printing process is necessary. One of the crucial process parameters is the temperature of the nozzle and the bed [4]. Our work aimed to describe the temperature distribution around these components, and especially around the printed material (printed with the layer thickness of 150 µm). Since measuring the temperature distribution during printing is nearly impossible, we created a two-dimensional FEM (finite element method) mathematical model.

2. Models and Inputs

There have been numerous works dealing with thermal simulation of the 3D printing, sometimes with a validation on a real print [5], sometimes with a focus on printing of specific structures and geometries [1], and often on studying the effect of interlayer interactions [6,7].

This work focused on the area that has not been sufficiently covered in the above-mentioned works, and that is the area between the extruder nozzle and the preheated bed. In contrast to [5], where the axial symmetry was anticipated, the temperature model was calculated with respect to real nozzle shape. Rotational symmetry of the nozzle would not be applicable for the nozzle-bed interface, as the bed is not a rotationally symmetrical object.

Two models, static and dynamic, were created in this work. The static model provided the temperature distribution in the given geometry; hence, the temperature in places that cannot be normally accessed (and measured) may be estimated. Such an example is the printed material in the area just below the printer nozzle. Furthermore, the distribution of spatial temperature changes in the printer material (filament) can be visualized as well. In the case of the dynamic model, these changes are also observed in time and movement. The heat conduction in both models is described by the Fourier-Kirchhoff Equation:

$$\rho \cdot c_p \cdot \left(\frac{\partial T}{\partial t} + v \cdot \nabla T \right) + \nabla \cdot q = Q_v \qquad (1)$$

where the heat flux q is defined as:

$$q = \lambda \cdot \nabla T \qquad (2)$$

This form is not suitable for numerical solutions; hence, it is typically transformed into individual equations for each axis of the used geometry (2D or 3D). The terms in parenthesis are often expanded, so the temperature variable can be separated from the rest. Nevertheless, the form of Equation (1) illustrates the issue most satisfactorily. The terms ρ and c_p are material parameters: density and specific heat capacity, respectively. Another material parameter is the thermal conductivity λ. In static and stationary problems, the time and spatial changes are zero, and the temperature distribution is described only by the divergence of heat flux q and by heat source Q_v. In dynamic situations, the full Equation (1) must be used to describe both time and spatial changes of temperature.

2.1. Static Model and Inputs

Before the realization of the dynamic model, it was deemed beneficial to first create a simplified description of the problem using a static model of temperature distribution in the area of printer head, bed, and part of the printed material. As was stated earlier, the time change of parameters in the static model is zero; hence the employed Equation (1) can be significantly reduced. Moreover, no mutual movement of any parts was present, and therefore the model was stationary (spatial change of parameters is zero) as well. For calculation of field distribution, MUMPS (MUltifrontal Massively Parallel sparse direct Solver) [8] was used. The model's areas of interest represent the choice of substantial thermal-sensitive parts of the printer: heat block, extruder, heat sink, nozzle, printed material, and heated printer bed.

Figure 1 illustrates the modelled geometry, in which the blue lines highlight the boundary conditions, where the temperature was fixed at 220 °C.

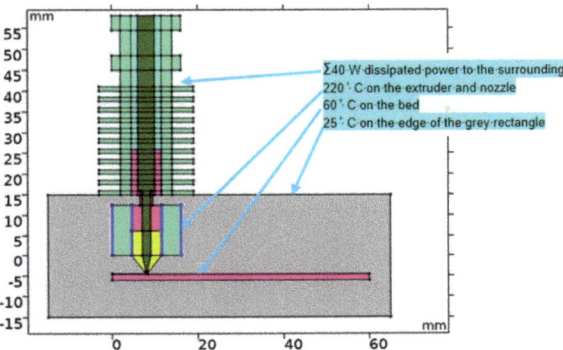

Figure 1. Geometry of the static model with boundary conditions.

On the edge of the heat sink (the upper part of the geometry where distinct fins are present), a boundary condition in the form of dissipated heating power into the ambient air was used. The total power dissipated from the surface was chosen as 40 W, as the resulting temperature distribution in the extruder axis corresponded to the one presented in a similar work [9] in which the airflow was modelled. No additional airflow was considered, as in this case, the printer was placed in a heated box in which the ambient temperature was practically constant. The grey area of Figure 1 represents the air domain, whose boundary condition was set to 25 °C (ambient temperature). This domain primarily serves to estimate the temperature distribution on the surface of the printed material and the nozzle. Lastly, the temperature of the printer bed was set at a constant 60 °C, which corresponds to the standard setting of PLA filament printing. All the employed boundary conditions can be seen in Table 1. Material parameters used in the simulation are listed in Table 2.

Table 1. Boundary conditions of the static model.

Edge Denotation	Boundary Condition
Problem boundary	25 °C
Inner part of the nozzle	220 °C
External part of the extruder	220 °C
Printer bed	60 °C
Outer part of the extruder (heat sink)	Power dissipated into surroundings (Σ40 W)

The model disregarded phase and temperature changes of these parameters for all materials. The reason is that the temperature of most parts does not practically change, except for the PLA filament and air. However, both thermal conductivity and specific heat capacity of air are very low, and hence their changes play virtually no role in the determination of temperature distribution. On the other hand, even though PLA's parameters are non-negligible, their changes in the range of the observed temperatures (60 to 220 °C) are marginal (only about 8%) [10], and hence only a small error in the calculation occurs by disregarding these changes.

Table 2. Material parameters of the static model.

Model Area	Material	Color	Thermal Conductivity $W \cdot m^{-1} \cdot K^{-1}$	Specific Heat Capacity $J \cdot kg^{-1} \cdot K^{-1}$
Printer bed	Steel	Purple	76	440
Printer nozzle	Brass	Yellow	120	400
Heat sink body	Aluminum	Darker turquoise	238	900
Extruder body	Aluminum	Darker turquoise	238	900
Printing material	PLA	Dark green	0.13	2100
Surroundings	Air	Grey	0.02	1.01
Teflon tube	Teflon	Lighter turquoise	0.24	1050

2.2. Dynamic Model and Inputs

During the printing process, the printer head performs various movements at different speeds. As a result, the temperature distribution in the printed material is different from the static case and changes according to the movement of the printer head. Hence, a dynamic model, which simulates the movement of the printer head and partially even the extrusion process, was created. Due to the dependence of parameters on both time and position, Equation (1) must be considered in its full form. For time-dependent calculation, the implicit BDF (Backward Differentiation Formula) solver with variable time stepping was used, while the MUMPS solver was employed for auxiliary static field distribution calculations.

Due to the complex nature of the simulation, only the geometry of the printer nozzle and bed was modelled (i.e., parts crucial in the extrusion and printing process). Modeling of the heat sink would increase the calculation time considerably, and its rather intricate surface could cause issues in mesh generation during the movement of the printer head. Moreover, the heat sink part does not influence the temperature distribution in the area of interest (nozzle and bed) and can thus be disregarded. The overall geometry used in the dynamic model is shown in Figure 2, which also includes the calculation mesh.

The mesh changes with each step of the calculation, and therefore the mesh in the area where the nozzle movement occurs was chosen as square shaped. The total number of elements remains constant; therefore, when the movement occurred and the nozzle in Figure 2 moved to the right, then the mesh before the nozzle (grid on the right side) condensed and the mesh located after the nozzle (in the figure on the left side) became loose.

The maximum mesh element size for the final calculation was set to 0.05 mm in the printed material (blue area on Figure 2). The calculation precision was of utmost priority in this area, as the required temperature distribution was obtained here. The impact of the mesh elements' volume in this area was evaluated carefully before the final solving; convergence of the results depending on the mesh quality was observed.

For this purpose, the temperature in the printed material under the nozzle was evaluated. The evaluation was carried out on the cut line in the middle of the height of printed material, and results for different mesh qualities can be seen in Figure 3.

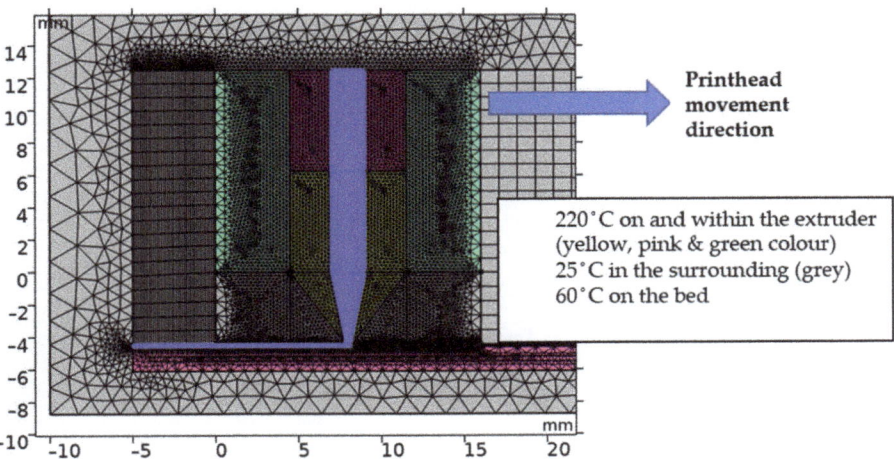

Figure 2. Geometry of the dynamic model with initial conditions; light turquoise—aluminum (heat block); blue—PLA (PLA material in the nozzle and on the bed); pink—steel (bed and heat break); yellow—brass (nozzle); gray—surroundings (air). The nozzle moves to the right.

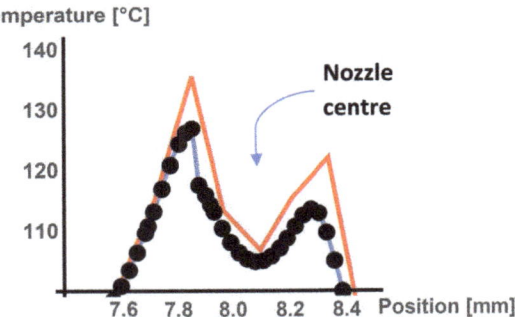

Figure 3. Mesh convergence evaluation under the nozzle. Black—extremely fine Mesh; blue—fine mesh; red—coarse mesh.

For results solved with coarse elements, we can see the different course (and peak value) of the temperature than for the results obtained with detailed meshing. When elements with a maximal dimension of 1 mm were used, the results differed by about 6%, whereas with 0.1 mm, the differences were not larger than 0.5% (both compared to results with max. element size 0.05 mm). Even though the latter maximal element size model returned satisfactorily accurate results, the triangle element maximal size of 0.05 mm was set for the final calculation for good measure. A summary of the mesh element size parameters is presented in Table 3 and includes the corresponding model computational time.

Table 3. Parameters for mesh quality check and computational time.

Name	Max. Triangle Element Size (mm)	Min. Triangle Element Size (mm)	Comp. Time (s)	Number of Mesh Elements	Number of Degrees of Freedom
Ex. Fine	0.05	0.025	288	101,740	214,341
Fine	0.1	0.05	246	73,386	157,449
Coarse	1	0.5	31	5536	21,297

The geometry includes an array of auxiliary construction parts, which are vital for the correct function of various parts of the model (movement simulation, suitable meshing). An example of these parts is the rectangular block on the right-hand side, which consists of quadrilateral elements. Such elements are much more suitable for horizontal movement than triangular elements, which would require continuous reconstruction during the movement.

The model simulated the movement of the printer head above the printer bed. Three different speeds were examined—5 mm/s, 10 mm/s, and 15 mm/s (see Table 4)—resulting in three different temperature distributions. The other parameters and conditions were identical to the static model.

Table 4. Parameters of time-dependent solver.

Speed (mm/s)	Stop Time (s)	Time Step (s)	Num. of Values (Time)
5	3	0.025	120
10	2.5	0.020	120
15	1.6	0.008	200

Since the model is time-dependent, initial conditions must be set as well. Naturally, they were based on the results of the static model. The temperature of the printer nozzle and bed was constant in time, as it is regulated, whereas the temperature distribution of the printed material changed due to the movement of the printer head.

In order to track the influence of the printer bed's heat capacity on the cooling process of the printed material, the constant temperature area (i.e., the heat source) was placed inside in bed rather than on its surface. In this manner, the local cooling processes of the printer bed can take place. Such setup corresponds to the actual configuration of a 3D printer, where a sheet is placed on the heated printer bed. Due to the static nature of air in this model, distinct areas with heightened temperature will be present in the movement path of the printer head.

Ultimately, the dynamic model should provide us with the distribution of temperature in time in the area inside and outside the extruder. Therefore, the thermal history of the printed material during the complete printing process can be estimated.

3. Results
3.1. Results from the Static Model

The temperature distribution obtained from the static model is shown in Figure 4. The temperature of the upper part of the heat sink reached about 60 °C, which corresponds to the temperature given in [9]. This value is determined by the dissipated power leaving the extruder surface. As stated earlier, this power was equal to 40 W.

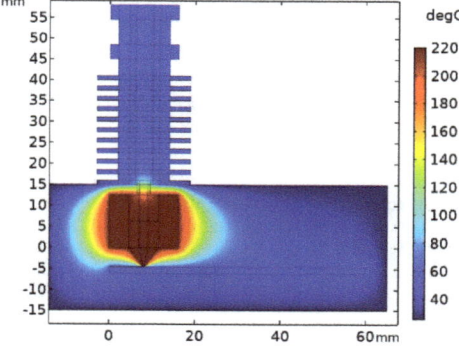

Figure 4. Static model output—temperature distribution in the examined area.

If we focus on the area near the heat block, it is evident that the temperature just below the heat break started to grow. The temperature started at about 100 °C in the upper part of the break and then reached a value of 220 °C as it entered the inner part of the heat block. On the other side of the heat block, the printed material exited via the nozzle and quickly cooled down to a temperature of 60 °C, i.e., the temperature of the printer bed.

This process is better illustrated in Figure 5, which shows the detail of the nozzle/bed interface (left) together with the temperature trend of the material along the deposited layer axis (right). However, the initial part of the axis is situated in the air (prior to the extrusion and deposition)—it is therefore shown as a dashed part of the resulting curve.

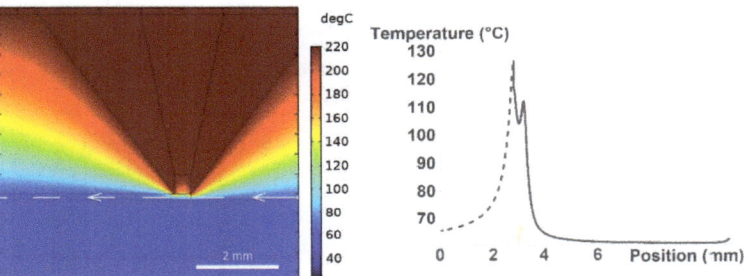

Figure 5. Detail of the printer nozzle/bed interface (left). Temperature distribution along the white line, where the white arrow denotes the direction of the x-axis (right). Printhead moves in the right direction.

3.2. Results from the Dynamic Model

As in the previous case, the dynamic model provides knowledge of temperature distribution in the given geometry. However, the distribution is now time-dependent, as the printer head performs the movement. Figure 6 shows the temperature distribution in and around the heat block at time $t = 5$ s with a simulated movement speed of 5 mm/s.

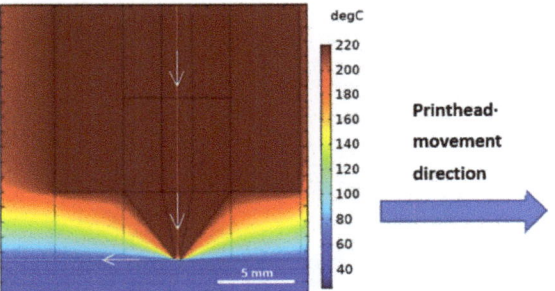

Figure 6. Temperature distribution at movement speed of 5 mm/s and at the time $t = 5$ s. Temperature distribution along the white line, where the white arrow denotes the direction of the x-axis is shown in Figure 7.

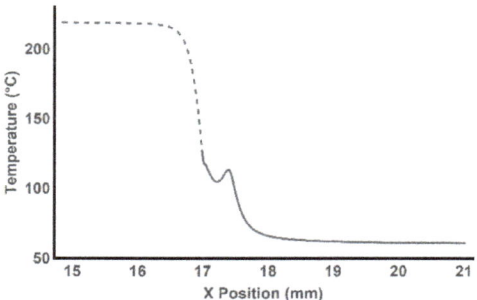

Figure 7. Temperature distribution in the material during printing (nozzle outlet is located on position 17.3 on the *x*-axis).

A thin white line representing the axis of the filament/deposited layer can be observed. Along this line, the thermal history of the material is examined, as illustrated in Figure 7.

Therein, the dashed line represents the movement of the printed material through the extruder. As the material exited the extruder via the nozzle, the line became solid.

A distinct "tooth" in the temperature trend was present—this phenomenon is caused by the subsequent movement of the printer head, during which the hot brass edge of the nozzle passes the material closely, increasing its temperature once again.

Furthermore, the temperature of the material on the printer bed can be examined for various times t. The temperature trend was obtained by placing a singular point above the printer bed in the middle of the imaginary layer that would be printed and 7 mm ahead of the nozzle. The temperature of this point was calculated for each step of the printer head movement. Figure 8 shows the results of these calculations for speeds 5 mm/s (blue line), 10 mm/s (green line), and 15 mm/s (red line).

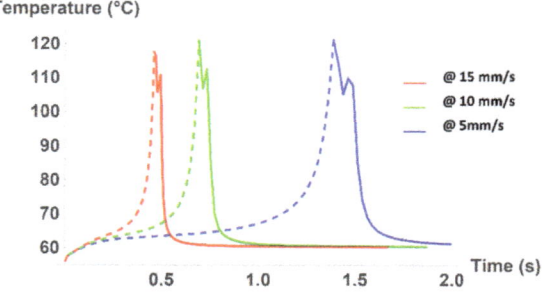

Figure 8. Temperature of the fixed-point during printing.

The slopes of the decreasing part of the temperature trends in Figure 8 were calculated in order to compare the individual cooling processes. The slope was considered from the second peak (i.e., from the "tooth") onwards, and was determined as the slope of a line passing through the curve intersection with 10% and 90% of the difference between the maximum and minimum temperatures. This is clearly visible from Table 5, where the temperature decline is numerically compared. This allows the evaluation of the movement speed effect. The temperature decline was initially linear with the increasing speed, but with faster speeds, the material could not absorb the heat energy and the deviation from the linear slope slightly increased. Therefore, it may be assumed that even higher speeds would further increase the temperature slope, which would significantly change the physical parameters of the print.

Table 5. Comparison of cooling processes of the material for various movement speeds of the printer head.

Movement Speed	Maximum "Second Tooth" Temperature (°C)	Minimum Temperature (°C)	Calculated Slope (°C/s)
5 mm/s	109.7	60.5	−201.7
10 mm/s	112.7	60.5	−404.1
15 mm/s	111.1	60.5	−620.2

4. Discussion

As a result of the static model, the temperature distribution in the extrusion head surroundings was presented. Remarkably, the influence of the nozzle on the material on the bed was apparent. The constantly heated printing nozzle with a different heat conductivity than the printed material and its geometry caused a typical "tooth" on the spatial temperature distribution.

In the dynamic model, a significant difference in the cooling process between different speeds of the extrusion head movement was observed.

The higher the speed of the head, the higher the slope of the temperature curve. When the speed is low, the material (in our case, semicrystalline PLA) absorbs more heat energy, which may affect the physical properties of the final print. The cooling rate directly affects the degree of crystallization of the polymer melt, which has a direct impact on the content of the crystalline phase [11]. Faster cooling results in lower crystalline content. Crystalline content directly affects the physical properties of the print, especially the mechanical properties, such as the Young's modulus of elasticity [12].

Furthermore, it is apparent from the dynamic model that the bed temperature remained constant even though the hot-end was present above the observed point. We attribute this fact to a large amount of mass of the printing bed with high specific heat capacity.

5. Conclusions

The presented FEM models describe the temperature distribution in the extrusion head of the polymer extrusion FFF printer and, more importantly, inside the printed material.

The results from the static model showed the temperature distribution along the path of the filament through the extrusion head. Additionally, the simulation revealed the influence of the extrusion nozzle on the layer of already printed material. The temperature inside the printed material under the nozzle could reach 110–120 °C. Such temperature exceeds the glass transition temperature of PLA and is approaching the melting point, which ranged from 130 to 220 °C [13]. It is consistent with the principle of additive manufacturing; the bottom layer is partially melted, and the top layer sticks to it.

The results from the dynamic model are represented by the time-temperature curves of the observed point lying on the printing bed over which the extrusion head passes. The simulation showed that with the increasing speed of the extrusion head, the cooling rate also significantly increased and influenced the thermal history of the printed material, which has an impact on the mechanical or thermal properties.

These simulations and results demonstrate the importance of the proper process settings and their influence on the thermal history of the material. Together with further practical experiments that would prove the relation between the thermal history of PLA and its physical properties, it would lead to a better understanding of how to set up the FFF process properly.

From the point of further improvement of the model, we will also focus on the influence of the bed temperature setting on the temperature distribution.

Author Contributions: Conceptualization, T.T.; Data curation, O.Š.; Investigation, P.V.; Project administration, D.B.; Validation, O.Š.; Writing—original draft, T.T.; Writing—review & editing, T.T. and D.B.; Funding acquisition, K.D. All authors have read and agreed to the published version of the manuscript.

Funding: This research was funded by the Grant Agency of the Czech Technical University in Prague, grant number SGS21/060/OHK3/1T/13.

Institutional Review Board Statement: Not applicable.

Informed Consent Statement: Not applicable.

Data Availability Statement: The data presented in this study are available on request from the corresponding author.

Conflicts of Interest: The authors declare no conflict of interest.

References

1. Vassilakos, A.; Giannatsis, J.; Dedoussis, V. Fabrication of parts with heterogeneous structure using material extrusion additive manufacturing. *Virtual Phys. Prototyp.* **2021**, *16*, 267–290. [CrossRef]
2. Canessa, E.; Fonda, C.; Zennaro, M. *Low-Cost 3D Printing for Science, Education and Sustainable Development*, 1st ed.; ICTP—The Abdus Salam International Centre for Theoretical Physics: Trieste, Italy, 2013.
3. Henning, C.; Schmid, A.; Hecht, S.; Rückmar, C.; Harre, K.; Bauer, R. Usability of biobased polymers for PCB. In Proceedings of the 42nd International Spring Seminar on Electronics Technology (ISSE), Wroclaw, Poland, 15–19 May 2019; pp. 1–7. [CrossRef]
4. Veselý, P. Nozzle temperature effect on 3D printed structure properties. In Proceedings of the Zborník príspevkov z medzinárodnej konferencie ELEKTROTECHNOLÓGIA 2019, Zuberec, Slovakia, 21–23 May 2019.
5. Moretti, M.; Rossi, A.; Senin, N. In-process simulation of the extrusion to support optimisation and real-time monitoring in fused filament fabrication. *Addit. Manuf.* **2021**, *38*, 101817. [CrossRef]
6. Vanaei, H.R.; Deligant, M.; Shirinbayan, M.; Raissi, K.; Fitoussi, J.; Khelladi, S.; Tcharkhtchi, A. A comparative in–process monitoring of temperature profile in fused filament fabrication. *Polym. Eng. Sci.* **2020**, *61*, 68–76. [CrossRef]
7. Vanaei, H.R.; Raissi, K.; Deligant, M.; Shirinbayan, M.; Fitoussi, J.; Khelladi, S.; Tcharkhtchi, A. Toward the Understanding of Temperature Effect on Bonding Strength, Dimensions and Geometry of 3D-Printed Parts. *J. Mater. Sci.* **2020**, *55*, 14677–14689. [CrossRef]
8. COMSOL AB. *MEMS Module User's Guide, COMSOL Multiphysics®V. 5.4.*; COMSOL AB: Stockholm, Sweden, 2018; pp. 54–57.
9. Jerez-Mesa, R.; Travieso-Rodríguez, J.A.; Corbella, X.; Busqué, R.; Gómez-Gras, G. Finite element analysis of the thermal behavior of a RepRap 3D printer liquefier. *Mechatronics* **2016**, *36*, 119–126. [CrossRef]
10. Pyda, M.; Bopp, R.C.; Wunderlich, B. Heat capacity of poly (lactic acid). *J. Chem. Thermodyn.* **2004**, *36*, 731–742. [CrossRef]
11. Jaques, N.G.; dos Silva, I.D.S.; da Barbosa Neto, M.C.; Ries, A.; Canedo, E.L.; Wellen, R.M.R. Effect of Heat Cycling on Melting and Crystallization of PHB/TiO_2 Compounds. *Polímeros* **2018**, *28*, 161–168. [CrossRef]
12. Batista, N.L.; Olivier, P.; Bernhart, G.; Rezende, M.C.; Botelho, E.C. Correlation between Degree of Crystallinity, Morphology and Mechanical Properties of PPS/Carbon Fiber Laminates. *Mater. Res.* **2016**, *19*, 195–201. [CrossRef]
13. Avinc, A.; Akbar, K. Overview of poly (lactic acid) fibres. Part I: Production, properties, performance, environmental impact, and end-use applications of poly (lactic acid) fibres. *Fiber Chem.* **2009**, *41*, 391–401. [CrossRef]

Article

Characterizations of Polymer Gears Fabricated by Differential Pressure Vacuum Casting and Fused Deposition Modeling

Chil-Chyuan Kuo [1,2,*], Ding-Yang Li [1], Zhe-Chi Lin [1] and Zhong-Fu Kang [1]

[1] Department of Mechanical Engineering, Ming Chi University of Technology, No. 84, Gungjuan Road, New Taipei City 243, Taiwan; M10118002@mail.mcut.edu.tw (D.-Y.L.); U09117212@mail.mcut.edu.tw (Z.-C.L.); U09117222@mail.mcut.edu.tw (Z.-F.K.)
[2] Research Center for Intelligent Medical Devices, Ming Chi University of Technology, No. 84, Gungjuan Road, New Taipei City 243, Taiwan
* Correspondence: jacksonk@mail.mcut.edu.tw

Citation: Kuo, C.-C.; Li, D.-Y.; Lin, Z.-C.; Kang, Z.-F. Characterizations of Polymer Gears Fabricated by Differential Pressure Vacuum Casting and Fused Deposition Modeling. *Polymers* **2021**, *13*, 4126. https://doi.org/10.3390/polym13234126

Academic Editors: Mohammadali Shirinbayan, Nader Zirak, Khaled Benfriha, Sedigheh Farzaneh and Joseph Fitoussi

Received: 28 October 2021
Accepted: 23 November 2021
Published: 26 November 2021

Publisher's Note: MDPI stays neutral with regard to jurisdictional claims in published maps and institutional affiliations.

Copyright: © 2021 by the authors. Licensee MDPI, Basel, Switzerland. This article is an open access article distributed under the terms and conditions of the Creative Commons Attribution (CC BY) license (https://creativecommons.org/licenses/by/4.0/).

Abstract: In recent years, polymer gears have gradually become more widely employed in medium or heavy-duty conditions based on weight reduction in transmission systems because of low costs and low noise compared to metal gears. In the current industry, proposing a cost-effective approach to the manufacture of polymer gears is an important research issue. This paper investigates the wear performance of polymer gears fabricated with eight different kinds of materials using differential pressure vacuum casting and additive manufacturing techniques. It was found that both additive manufacturing and differential pressure vacuum casting seem to be an effective and cost-effective method for low-volume production of polymer gears for industrial applications. The gate number of one is the optimal design to manufacture a silicone rubber mold for differential pressure vacuum casting since the weld line of the polymer is only one. Polyurethane resin, 10 wt.% glass fiber-reinforced polylatic acid (PLA), or 10 wt.% carbon fiber-reinforced PLA are suggested for manufacturing gears for small quantity demand based on the deformation and abrasion weight percentage under process conditions of 3000 rpm for 120 min; epoxy resin is not suitable for making gears because part of the teeth will be broken during abrasion testing.

Keywords: polymer gear; additive manufacturing; differential pressure vacuum casting; polyurethane resin; abrasion

1. Introduction

In practice, product developers need to overcome a tricky issue by making a small batch of prototypes for testing economy and feasibility. A gear is a rotating circular machine part, which can change the torque, speed, and direction of a power source in industrial applications. The polymer gear has some distinct advantages compared to the metal gear, including low weight, quietness of operation, and no need for external lubrication [1], and has been widely used in the automotive industry and consumer electronics. Additive manufacturing (AM) [2,3] has been defined as the process of building physical models by joining materials layer upon layer using computer numerical control data. The application of AM processes has increased in fabricating physical models across various industries because of its capability in manufacturing functional parts with complex geometries. Thus, the AM technology has been widely used to produce prototypes or physical models since it has the capacity to manufacture components with sophisticated geometric shapes. Ghelloudj et al. [4] developed an engineering model to express the evolution of tooth flank wear in polyamide spur gears as a function of the number of cycles. It was found that a wear correction parameter was added to compensate for the measuring errors when plotting the wear profile curves. The simulation results are in good agreement with those obtained from experimental measurements. Lu et al. [5] detected the injection molding lunker defects by X-ray computed tomography. Results showed that the lunker defect jeopardizes the loading capacity of the tooth root under medium or heavy loading conditions, while the tooth flank failure is significantly influenced by the loading condition. Zhang et al. [6] optimized the performance of 3D-printed gears using a machine learning process using a genetic algorithm-based artificial neural network multi-parameter regression model; the authors found that the wear performance of 3D-printed gears was increased by three times. Vacuum casting (VC) [7,8] is a promising technique used for the production of functional parts due to its fast production of high-quality prototypes. Oleksy et al. [9] manufactured the gear wheels with epoxy composites using VC technology and found that developed multi-stage homogenized hybrid-filled epoxy resin had a regular layered morphology. Furthermore, the tensile strength was increased by up to 44 %. Kai et al. [10] integrated VC and AM as well as rapid tooling for fabricating connectors. It was found that a stereolithography apparatus mold cannot be used directly in the VC process since the stereolithography apparatus mold must be broken into pieces for extracting the molded parts. Puerta et al. [11] proposed a new approach to determine the suitability of the usage of standard tensile test specimens fabricated by VC and fused deposition modeling (FDM). The results revealed that the surface quality of the model used for the creation of the silicone rubber mold is an important issue in the VC. Zhang et al. [12] proposed a differential pressure technology to improve the quality of resin parts using VC technology through the optimization method. The results revealed that the artificial fish-swarm algorithm optimized the response surface model of the warpage via the optimized process parameters. Zhao et al. [13] manufactured an accurate shark-skin surface in a large area to overcome some difficulties in the replication process via VC technology. It was found that process parameters played an important role in eliminating air bubbles on the surface of the resin parts. Frankiewicz et al. [14] demonstrated the results of analyses performed for the process of replicating mechanoscopic marks with the use of three vacuum-casting variants, including a hybrid vacuum-pressure casting process developed in particular for the purposes of replication. It was found that the proposed method not only allowed the tool preparation to be simplified and shortened, but also caused the entire process time to be shortened from 10 to 1.5 h.

Injection molding and machine cutting are normally used to fabricate polymer gears. However, the use of plastic injection molding to manufacture polymer gears requires a set of steel injection molds, which does not seem to be a good approach during the research and development stage of a new polymer gear. A set of cutting tools is required for machining polymer gears by machine cutting. Note that these methods are suitable for mass production of polymer gears based on cost-effectiveness. Therefore, developing a cost-effective method for batch production of polymer gears in the research and development stage is an important research issue. In general, the integration of silicone rubber mold and vacuum casting technology [15] is widely used for rapid manufacturing prototypes since the silicone rubber mold has elastic and flexible characteristics. Accordingly, a prototype with complex geometries can be fabricated easily [16]. Chu et al. [17] proposed an efficient generation grinding method for a spur face gear along the contact trace using a disk CBN wheel. Results demonstrated that the proposed method breaks new ground for the engineering application of face gears.

Vacuum casting is a cost-effective method used for the low-volume production of physical models. However, conventional vacuum casting employs the gravity of molding material to fill the mold cavity, resulting in some common defects, such as insufficient filling, shrink marks, or trapped air observed in the cast. Especially, these defects can be eliminated using differential pressure vacuum casting (DPVC) [18]. Therefore, the end-use prototypes can fundamentally be formed by silicone rubber mold using DPVC. The advantages of manufacturing polymer gears using AM techniques include design freedom and less waste of materials. However, not much work has been conducted to characterize the differences in polymer gears fabricated by AM and DPVC. The goal of this investigation is to investigate the characterizations of polymer gears fabricated by AM and DPVC techniques using eight different kinds of polymers. In addition, in-house abrasion testing equipment was designed and implemented to evaluate spur gear life. Finally, an effective and cost-effective method for the low-volume production of polymer gears was proposed.

2. Materials and Methods

Figure 1 shows the research process of this study. The gear type selected in this study is a spur gear since this is the simplest type of gear. Firstly, two spur gears were designed using computer-aided design (CAD) software (Cero, parametric technology corporation Inc.. Taipei, Taiwan), i.e., driving gear and passive gear.

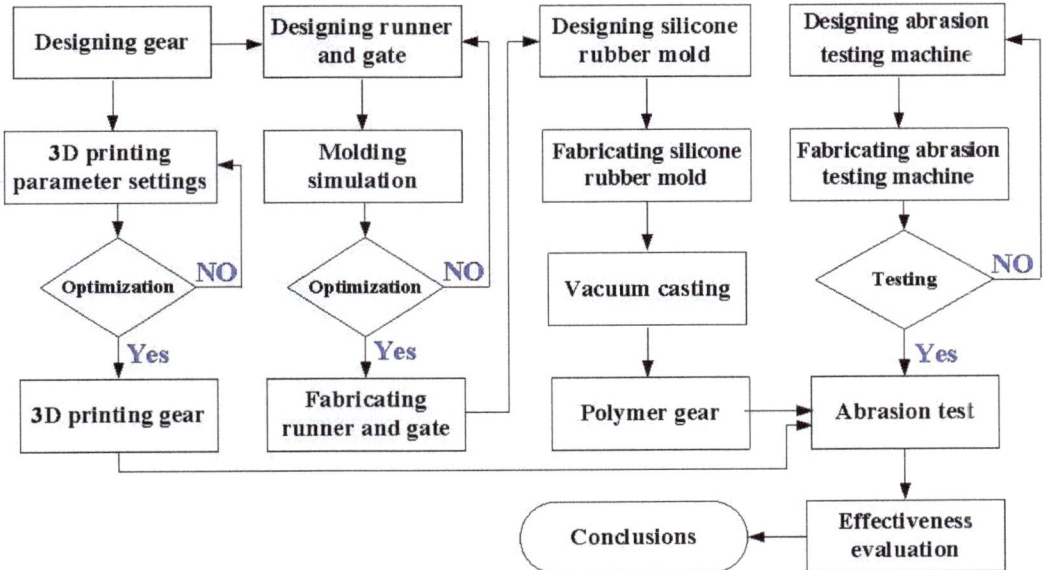

Figure 1. Research process of this study.

Figure 2 shows a three-dimensional (3D) CAD model and the dimensions of the driving gear and the passive gear. The number of teeth, pitch diameter, tooth module, pressure angle, and thickness of the gear are 30, 60 mm, 2 mm, 20° and 5 mm. respectively.

Figure 3 shows the 3D printing software interface of the driving gear and the passive gear. Designing the runner system for the silicone rubber mold is crucial to the mold design. Conventionally, designing the runner system significantly depends on the mold designer's experiences. To address these issues, the filling system of the silicone rubber mold is investigated using numerical simulation software. To investigate the optimum filling system of the silicone rubber mold, the 3D CAD models of spur gear, runner, and gate were imported to the Moldex3D simulation software (R16SP3OR, CoreTech System Inc., Hsinchu, Taiwan) via a data exchange STEP format. Table 1 shows the main numerical modeling parameters used in the numerical analysis.

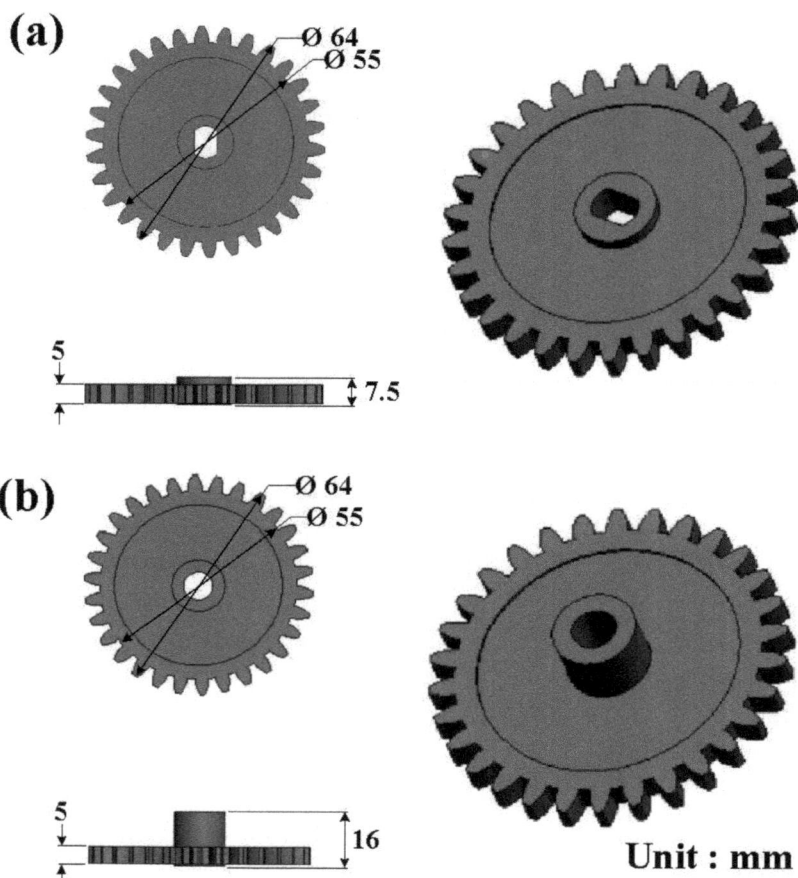

Figure 2. 3D CAD model and dimensions of (**a**) driving gear and (**b**) passive gear.

Table 1. Main numerical modeling parameters used in the numerical analysis.

Properties	Value
Filling time (s)	10
Material temperature (°C)	27
Mold temperature (°C)	27
Maximum injection pressure (kPa)	30

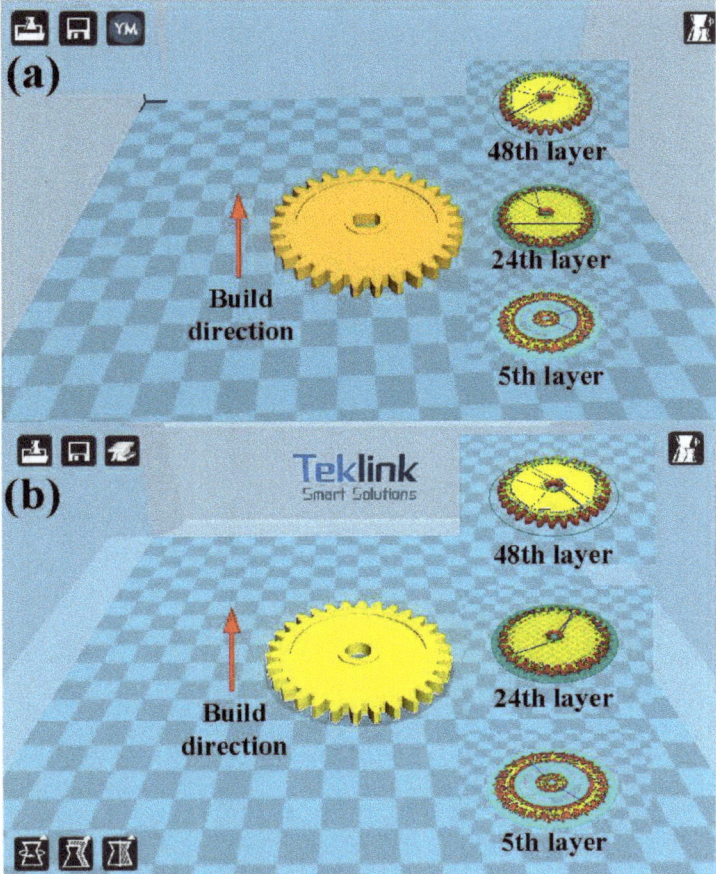

Figure 3. 3D printing software interface of (**a**) driving gear and (**b**) passive gear.

Figure 4 shows the viscosity as a function of the temperature of the epoxy molding material. Q stands for temperature ramping rate of the mixture. Figure 5 shows the viscosity as a function of the temperature of the polyurethane (PU) molding material. In this study, a standard sprue–runner–gate system was used due to the low pressure drop during DPVC. Thus, the pouring materials can flow directly into the silicone rubber mold cavity without passing through the intricate runner system. Figure 6 shows the relationship between the filling system, cast part, and the silicone rubber mold.

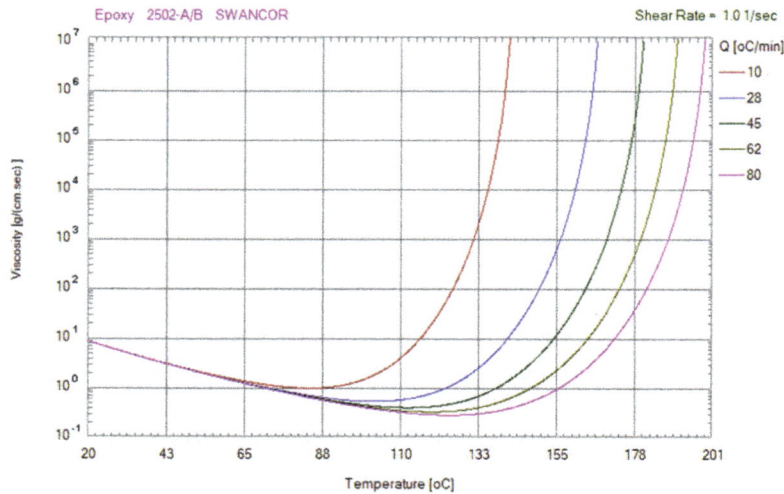

Figure 4. Viscosity as a function of the temperature of the epoxy molding material.

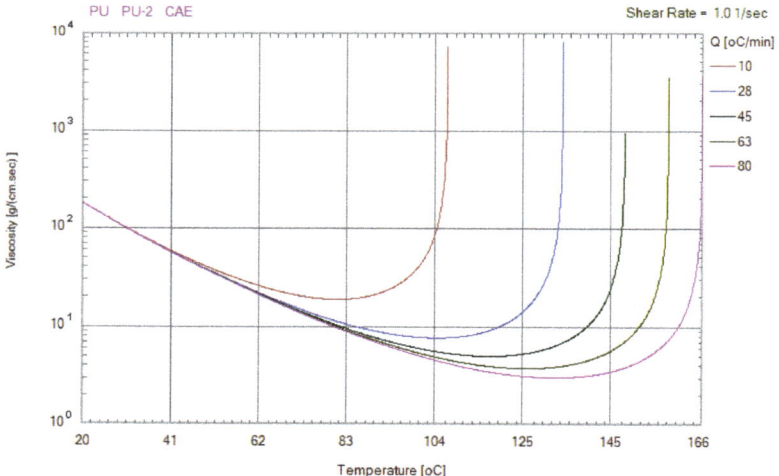

Figure 5. Viscosity as a function of the temperature of the polyurethane molding material.

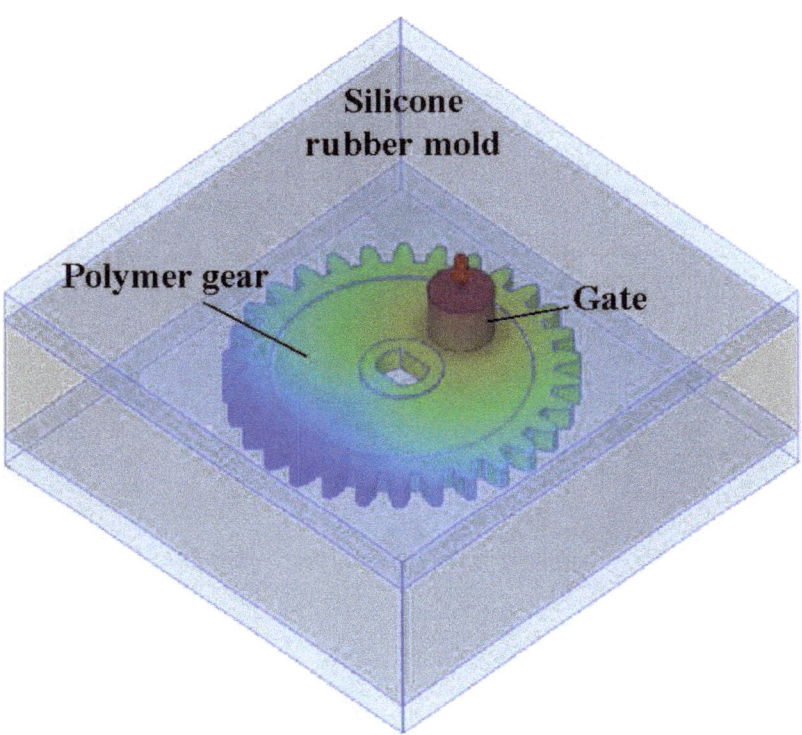

Figure 6. Relationship between filling system, cast part, and silicone rubber mold.

Figure 7 shows the five stages of the VC and information about ball valve and intake area. In general, the VC process involves five distinct stages: preliminary, vacuuming, casting, vacuum relief, and post-processing stages. The P1, P2, and P3 stand for mixing chamber pressure, casting chamber pressure, and atmospheric pressure, respectively. The preliminary stage is the preparation of the silicone rubber mold based on the size of the gear prototype. The radii of ball valve, ball, and seat are 15 mm, 7.5 mm, and 6.25 mm, respectively. In this study, a room temperature vulcanization liquid silicone rubber (KE-1310ST, Shin Etsu Inc, Hsinchu, Taiwan) was used to fabricate the silicone rubber mold. The base compound and hardener (CAT-1310S, Shin Etsu Inc.) were mixed in a weight ratio of 10: 1. A vacuum casting machine (F-600, Feiling Inc., Taoyuan, Taiwan) was used to remove air bubbles in the mixture resulting from the mixing process under vacuum conditions. The epoxy and polyurethane resins were selected as casting materials to fabricate spur gears by silicone rubber mold using differential pressure vacuum casting technology. The process parameters for manufacturing gears include a ball valve angle of 60 °, a silicone rubber mold preheating temperature of 27 °C, a molding material mixing time of 30 s, a pouring time of 40 s, and a differential pressure time of 20 s. The spur gears were also manufactured using an FDM machine (Infinity X1E, Photonier Inc., Taipei, Taiwan) with a nozzle diameter of 0.4 mm. In this study, the six different kinds of filaments, i.e., virgin polylactic acid (PLA) (Thunder 3D Inc., Taipei, Taiwan), PLA filled with 10 wt.% glass fiber (Thunder 3D Inc.), PLA filled with 10 wt.% carbon fiber (Thunder 3D Inc.), acrylonitrile butadiene styrene (ABS) (Thunder 3D Inc.), polycarbonate (PC), and polyamide (PA) were used to print polymer gears using the FDM technique according to the standard of ASTM52900. The process parameters for printing spur gears with a PLA filament are printing temperature of 200 °C, hot bed temperature of 60 °C, printing speed of 50 mm/s, and layer thickness of 0.1 mm. The process parameters for printing spur gears with both PLA filled with 10 wt.%

glass fiber and 10 wt.% carbon fiber filaments are printing temperature of 200 °C, hot bed temperature of 70 °C, printing speed of 50 mm/s, and layer thickness of 0.1 mm. The process parameters for printing spur gears with ABS, PC, and PA filaments are printing temperature of 100 °C, hot bed temperature of 60 °C, printing speed of 50 mm/s, and layer thickness of 0.1 mm. The infill density was set as 100%. The Ultimaker Cura software (New Taipei, Taiwan) was used to generate the program for the FDM machine. Chemical compositions of six different kinds of filaments were characterized using energy-dispersive x-ray spectroscopy (EDS) (D8 ADVANCE, Bruker Inc., Karlsruhe, Germany) and field-emission-scanning electron microscopy (FE-SEM) (JEC3000-FC, JEOL Inc., Tokyo, Japan).

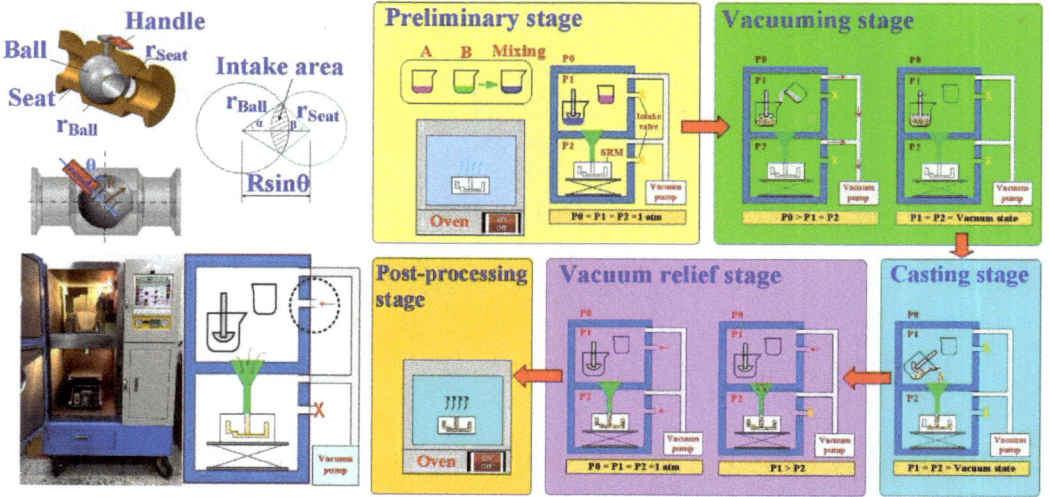

Figure 7. Five stages of the VC and information about ball value and intake area.

Tool wear is the main factor contributing to tool failure in cutting difficult-to-machine materials [19]. Similarly, the abrasion rate is the main factor causing spur gear failure. Various methods, including cylinder-on-plate [20], block-on-wheel, pin-on-disk [21], block-on-ring, pin-on-plate, or flat-on-flat can be used to investigate the wear rate. However, these methods require several testing conditions. In this study, a simple gear abrasion testing equipment was designed and implemented for investigating the wear performance of the fabricated polymer gears. Figure 8 shows a gear abrasion testing machine developed in this study. The tooth flank wear of spur gears as a function of the number of cycles was investigated. Corner wear evolution of gears fabricated with eight different materials was investigated using an OM (M835, Microtech, Inc., Dresden, Germany). The deformation angles of the printed spur gears were measured using a vision measuring system (Quick Vision 404, Mitutoyo Inc., Gunpo, Korea).

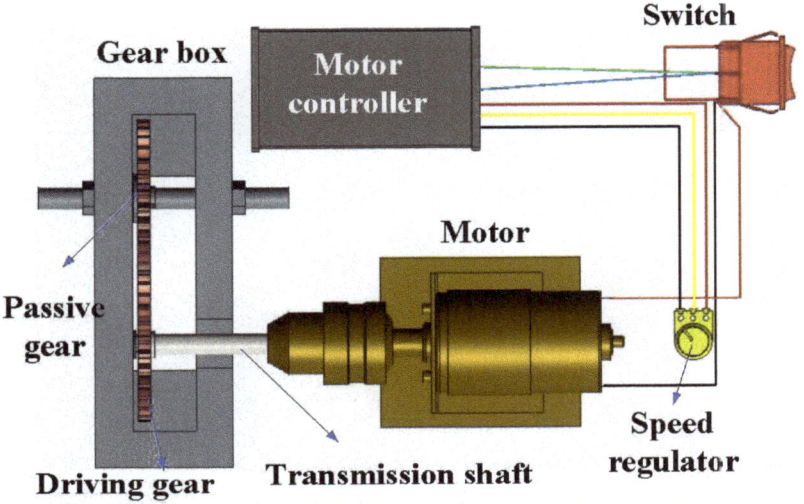
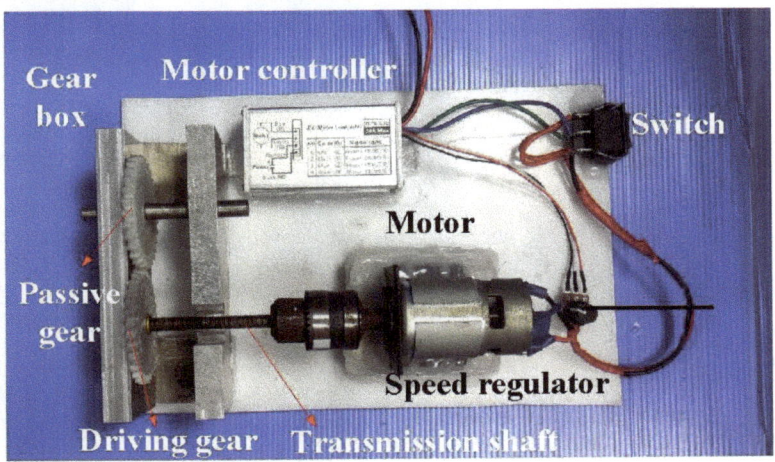

Figure 8. A gear abrasion testing machine developed in this study.

3. Results and Discussion

The efficiency, yield, or product quality of the vacuum casting was affected by the design of the pouring gate. The most common defects such as air-traps or short shot will occur due to poor filling in the vacuum casting. The shrinkage or warpage of the cast part will occur due to unbalanced flow. The post-processing time and costs will increase due to incorrect gate size or location. To avoid these disadvantages described above, the Moldex3D molding simulation software was utilized to investigate the most suitable gate for vacuum casting. There are four different gate types: single point, two points, three points and four points. These gate types were investigated for the gear design in vacuum casting. Figure 9 shows the filling results of different gate numbers. It was found that the gears can be filled completely for four different gate numbers. The fill times for gate numbers of one, two, three, and four are all about 10 s.

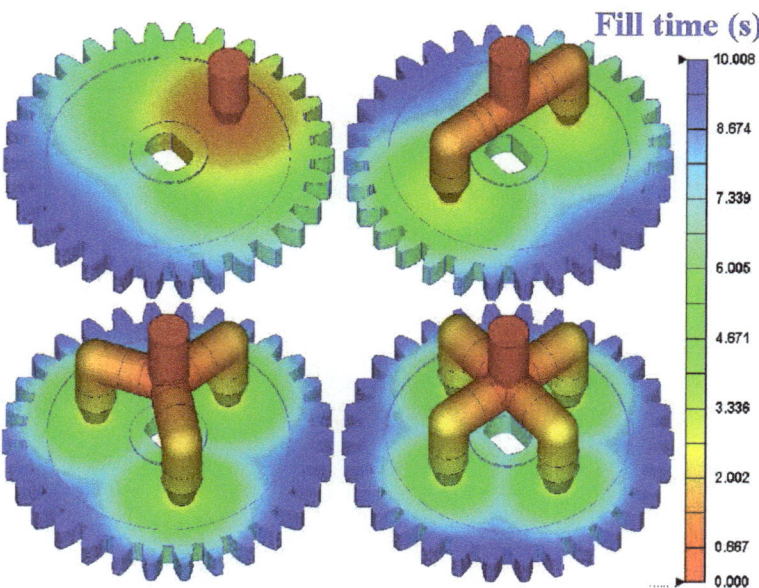

Figure 9. Filling results of different gate numbers.

Figure 10 shows the weld line results for different gate numbers. The weld lines are formed by two different melt fronts joining together during the filling stage, which significantly reduces the strength of the molded part. Figure 11 shows the filling maximum pressures for different gate numbers. The filling maximum pressures for gate numbers of a single point, two points, three points, and four points are 1.439×10^{-4} MPa, 1.035×10^{-4} MPa, 8.441×10^{-5} MPa, and 4.272×10^{-5} MPa, respectively. The maximum filling pressure decreases with as the number of gates increases. It should be noted that the differences in filling pressure can be ignored since the material was poured in a vacuume nvironment.

Figure 12 shows the silicone rubber molds with different gate numbers. As can be seen, the number of weld lines for gate numbers of one, two, three, and four are one, two, three, and four, respectively. Based on practical experience, fewer weld lines represent a better the quality of gears. In addition, the post-processing time and costs of the cast parts for the gate number of one were less than those of the cast parts made with gate numbers of two, three, and four. According to the results described above, the single-point gate seems to be the optimal gate number to fabricate a silicone rubber mold for DPVC.

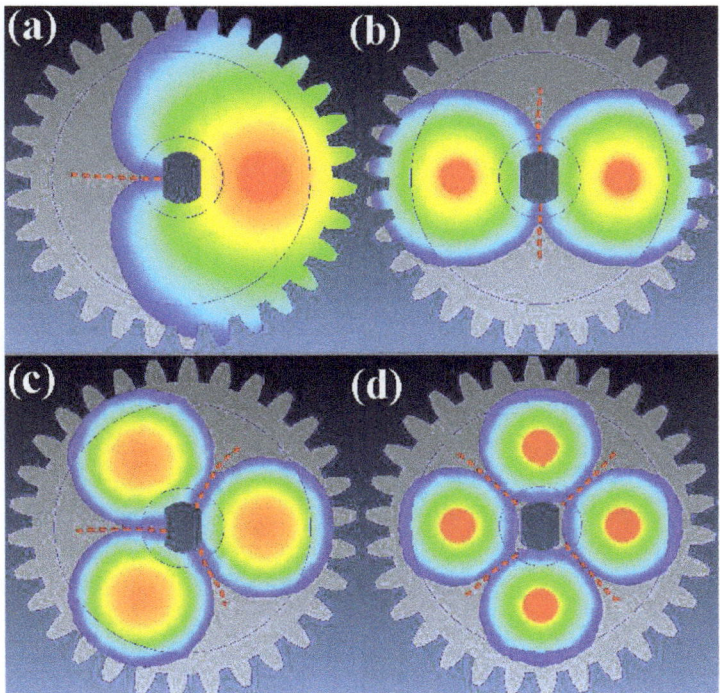

Figure 10. Weld line results for different gate numbers of (**a**) single point, (**b**) two points, (**c**) three points, and (**d**) four points.

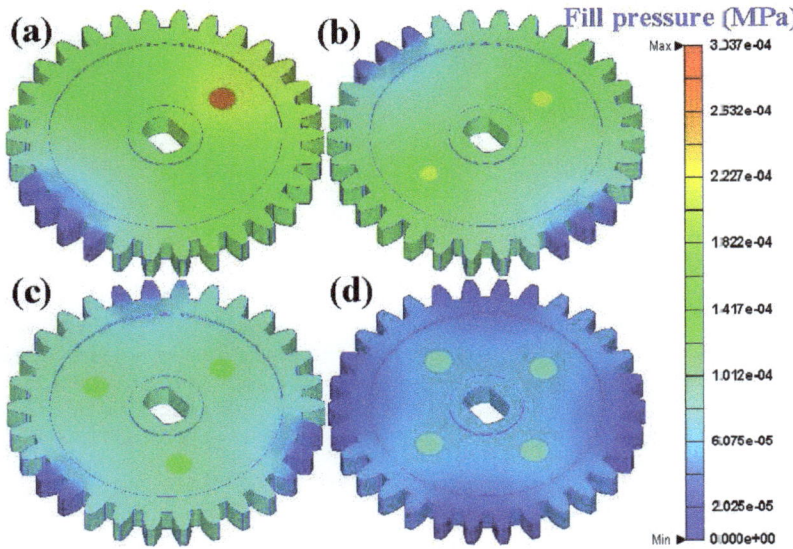

Figure 11. Filling maximum pressures for gate numbers of (**a**) single point, (**b**) two points, (**c**) three points, and (**d**) four points.

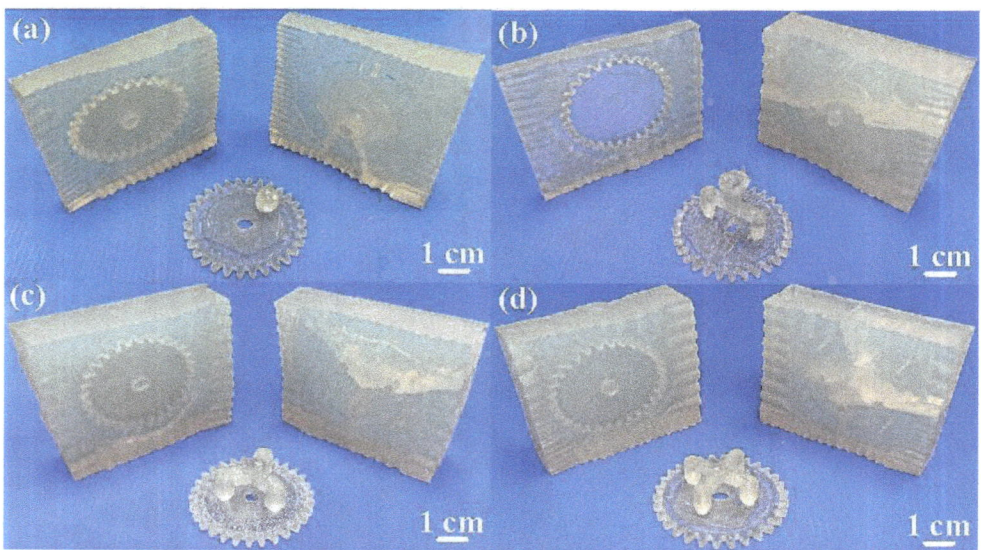

Figure 12. Silicone rubber molds for gate numbers of (**a**) single point, (**b**) two points, (**c**) three points, and (**d**) four points.

Figure 13 shows FE-SEM images of 10 wt.% glass fiber-reinforced PLA and 10 wt.% carbon fiber-reinforced PLA. This result indicates that glass fiber or carbon fiber was observed in the filaments applied to fabricate polymer gears using the FDM technique. Impurity was not observed, which was also confirmed by EDS element mapping analysis. Figure 14 shows EDS analysis of PLA, ABS, 10 wt.% glass fiber-reinforced PLA, 10 wt.% carbon fiber-reinforced PLA, PA, and PC filaments. The major compositions of PLA, ABS, 10 wt.% carbon fiber-reinforced PLA, PA, and PC filaments are C and O. In particular, components of 10 wt.% glass fiber-reinforced PLA are Si, C, O, Ca, and Al. Figure 15 shows the spur gears fabricated with filaments of PLA, ABS, 10 wt.% glass fiber-reinforced PLA, 10 wt.% carbon fiber-reinforced PLA, PA, and PC using the FDM technique.

Figure 16 shows typical spur gears printed with PLA, ABS, PC, and PA filaments. The distinct warpage of the printed gear was found due to uneven shrinkage [22]. It should be noted that two phenomena were found. One is that the deformation of the gear printed with the PC filament is the largest, followed by PA and ABS; the deformation angles are about 5.7°, 2.2°, and 1.8°, respectively. Note that this drawback can be resolved by mounting an auxiliary heating plate on the printing head [23]. The other phenomenon observed is that the flatness of gears printed with PLA filament is better. Small batch production of prototypes via vacuum casting seems to be a good solution, since the cost of silicone rubber mold is at least ten times less than a conventional steel injection mold. In addition, the fatigue life of the polymer gear was greatly influenced by the lunker defects generated during the injection molding process. Note that no lunker defects were observed, which is widely observed with the polymer gears fabricated by plastic injection molding. Figure 17 shows the spur gears fabricated by epoxy and polyurethane resins using the DPVC technique. The results clearly show that the gears fabricated by DPVC have excellentf latness.

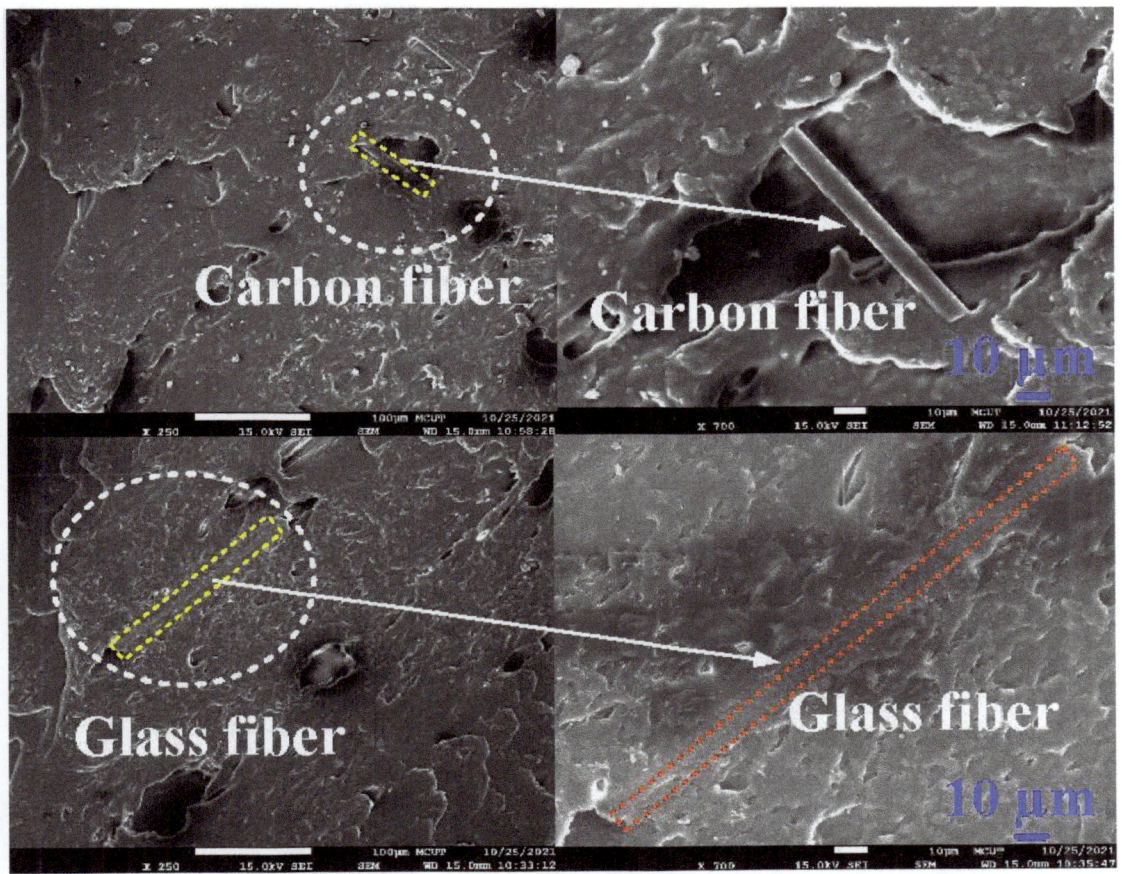

Figure 13. FE-SEM images of 10 wt.% glass fiber-reinforced PLA and 10 wt.% carbon fiber-reinforced PLA.

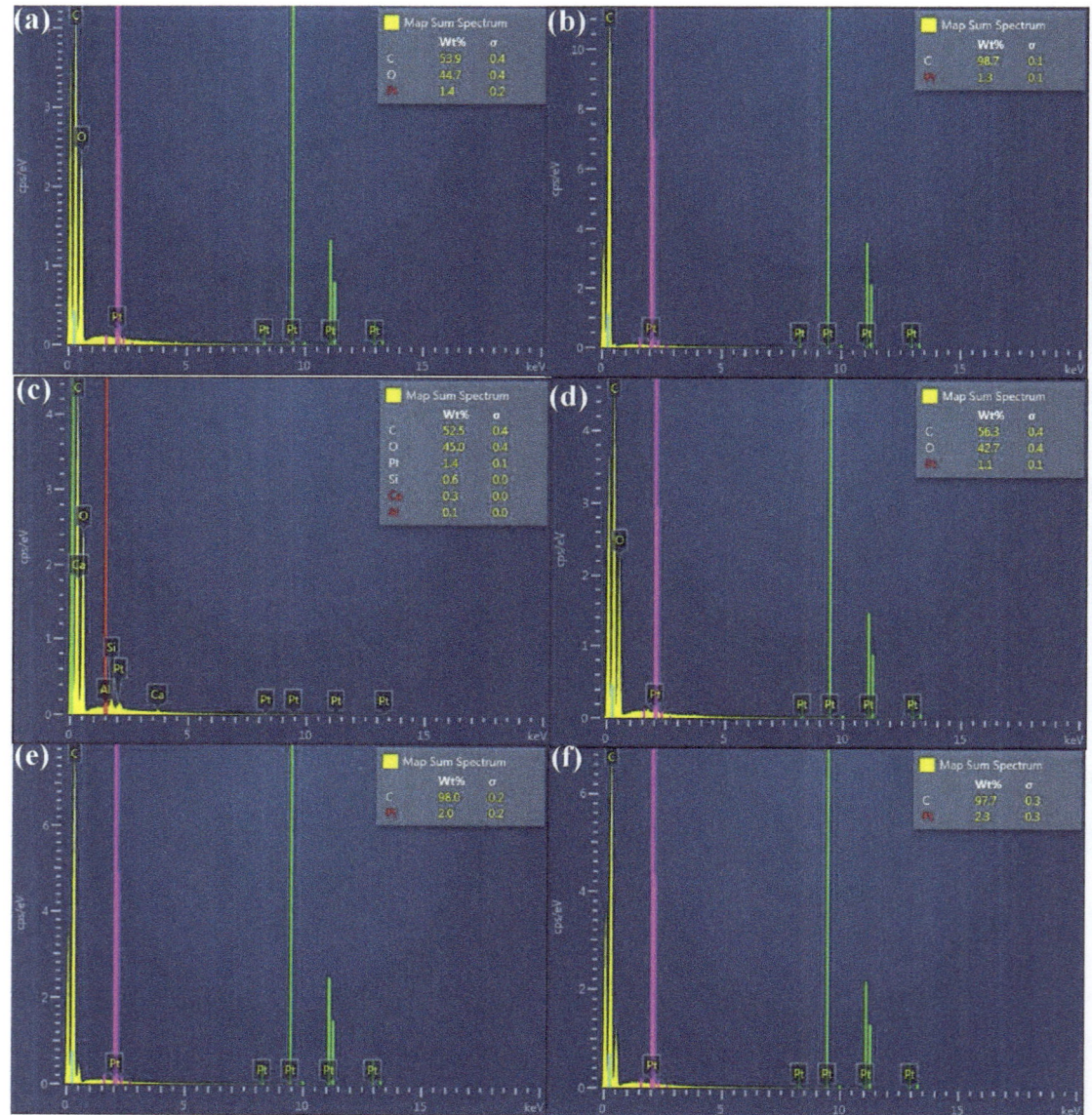

Figure 14. EDS analysis of (a) PLA, (b) ABS, (c) 10 wt.% glass fiber-reinforced PLA, (d) 10 wt.% carbon fiber-reinforced PLA, (e) PA, and (f) PC filaments.

Figure 15. Typical spur gears fabricated with six different filaments of (**a**) PLA, (**b**) ABS, (**c**) 10 wt.% glass fiber-reinforced PLA, (**d**) 10 wt.% carbon fiber-reinforced PLA, (**e**) PA, and (**f**) PC using FDM technique.

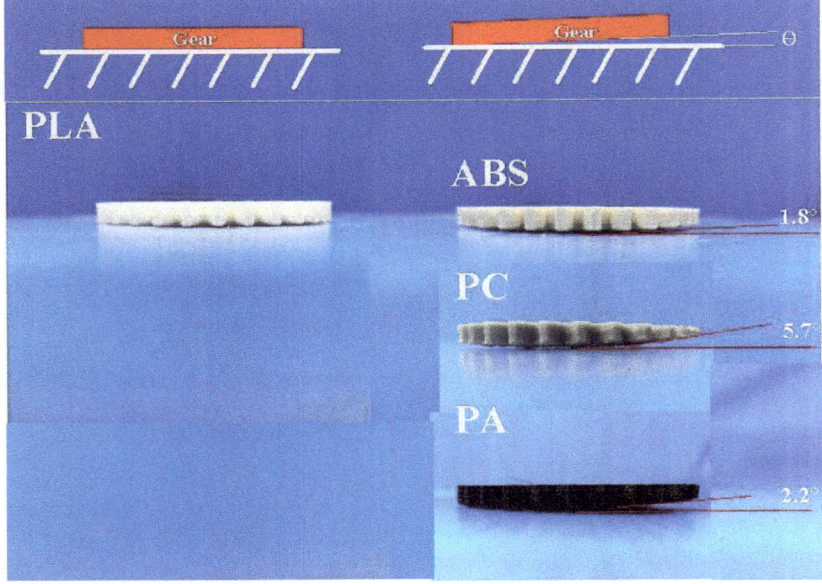

Figure 16. Spur gears printed with PLA, ABS, PC, and PA filaments.

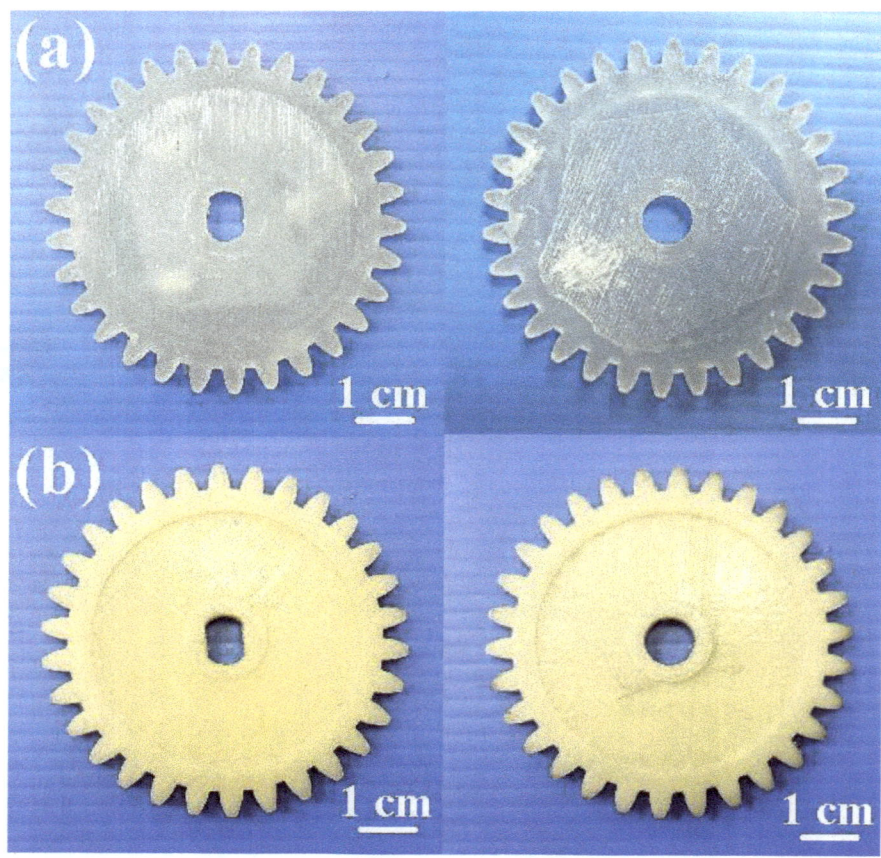

Figure 17. Spur gears fabricated by (**a**) epoxy and (**b**) polyurethane resins using DPVC technique. Driving gear (**left**) and passive gear (**right**).

Polymer gears are usually designed with small tooth modules and operated in dry contact conditions for light loading transmissions [24]. Polymer gears involve three obvious failure types, including tooth root breakage, tooth wear, and tooth flank failure. In general, wear and thermal damages are widely observed in polymer gears in light loading conditions. To evaluate the wear resistance characteristics of gears fabricated by DPVC and AM technologies, an in-house abrasion testing machine was applied to investigate wear loss of the gear under 3000 rpm and an operating time of 120 min. The wear losses were discovered from the changes in the weight of gears before and after abrasion testing using a precision electronic scale.

Figure 18 shows the abrasion weight percentage of gears fabricated with eight different materials for driving and passive gears. The average abrasion weight percentages of driving gears fabricated by filaments of PLA, ABS, 10 wt.% glass fiber-reinforced PLA, 10 wt.% carbon fiber-reinforced PLA, PA, PC, epoxy, and polyurethane resins are 0.173%, 0.182%, 0.192%, 0.155%, 0.485%, 0.524%, 2.379%, and 0.373%, respectively. In addition, the average abrasion weight percentages of passive gears fabricated by filaments of PLA, ABS, 10 wt.% glass fiber-reinforced PLA, 10 wt.% carbon fiber-reinforced PLA, PA, PC, epoxy, and polyurethane resins are 0.325%, 0.302%, 0.192%, 0.287%, 0.418%, 0.696%, 5.039%, and 0.761%, respectively.

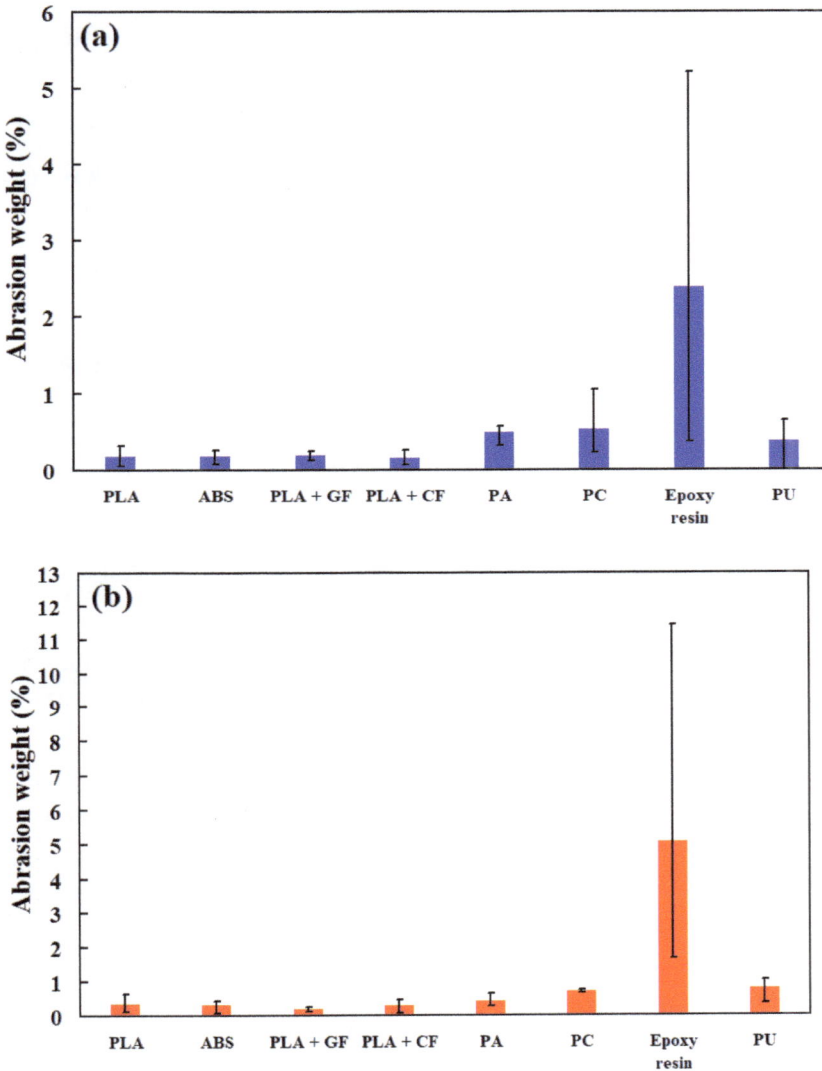

Figure 18. Abrasion weight percentage of gears fabricated with eight different materials (**a**) driving gear and (**b**) passive gear.

Figure 19 shows the corner wear evolution of gears fabricated with eight different materials. It is evident that there is significant wear of the tooth surface. However, some common defects of gears (fisheye defects, debris frosting, pitting, or moderate pitting) were not found on the surface of the failed spur gears.

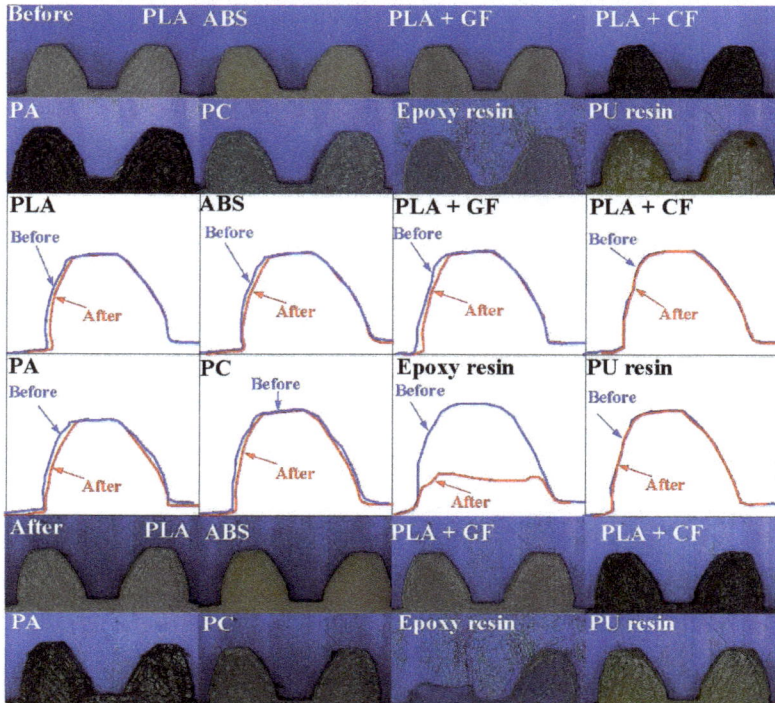

Figure 19. Corner wear evolution of gears fabricated with eight different materials.

Figure 20 shows the cost of materials and manufacturing time for gears fabricated with eight different materials. The results show that manufacturing times for gears fabricated with PLA, ABS, 10 wt.% glass fiber-reinforced PLA, 10 wt.% carbon fiber-reinforced PLA, PA, PC, epoxy resin, and polyurethane resin are 169, 208, 173, 185, 212, 206, 305, and 134 min, respectively. The costs of materials for gears fabricated with PLA, ABS, 10 wt.% glass fiber-reinforced PLA, 10 wt.% carbon fiber-reinforced PLA, PA, PC, epoxy resin, and polyurethane resin are 4.16, 12.13, 22.64, 23.75, 18.75, 31.62, 19.28, and 37.5 in new Taiwan dollars (NTD), respectively.

Based on wear resistance, flatness, production time, and the materials cost of gears, four suggestions are proposed: (a) epoxy resin is not suitable for making gears since part of the teeth will be broken during abrasion test. The underlying reason for gear failure is that the material of polymer gears is fragile; (b) 10 wt.% glass fiber-reinforced PLA or 10 wt.% carbon fiber-reinforced PLA are recommended for making a small batch of gears for functional testing; (c) ABS, PA, or PC are not suitable for making gears because of the larger amount of deformation produced, and (d) polyurethane resin is also suitable to make gears for small quantity demand based on the inconspicuous deformation and abrasion weight percentage. In addition, the wear resistance of gears fabricated with polyurethane resin can be further enhanced by adding reinforcing fillers into base materials.

According to the aforementioned results, the findings of this study are very practical and provide potential applications in consumer electronics, automotive, aerospace engineering, medical, or architectural industries because this technique can be used to fabricate small batch production of polymer gears for functional testing at the research and development stage. The fabricated polymer gears can be further machined, such as by polishing, grinding, cutting, tapping, or drilling. In practice, pressure and temperature are the most significant variables in the differential pressure vacuum casting process. To achieve intelligent manufacturing during mass production of transmission components using VC technology, it is recommended that both pressure and temperature sensors are embedded in the cavity of the silicone rubber mold to monitor operational parameters during the differential pressure vacuum casting process. In this study, both epoxy resin and polyurethane resin were employed to manufacture polymer gears. Alternative polymers, such as polycarbonate, nylon, acrylonitrile butadiene styrene, or polypropylene were recommended for the manufacture of polymer gears. In addition, the mechanical properties of the fabricated polymer gears were dramatically affected by the intrinsic material properties of the molding material. Hence, the mechanical properties of the fabricated polymer gears can be further improved by adding reinforcing fillers, such as bentonite [25], silsesquioxanes, silica, alumina [26], zirconium dioxide, silicon dioxide [27], silicon carbide [28], silicon nitride [29], or molybdenum disulfide [30] into the matrix materials. These issues are currently being investigated and the results will be presented in a later study.

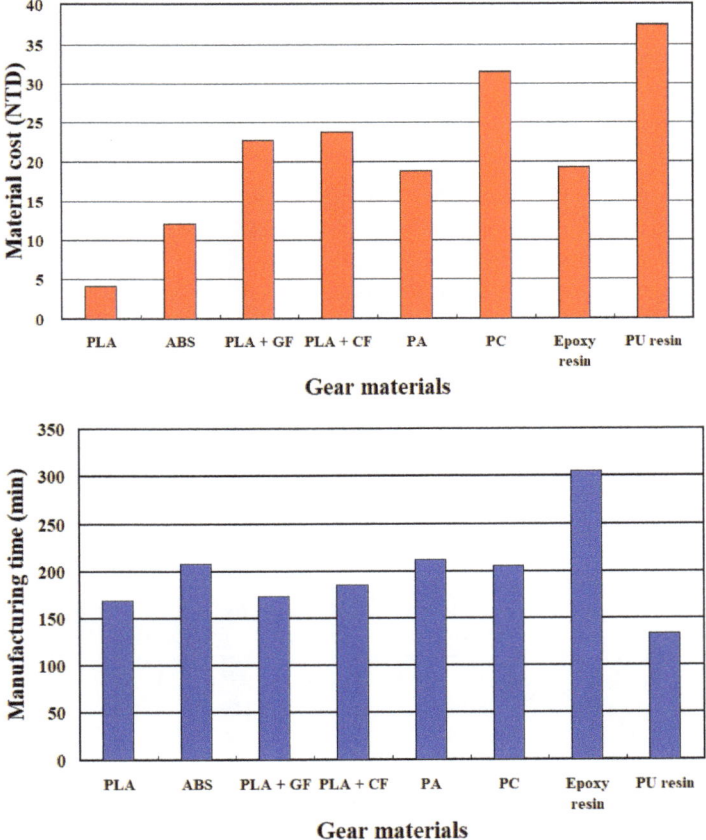

Figure 20. Materials cost and manufacturing time for gears fabricated with eight different materials.

4. Conclusions

Polymer gears have been widely applied in transmission systems due to low noise and low costs compared to metal gears. The main purpose of this study was to characterize polymer gears fabricated by both DPVC and AM. The filling system of the silicone rubber mold was optimized by utilizing the numerical simulation software. Abrasion test equipment for evaluating spur gear life was designed and implemented. The main conclusions from the experimental work in this study are as follows:

1. The remarkable findings in this study are very practical and provide potential applications in the research and development stage because this technique can be used to fabricate small batch production of polymer gears for functional testing.
2. Notably, 10 wt.% glass fiber-reinforced PLA or 10 wt.% carbon fiber-reinforced PLA are suggested for the small batch production of gears for functional testing. ABS, PA, or PC are not suitable for making gears because they produce a larger amount of deformation.
3. Polyurethane resin is suitable for manufacturing polymer gears for small quantity demand based on the inconspicuous deformation and abrasion weight percentage. In addition, the wear resistance of gears fabricated with polyurethane resin can be further enhanced by adding reinforcing fillers into base materials.

Author Contributions: C.-C.K.; wrote the paper/conceived and designed the analysis/performed the analysis/devised the conceptualization; D.-Y.L., Z.-C.L., Z.-F.K.; collected the data/contributed data or analysis tools. All authors have read and agreed to the published version of the manuscript.

Funding: This study received financial support by the Ministry of Science and Technology of Taiwan under contract nos. MOST 110-2221-E-131-023 and MOST 109-2637-E-131-004.

Institutional Review Board Statement: Not applicable.

Informed Consent Statement: Not applicable.

Data Availability Statement: The data presented in this study are available on request from the corresponding author.

Conflicts of Interest: The authors declare no conflict of interest.

References

1. Polanec, B.; Zupanič, F.; Bončina, T.; Tašner, F.; Glodež, S. Experimental Investigation of the Wear Behaviour of Coated Polymer Gears. *Polymers* **2021**, *13*, 3588. [CrossRef]
2. Sutton, J.T.; Rajan, K.; Harper, D.P.; Chmely, S.C. Improving UV Curing in Organosolv Lignin-Containing Photopolymers for Stereolithography by Reduction and Acylation. *Polymers* **2021**, *13*, 3473. [CrossRef] [PubMed]
3. Buschmann, B.; Henke, K.; Talke, D.; Saile, B.; Asshoff, C.; Bunzel, F. Additive Manufacturing of Wood Composite Panels for Individual Layer Fabrication (ILF). *Polymers* **2021**, *13*, 3423. [CrossRef] [PubMed]
4. Ghelloudj, O.; Zelmati, D.; Amirat, A. Engineering Modeling of Wear Profiles in Tooth Flank of Polyamide Spur Gears. *Int. J. Adv. Manuf. Technol.* **2017**, *93*, 3531–3540. [CrossRef]
5. Lu, Z.; Liu, H.; Wei, P.; Zhu, C.; Xin, D.; Shen, Y. The Effect of Injection Molding Lunker Defect on the Durability Performance of Polymer Gears. *Int. J. Mech. Sci.* **2020**, *180*, 105665. [CrossRef]
6. Zhang, Y.; Mao, K.; Leigh, S.; Shah, A.; Chao, Z.; Ma, G. A Parametric Study of 3D Printed Polymer Gears. *Int. J. Adv. Manuf. Technol.* **2020**, *107*, 4481–4492. [CrossRef]
7. Kuo, C.C.; Qiu, W.K.; Liu, H.A.; Chang, C.M. Filling Mechanism for Prototype Parts Produced by Vacuum Differential Pressure Casting Technology. *Int. J. Adv. Manuf. Technol.* **2019**, *105*, 1469–1481. [CrossRef]
8. Kuo, C.C.; Wu, M.X. Evaluation of Service Life of Silicone Rubber Molds Using Vacuum Casting. *Int. J. Adv. Manuf. Technol.* **2017**, *90*, 3775–3781. [CrossRef]
9. Oleksy, M.; Heneczkowski, M.; Oliwa, R.; Budzik, G.; Dziubek, T.; Markowska, O.; Szwarc-Rzepka, K.; Jesionowski, T. Hybrid Composites with Epoxy Resin Matrix Manufactured with Vacuum Casting Technology. *Polimery* **2014**, *59*, 677–681. [CrossRef]
10. Kai, C.C.; Howe, C.T.; Hoe, E.K. Integrating Rapid Prototyping and Tooling with Vacuum Casting for Connectors. *Int. J. Adv. Manuf. Technol.* **1998**, *14*, 617–623. [CrossRef]
11. Puerta, P.V.; Sanchez, D.M.; Batista, M.; Salguero, J. Criteria Selection for a Comparative Study of Functional Performance of Fused Deposition Modelling and Vacuum Casting Processes. *J. Manuf. Process.* **2018**, *35*, 721–727. [CrossRef]
12. Zhang, H.G.; Hu, Q.X. Study of the Filling Mechanism and Parameter Optimization Method for Vacuum Casting. *Int. J. Adv. Manuf. Technol.* **2016**, *83*, 711–720. [CrossRef]
13. Zhao, D.Y.; Huang, Z.P.; Wang, M.J.; Wang, T.; Jin, Y. Vacuum Casting Replication of Micro-Riblets on Shark Skin for Drag-Reducing Applications. *J. Mater. Process. Technol.* **2012**, *212*, 198–202. [CrossRef]
14. Frankiewicz, M.; Kobiela, K.; Kurzynowski, T. Possibility for Replicating Mechanoscopic Surface Marks in the Hybrid Vacuum-Pressure Casting Process. *Polymers* **2021**, *13*, 874. [CrossRef]

15. Zhan, S.A.; Song, J.T.; Ding, M.H.; Guo, J.; Liu, H.H. A Study of Thin-Walled ZL105A Casting Manufactured by Vacuum Differential Pressure Casting. *Adv. Eng. Res.* **2017**, *135*, 574–582.
16. Enemuoh, E.U.; Duginski, S.; Feyen, C.; Menta, V.G. Effect of Process Parameters on Energy Consumption, Physical, and Mechanical Properties of Fused Deposition Modeling. *Polymers* **2021**, *13*, 2406. [CrossRef]
17. Chu, X.; Wang, Y.; Du, S.; Huang, Y.; Su, G.; Liu, D.; Zang, L. An Efficient Generation Grinding Method for Spur Face Gear Along Contact Trace Using Disk CBN Wheel. *Int. J. Adv. Manuf. Technol.* **2020**, *110*, 1179–1187. [CrossRef]
18. Kuo, C.C.; Qiu, W.K. Effect of Differential Pressure on the Transcription Rate of Micro-Featured Components. *Int. J. Adv. Manuf. Technol.* **2019**, *104*, 1229–1237. [CrossRef]
19. Zheng, G.; Lin, Y. Tribological Properties of Micro-Groove Cemented Carbide by Laser Processing. *Micromachines* **2021**, *12*, 486. [CrossRef] [PubMed]
20. Hanon, M.M.; Alshammas, Y.; Zsidai, L. Effect of Print Orientation and Bronze Existence on Tribological and Mechanical Properties of 3D-Printed Bronze/PLA Composite. *Int. J. Adv. Manuf. Technol.* **2020**, *108*, 553–570. [CrossRef]
21. Singh, P.K.; Singh, A.K. An Investigation on the Thermal and Wear Behavior of Polymer Based Spur Gears. *Tribol Int.* **2018**, *118*, 264–272. [CrossRef]
22. Kuo, C.C.; Wu, Y.R.; Li, M.H.; Wu, H.W. Minimizing Warpage of ABS Prototypes Built with Low-Cost Fused Deposition Modeling Machine Using Developed Closed-Chamber and Optimal Process Parameters. *Int. J. Adv. Manuf. Technol.* **2019**, *101*, 593–602. [CrossRef]
23. Kurahashi, Y.; Tanaka, H.; Terayama, M.; Sugimura, J. Effects of Environmental Gas and Trace Water on the Friction of DLC Sliding with Metals. *Micromachines* **2017**, *8*, 217. [CrossRef]
24. Yu, N.; Sun, X.; Wang, Z.; Zhang, D.; Li, J. Effects of Auxiliary Heat on Warpage and Mechanical Properties in Carbon Fiber/ABS Composite Manufactured by Fused Deposition Modeling. *Mater. Des.* **2020**, *195*, 108978. [CrossRef]
25. Fang, M.; Yu, T.; Xi, F. An Experimental Investigation of Abrasive Suspension Flow Machining of Injector Nozzle Based on Orthogonal Test Design. *Int. J. Adv. Manuf. Technol.* **2020**, *110*, 1071–1082. [CrossRef]
26. Ribeiro, F.S.F.; Lopes, J.C.; Garcia, M.V.; de Angelo Sanchez, L.E.; de Mello, H.J.; de Aguiar, P.R.; Bianchi, E.C. Grinding Performance by Applying MQL Technique: An Approach of the Wheel Cleaning Jet Compared with Wheel Cleaning Teflon and Alumina Block. *Int. J. Adv. Manuf. Technol.* **2020**, *107*, 4415–4426. [CrossRef]
27. Meshram, T.; Yan, J. Generation of Microcones on Reaction-Bonded Silicon Carbide by Nanosecond Pulsed Laser Irradiation. *Int. J. Adv. Manuf. Technol.* **2020**, *108*, 1039–1048. [CrossRef]
28. Barbouche, M.; Zaghouani, R.B.; Benammar, N.; Khirouni, K.; Turan, R.; Ezzaouia, H. Impact of rapid thermal annealing on impurities removal efficiency from silicon carbide for optoelectronic applications. *Int. J. Adv. Manuf. Technol.* **2019**, *106*, 731–739. [CrossRef]
29. Yao, J.; Wu, Y.; Sun, J.; Xu, Y.; Wang, H.; Zhou, P. Research on the Metamorphic Layer of Silicon Nitride Ceramic under High Temperature Based on Molecular Dynamics. *Int. J. Adv. Manuf. Technol.* **2020**, *109*, 1249–1260. [CrossRef]
30. Darzi, S.; Mirnia, M.J.; Elyasi, M. Single-Point Incremental Forming of AA6061 Aluminum Alloy at Elevated Temperatures. *Int. J. Adv. Manuf. Technol.* **2021**, *116*, 1023–1039. [CrossRef]

Article

Mathematical Modeling and Optimization of Fused Filament Fabrication (FFF) Process Parameters for Shape Deviation Control of Polyamide 6 Using Taguchi Method

Zohreh Shakeri [1,*], Khaled Benfriha [1], Mohammadali Shirinbayan [2], Mohammad Ahmadifar [1,2] and Abbas Tcharkhtchi [2]

1. Arts et Metiers Institute of Technology, CNAM, PIMM, HESAM University, F-75013 Paris, France; khaled.benfriha@ensam.eu (K.B.); mohammad.ahmadifar@ensam.eu (M.A.)
2. Arts et Metiers Institute of Technology, CNAM, LCPI, HESAM University, F-75013 Paris, France; Mohammadali.shirinbayan@ensam.eu (M.S.); abbas.tcharkhtchi@ensam.eu (A.T.)
* Correspondence: zohreh.shakeri@ensam.eu

Citation: Shakeri, Z.; Benfriha, K.; Shirinbayan, M.; Ahmadifar, M.; Tcharkhtchi, A. Mathematical Modeling and Optimization of Fused Filament Fabrication (FFF) Process Parameters for Shape Deviation Control of Polyamide 6 Using Taguchi Method. *Polymers* **2021**, *13*, 3697. https://doi.org/10.3390/polym13213697

Academic Editors: Antonio Gloria and Roberto De Santis

Received: 17 September 2021
Accepted: 22 October 2021
Published: 27 October 2021

Publisher's Note: MDPI stays neutral with regard to jurisdictional claims in published maps and institutional affiliations.

Copyright: © 2021 by the authors. Licensee MDPI, Basel, Switzerland. This article is an open access article distributed under the terms and conditions of the Creative Commons Attribution (CC BY) license (https://creativecommons.org/licenses/by/4.0/).

Abstract: Fused filament fabrication (FFF) is a layer-by-layer additive manufacturing (AM) process for producing parts. For industries to gain a competitive advantage, reducing product development cycle time is a basic goal. As a result, industries' attention has turned away from traditional product development processes toward rapid prototyping techniques. Because different process parameters employed in this method significantly impact the quality of FFF manufactured parts, it is essential to optimize FFF process parameters to enhance component quality. The paper presents optimization of fused filament fabrication process parameters to improve the shape deviation such as cylindricity and circularity of 3D printed parts with the Taguchi optimization method. The effect of thickness, infill pattern, number of walls, and layer height was investigated as variable parameters for experiments on cylindricity and circularity. The MarkForged® used Nylon White (PA6) to create the parts. ANOVA and the S/N ratio are also used to evaluate and optimize the influence of chosen factors. As a result, it was concluded that the hexagonal infill pattern, the thickness of 5 mm, wall layer of 2, and a layer height of 1.125 mm were known to be the optimal process parameters for circularity and cylindricity in experiments. Then a linear regression model was created to observe the relationship between the control variables with cylindricity and circularity. The results were confirmed by a confirmation test.

Keywords: Taguchi design; ANOVA; FFF; response surface; cylindricity; process optimization

1. Introduction

In the additive manufacturing process (AM), one of the fast prototyping methods, the CAD model is designed first and then made in 3D. The AM process is a layer-by-layer production process. Other names for this process are layer manufacturing, additive processes, free-form fabrication, and solid free-form fabrication [1]. Three-dimensional printed parts show different properties depending on other AM techniques [2]. Fused deposition modeling (FDM), selective laser melting (SLM), multi-jet modeling (MJM), laminated object manufacturing (LOM), and stereolithography (SLA) and selective laser sintering (SLS) are various additive manufacturing (AM) methods [3–8]. Additive manufacturing provides the ability to produce complex geometries that are difficult to produce by conventional methods without complex tooling. The usage of AM technology has risen in recent years. Today, the use of AM method has increased because it reduces post-processing, material wastes, lower costs, creates high customization manufacturing parts, and greatly reduces overall product development [9,10].

Fused filament fabrication (FFF) is the most common process, and it is a 3D printing process that has been extensively investigated to produce metal and thermoplastic structures [11]. Due to less waste of material, high quality, and low manufacturing cost it is a common extrusion-based process [12]. FFF is a material extrusion process, according to ISO/ASTM terminology [13]. Thermoplastics are the base material in the form of filament that is selectively deposited through a nozzle over a movable bed. Among the different thermoplastics, we can mention PLA, ABS, ASA, and Nylon, which are more common in AM [14]. One can note that the limitation of this method is dimensional and geometrical accuracy. The Stratasys Company introduced this technology, and the proprietary term fused deposition modeling (FDM) was established [13].

Polyamide is a semi-crystalline thermoplastic with repeated amide sequences in the polymer backbone and H bonds between neighboring polymer chains. It has good mechanical, electrical, and thermal properties. PA6, PA12, and PA66 are all varieties of this thermoplastic, depending on the monomers that make it up. They are employed in wind turbines and oil and gas. PA6 is a low-cost, widely used synthetic polymer with wide applications. In the automobile industry, these polymers are commonly used. To improve some of its characteristics, extensive research is continuously being done [15,16].

Various process parameters utilized in this technique have an impact on the quality of FFF produced components [17]. On the other hand, the geometrical tolerance of AM-3D printed parts is mostly determined by the process parameters' setting. The process parameter could achieve improvement of the quality of prototypes by making appropriate adjustments to manufacturing parameters [18–20].

Fisher [21] proposed the concept of design of experiments (DOE) in the 1920s. DOEs are a structured and systematic method of running and evaluating controlled experiments to identify the factors that influence output variables. Because each component is independent of the others, this is a multivariable testing technique that varies them all simultaneously. DOEs define the specific setting levels of a couple of variables at which each run of the experiment will be carried out. For experimental planning, the Taguchi design method is a crucial tool. It provides a methodical and effective approach to cost, quality, and performance optimization. Taguchi [22] is the developer of the Taguchi design. A greater number of parameters may be evaluated at once in the Taguchi design method, and the optimal configuration can be reached with fewer resources than in the classic DOE approach. In fact, the main advantages of adopting Taguchi's approach to design experiments with a simplicity of the experimental plan and the capability of studying interactions between multiple process parameters. The Taguchi orthogonal array (OA) is a basic fractional factorial design. It is a fractional orthogonal design based on a design matrix that lets users evaluate a set of many factor combinations at several levels. The Taguchi L9 [23,24] orthogonal array is a good experimental design approach with a small number of tests.

The response surface method is also one of the DOEs that examines the effect of different parameters on responses. This method helps to improve responses through a set including regression analysis and parameter optimization [25]. Many researchers have studied the effect and optimization of FDM process parameters on the shape deviation in FFF. Also, simple specimens and experimental designs such as Taguchi, ANOVA, and others are used in experimental investigations [26].

For example, Lee and Abdulla [27], using the Taguchi method, investigated the optimal elastic performance of a piece of ABS produced by 3D printing to achieve the maximum throwing distance from the prototype. They concluded that FDM variables such as raster angle, air gap and layer thickness have a major effect on the compliant ABS prototype's elastic behavior.

Alafaghani and Qattawi [28], utilized the Taguchi technique to study the effects of infill density, infill pattern, layer height, and extrusion temperature in terms of the mechanical properties and dimensional accuracy of the FDM process with PLA filament. The results indicate that a lower extrusion temperature, smaller layer thickness, lower infill density, and hexagonal infill pattern will improve the dimensional accuracy. Also, with optimal layer thickness, higher extrusion temperature, and a larger infill density and triangle infill pattern, the strength of FDM parts is at a maximum.

Anitha and Arunachalam [18], using the Taguchi method (L18 orthogonal array), examined parameters such as road width, speed deposition, layer height on surface roughness. They observed that layer height, followed by road width and deposition speed, had the greatest influence on surface roughness.

Rizea and Anghel [29] investigated the effect of three critical parameters, including layer thickness, infill density, and orientation, on flatness and dimensional accuracy of Z-ULTRAT parts produced under FDM using the Taguchi method and L9 orthogonal array. Finally, they found that the effect of layer thickness and infill pattern was more significant than the orientation on shape deviation.

Sood and Ohdar [30], used the central composite design method, which is one of the DOE methods, and ANOVA analysis to optimize the parameters to investigate the impact of process parameters on specimen mechanical strength. Five basic process settings were studied, including raster angle, orientation, raster width, layer thickness and air gap and their impact on specimen responses, including tensile, flexural, and impact strength. Small raster angle, thick raster, lower number of layers and zero air gap will increase the mechanical strength.

Sheth and George [31] comprehended that spindle speed, feed rate, and the interaction between them have significant effects on cylindricity. They concluded that at lower spindle speeds, cylindricity is minimum.

Das and Mhapsekar [32], have evaluated the effect of FDM process parameters on optimization for the cylindricity tolerances with build orientation which minimize the support contact area. The circularity error is reduced with lower infill density. Also when the circular object is oriented with the vertical axis as the center and the base with the horizontal axis, the circularity error is minimal.

Aslani and Chaidas [33], applied the Taguchi method (L9) to estimate the effect of wall thickness and extraction temperature on dimensional accuracy and the surface quality of PLA parts under the FFF process. According to the results obtained, they understood that the dimensional accuracy and surface quality of PLA is improved and optimized using high-temperature extraction and average wall thickness values.

Chang and Huang [34] have worked on the optimization of raster width, contour width, raster angle, and contour depth variables for the optimization of flatness and cylindricity in FDM parts. The contour depth has the greatest influence, according to the ANOVA analysis of individual process parameters. Contour width is the second most important parameter, and raster width and raster angle have the least values.

Prakash and Sivakumar [35], investigated the effect of three parameters: filling density, horizontal and vertical orientation on the circularity of ABS parts 3D printed by the FDM method. It should be noted that they used the Taguchi method. They found that when a circular object is oriented with the vertical axis as the center and the horizontal axis as the base, the error on circularity is reduced compared to other orientations. The circularity of the parts is more influenced by the horizontal orientation.

Doloi and Kumar [36], applied the Taguchi approach to determine the impact of several process parameters on circularity error and diametrical deviation of ABS parts produced by FDM, including layer thickness, bed temperature, extruder temperature, infill density, and speed. From the results, it was found that in low bed temperature and moderate layer thickness, diametrical variation decreased. Also, the lowest circularity error was obtained at high speed, lower layer height, lower extruder temperature, and moderate infill density.

Using the Taguchi approach, Nagendra and Vikas [37], investigate how infill pattern, layer height, build orientation, and infill density impacted dimensional accuracy (DA), flatness, and cylindricity. An analysis of variance was used to assess the influence of process factors.

Saqib and Urbanic [38] investigated the influence on component accuracy on the FDM process by process variables with geometric shapes. In fact, to investigate the most influential process variables on the deformation of printed 3D parts, they designed experiments on flatness and circularity. The layer height, work envelope, and orientation were all investigated. The variation in cylindricity was maximum at 90° orientations, according to the results.

One of the most critical problems in the manufacturing of polymer parts is deformation during the process, which depends on the material's rheological behavior. Also, 3D printing as one of the manufacturing methods is deformed due to the presence of micro voids in the interfaces of the deposited layers and layered structure of the 3D printing. Therefore dimensional stability is an essential factor for the designer. One of the methods to control geometric accuracy and reduce the resulting errors is to optimize the process parameters. In this paper, the cylindricity and the circularity as geometric tolerances of PA6 parts fabricated by FFF with variable parameters such as infill pattern, wall layers, layer height, and thickness were analyzed. The S/N ratios and ANOVA were employed to analyze the significant impacts and find the optimal parameter for minimum cylindricity and circularity simultaneously. Regression model of cylindricity and circularity was developed to predict them and to examine the correlation between different variables, which determines the relationship between each response and process parameters.

2. Materials and Methods

In this research, the Markforged printer, one of the advanced FFF desktop 3D printers, has been used to 3D print the parts from polyamide 6 (Figure 1). Nylon White is a commercial material that Markforged Company (Watertown, MA, USA) developed. Nylon 6 or polycaprolactam are other names for polyamide 6 (PA6). It is one of the most widely used polyamides in the world due to its versatility. It also outperforms other polymers such as PLA and ABS in terms of mechanical properties, and the surface quality of PA 6 is excellent. Also, the 3D printer used is capable of printing limited materials such as PA 6, so this material was selected for experiments. As shown in Figure 2 for the experiments, hollow cylindrical parts with a fixed height of 40 mm and inner diameter of 10 mm, but variable outer diameter with the amounts of 20, 30, and 40 mm, were designed by CATIA-V5 software (V5, Dassault Systèmes, Paris, France), and STL files have been exported from it. The parts were printed using thermoplastic (polyamide 6) at a temperature of 273 °C. Table 1 represents the levels, and the process parameters that will be employed in the experiment. In this table the column of thickness shows the difference between the outer and inner diameter, and infill pattern identifies the structure and shape of the material inside of a part. Also, the thickness of each layer of deposited material is given by layer height, and wall layer indicates the thickness of the part's walls. In this printer, some parameters have limitations, for example, for the layer height variable, the numbers 1, 1.125 and 2 can be selected. The selection of these parameters and the selection of different levels of each parameter primarily were performed based on the limitation of the variable parameters of the markforged 3D printer. Selected thicknesses were considered to observe both small and large dimensions of the responses. The number of wall layers was also selected based on the dimensions of the cylinder. Considering the volume required to apply different infill patterns, the number of one wall created an unsuitable surface in the cylinder. Table 2 shows the Taguchi orthogonal array that controls the parameter combinations for each experiment.

Figure 1. Markforged 3D printer.

Figure 2. Hollow cylindrical parts.

Table 1. The process parameters and their levels.

Level	Thickness (mm)	Infill Pattern	Layer Height (mm)	Wall Layer
1	5	Hexagonal	1	2
2	10	Rectangular	1.125	3
3	15	Triangular	2	4

The standard modeling software CATIA-V5™ was used for 3D modeling. An STL file is extracted from the designed CAD model. Then, pieces were 3D printed, and all the parts were scanned with a 3D laser scanner named Solutionix D500 (Medit, Seoul, Korea). This professional scanner specializes in scanning small and detailed things and the most complex products. It has a resolution of 0.055 mm and an accuracy of 0.01 mm. Also, the scan speed in this scanner is high because of more powerful engines and enhanced algorithms. The 3D scanner scans the product from multiple angles automatically. A blue light reflects objects and reaches the camera lens, which obtains point-by-point coordinate measurements and geometry of the parts. The STL files were exported from the Solutionix ezScan which controls the Solutionix D500 scanner. The evaluation of shape deviation errors was undertaken with Geomagic® Control X™ software (based on ASME Y14.5M standard, v2020.1.1, 3D Systems, Research Triangle, SC, USA) by comparing the CAD model whit STL files were exported from Solutionix ezScan™. On the other hand, The STL scanned file must be aligned with CAD using component alignment to obtain consistent results. At least four points of cloud data must be matched to the CAD model during the alignment process. The circularity and cylindricity were evaluated by using the standard ASME Y14.5M 19,941 The measured values of cylindricity and circularity are shown in Table 3. In Figure 3, each step in the flowchart will be used in the rest of the article.

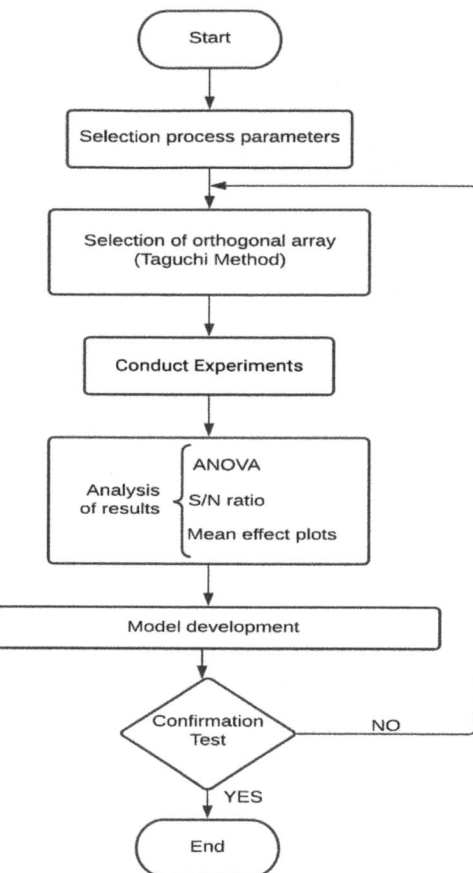

Figure 3. Flowchart of implementing the steps.

Table 2. L9 orthogonal array.

No. of Trial	Thickness (mm)	Infill Pattern	Layer Height (mm)	Wall Layer
1	5	Hexagonal	1	2
2	5	Rectangular	1.125	3
3	5	Triangular	2	4
4	10	Hexagonal	1.125	4
5	10	Rectangular	2	2
6	10	Triangular	1	3
7	15	Hexagonal	2	3
8	15	Rectangular	1	4
9	15	Triangular	1.125	2

Table 3. Measured values of cylindricity and circularity.

No. of Trial	cyl 1 (mm)	cyl 2 (mm)	Mean cyl (mm) (±St Dev)	cir 1 (mm)	cir 2 (mm)	Mean cir (mm) (±St Dev)
1	0.1112	0.1101	0.11065 (±0.0007778)	0.0864	0.0878	0.0871 (±0.0009899)
2	0.1047	0.1024	0.10355 (±0.0016263)	0.0853	0.0741	0.0797 (±0.0079196)
3	0.1648	0.1612	0.163 (±0.0025456)	0.1463	0.01376	0.14195 (±0.0937199)
4	0.1309	0.1294	0.13015 (±0.0010607)	0.1098	0.0975	0.10365 (±0.0086974)
5	0.1571	0.1528	0.15445 (±0.0030406)	0.1192	0.1204	0.1198 (=0.0008485)
6	0.1646	0.1644	0.1645 (±0.0001414)	0.1267	0.1262	0.12645 (±0.0003536)
7	0.2423	0.2444	0.24335 (±0.0014849)	0.2018	0.2015	0.20165 (±0.0002121)
8	0.26	0.263	0.2615 (±0.0021213)	0.1897	0.1943	0.192 (±0.0032527)
9	0.2287	0.2267	0.2277 (±0.0014142)	0.1798	0.176	0.1779 (=0.0026870)

3. Results and Discussion

3.1. Analysis of Experimental Data

S/N ratio and ANOVA were used to analyze data from the experiments. Experimental measured data of Table 3 for the cylindricity and circularity were analyzed by using the statistical software MINITAB 19.0® (LLC, State College, PA, USA).

3.2. Analysis Using S/N Ratio

To analyze the effect of process variables on each response (cylindricity), the S/N ratio was utilized. When the experimental results were presented as S/N ratios, it was discovered that they varied linearly. For the data analysis, out of the different quality characteristics of S/N ratio, the 'smaller is better' was considered. S/N ratio (η) can be obtained by using Equation (1), where MSD stands for mean-square deviation, the average of the data points is indicated by the Y, and Y_0 represents the target value, and σ^2 is the variance. Equation (2) calculates the MSD value [30].

$$\eta = -10\ log(MSD) \quad (1)$$

$$MSD = \sigma^2 - (Y - Y_0)^2 \quad (2)$$

3.3. Response Table for S/N Ratio for Cylindricity and Circularity

These S/N ratios are added for each level of each parameter according to Taguchi's procedures, and then their average is determined. The highest minus the lowest average is the delta statistic and the individual contribution of the parameters (Rank) is shown in Tables 4 and 5.

Table 4. Response of variable parameters for cylindricity.

Level	Thickness (mm)	Infill Pattern	Layer Height (mm)	Wall Layer
1	18.19	16.37	15.48	16.06
2	16.53	15.85	16.75	15.88
3	12.26	14.76	14.74	15.04
Delta	5.93	1.61	2.01	1.02
Rank	1	4	2	3

Table 5. Response of variable parameters for circularity.

Level	Thickness (mm)	Infill Pattern	Layer Height (mm)	Wall Layer
1	20.04	18.27	17.83	18.21
2	18.69	18.25	18.89	17.95
3	14.41	16.64	16.43	16.99
Delta	5.63	1.63	2.45	1.22
Rank	1	3	2	4

3.4. Mean Affects Plots for S/N Ratios

In Figures 4 and 5, The S/N graphs were used to determine the optimal parameters in the form of average S/N ratios for cylindricity and circularity. As minimization of the output parameters is required, the smaller is better is selected to maximize mathematical expression for the S/N ratio for cylindricity and circularity. In Figures 4A–D and 5A–D correspond to thickness, infill pattern, layer height, and wall layer, respectively. The horizontal axis shows the different levels for each parameter. According to the respective average S/N ratios, all phases of the given graphs show that in the 5 cm thickness, hexagonal infill pattern, 1.125 mm layer height, and two wall layers are the required 3D printing parameters for the best cylindricity and circularity values.

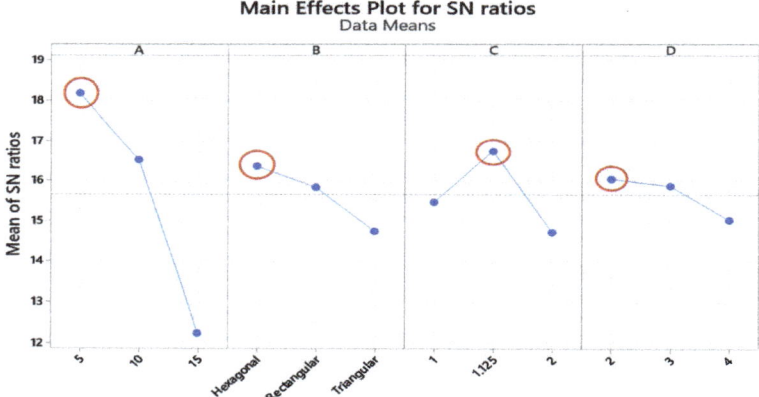

Signal-to-noise: Smaller is better

Figure 4. Main effect plot cylindricity. A, B, C, and D represent the Thickness, Infill Pattern, Layer height, and Wall Layer, respectively.

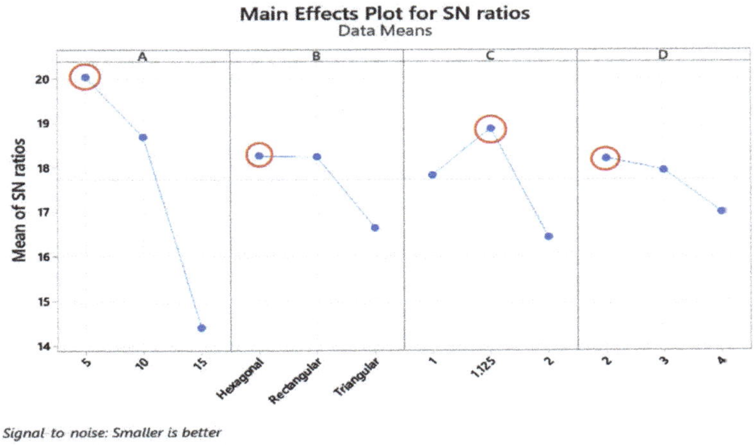

Figure 5. Main effect plot circularity. A, B, C, and D represent the Thickness, Infill Pattern, Layer height, and Wall Layer, respectively.

3.5. Analysis of Variance

The ANOVA method was used to determine the significance and contribution of each process parameter to the response variables. The results are shown in Tables 6 and 7, where DF represents degrees of freedom, and Adj SS shows the adjusted sum of squares and can be calculated as shown in Equation (3) where η_i shows the mean S/N ratio and η_j shows the total mean of S/N ratio and n is total number of experiments. Adj MS is the adjusted mean sum of squares, F-Value and *p*-Value are the variance of the group means and probability value respectively. *p*-Value describes the significance level of each parameter, and it can be calculated as shown in the 4, where SS_D is sum of squared deviations each process parameter and SS_T is Total sum of squared deviations [30]. The findings reveal that the F-Value of thickness and infill pattern is greater than the F-Value of the wall layer and layer height. As a result, they have the greatest influence on cylindricity and circularity values.

$$ss_T = \sum_{i=1}^{n}(\eta_i - \eta_j)^2 \quad (3)$$

$$P = \frac{S_{eq}ss_D}{S_{eq}ss_T} \quad (4)$$

Table 6. Response of variable parameters of cylindricity.

Source	DF	Adj SS	Adj MS	F-Value	*p*-Value
A	2	0.023508	0.011754	35.49	0.027
B	2	0.000841	0.000421	1.27	0.441
C	2	0.001806	0.000903	0.72	0.268
D	2	0.000662	0.000331	-	-
Pulled Error	2	0.000662	0.000331	-	0.559
Total	8	0.026818	0.01374	-	-

Table 7. Response of variable parameters of circularity.

Source	DF	Adj SS	Adj MS	F-Value	p-Value
A	2	0.013321	0.00666	28.51	0.034
B	2	0.000657	0.000328	1.41	0.416
C	2	0.0001749	0.000875	3.74	0.211
D	2	0.000467	0.000234	-	-
Pulled Error	2	0.000467	0.000234	-	0.584
Total	8	0.016194	0.008331	-	-

3.6. Response Surface Regression Model

Since the Taguchi approach just analyzes the major factors which influence variables, without taking account of the correlation between them, response surface regression was used to determine the relationship between the control variables and response variables for polyamide 6. The purpose of the response surface method is to formulate the response as a function of contributing variables and to discover the best set of factor levels that provides the best response value depending on the research goals.

Constants and predictor coefficients made up the regression model. The linear response surface regression model is represented by Equation (5) [39].

$$y = \beta_0 + \beta_1 x_1 + \beta_2 x_2 + \ldots + \beta_j x_j + \varepsilon \tag{5}$$

where x_i is the process parameter and β is the coefficient to be determined based on the experimental data (β_0 = constant coefficients, $\beta_1, \beta_2, \ldots, \beta_j$ = linear coefficients) and ε describes the measurement error. The response y can be any of the output parameters. Models were developed by using the software MINITAB 19.0®.

Linear regression equation used in the estimation of cylindricity values:

$$\text{cylindricity} = -0.0260 + 0.01185\ A + 0.0118\ B + 0.0192\ C + 0.0102\ D \tag{6}$$

Linear regression equation used in the estimation of circularity values:

$$\text{circularity} = -0.0317 + 0.00876\ A + 0.0090\ B + 0.0365\ C + 0.0088\ D \tag{7}$$

Here A, B, C and D are the factors that represent the thickness, infill pattern, layer height and wall layer, respectively. The above empirical model predicts the cylindricity and circularity for any combination of process parameters. The correlation coefficient or R-squared is a statistical measure that represents a dependent variable's proportion of variation and usually is between 0 to 100%. In the preceding model of cylindricity and circularity, the values of R-squared is 86.40% and 84.67%, respectively, which demonstrates that the actual and predicted values have a good correlation.

In a DoE study, response surface plots are extremely beneficial for evaluating the interaction effects between two parameters simultaneously on the responses. The contours of a response surface will be plotted to help visualize the shape of the response surface, and each contour corresponds to a certain response surface height. Response surface plots and contour plots of each two parameter combination on cylindricity (Figures 6 and 7) and circularity (Figures 8 and 9) are plotted when the other two parameters are held constant at their default values, as shown in the upper right area. As is shown in Figures 6 and 7, cylindricity is plotted versus different levels. Contour plots show that cylindricity is minimal at low thickness values and hexagonal infill patterns (B). Similarly, from the interaction plot of the thickness and layer height on cylindricity at low values of thickness and layer height levels, cylindricity is minimum. Also, at low levels of thickness and levels of wall layer, cylindricity is low. Cylindricity is minimum at low layer height levels and hexagonal infill pattern. At low wall layer levels and hexagonal infill pattern, cylindricity is low. Also, cylindricity is minimal in low wall layers and low layer height values. The 3D surface plots show the same combination of process variables.

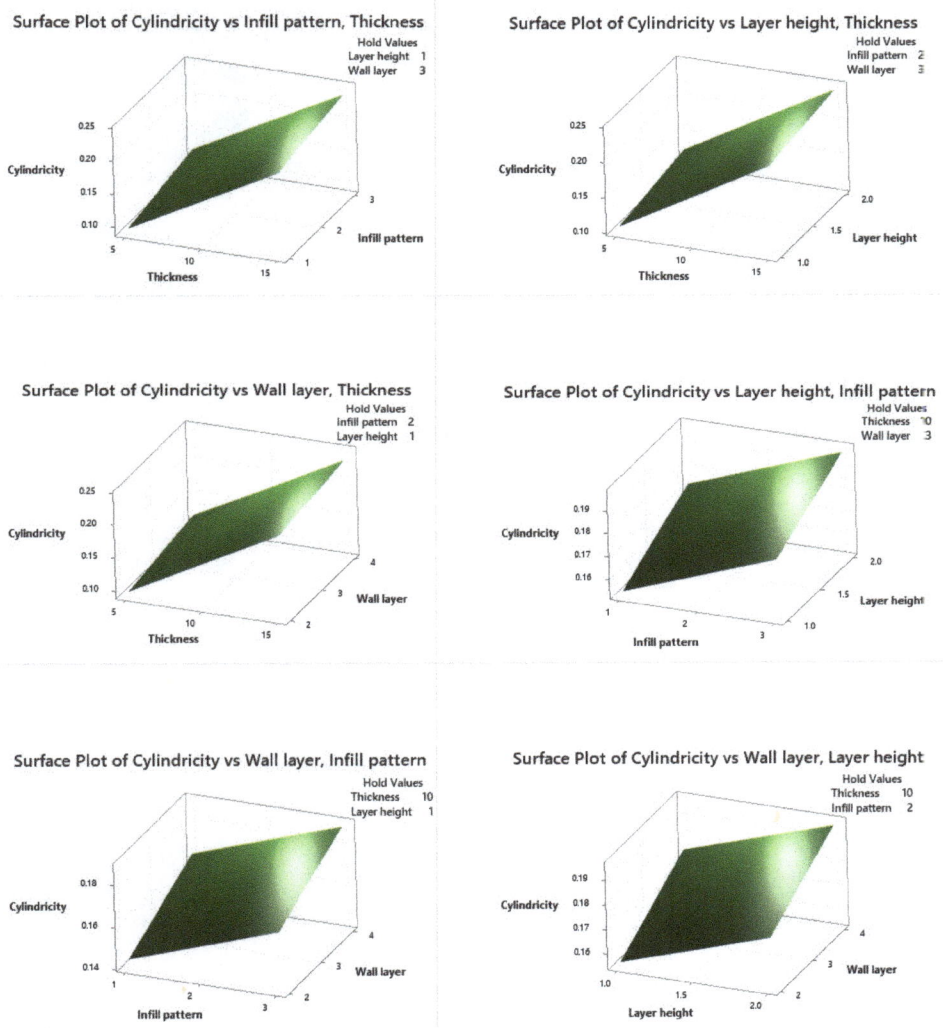

Figure 6. Cylindricity surface plots.

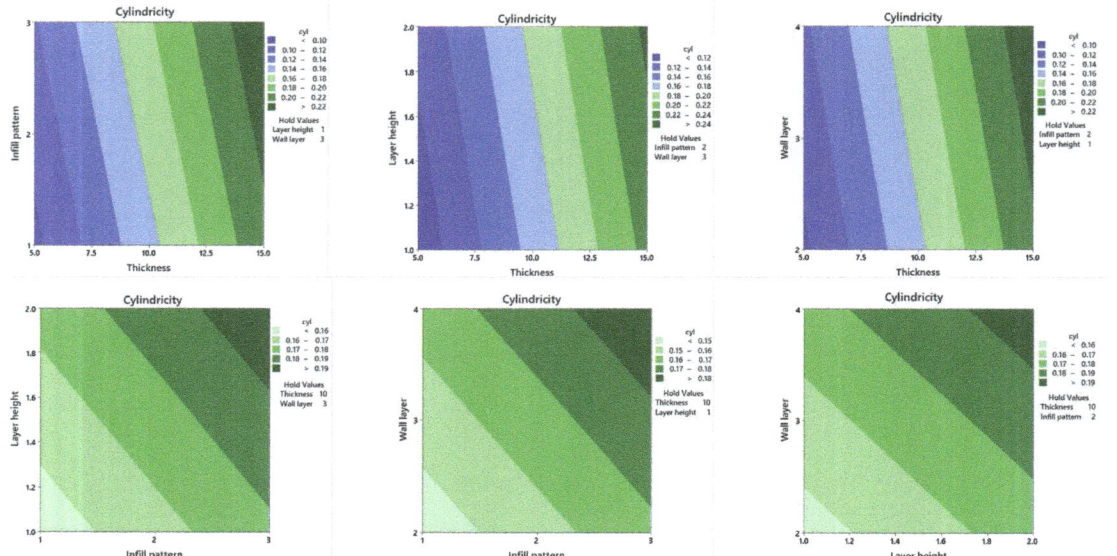

Figure 7. Cylindricity contour plots.

Contour and surface plots for circularity are shown against various levels in Figures 8 and 9. From the interaction plot of the thickness and infill pattern, low values of the thickness, and hexagonal infill patterns, circularity is low. At lower thickness and layer height levels, circularity is also minimal. Circularity is low at low thickness levels and wall layers values. Circularity is minimal in lower layer height levels and hexagonal infill pattern. Also, lower wall layer levels with a hexagonal infill pattern provide lower circularity. Cylindricity is minimal in low wall layers levels and low layer height values. The 3D contour plots show the same combination of process variables.

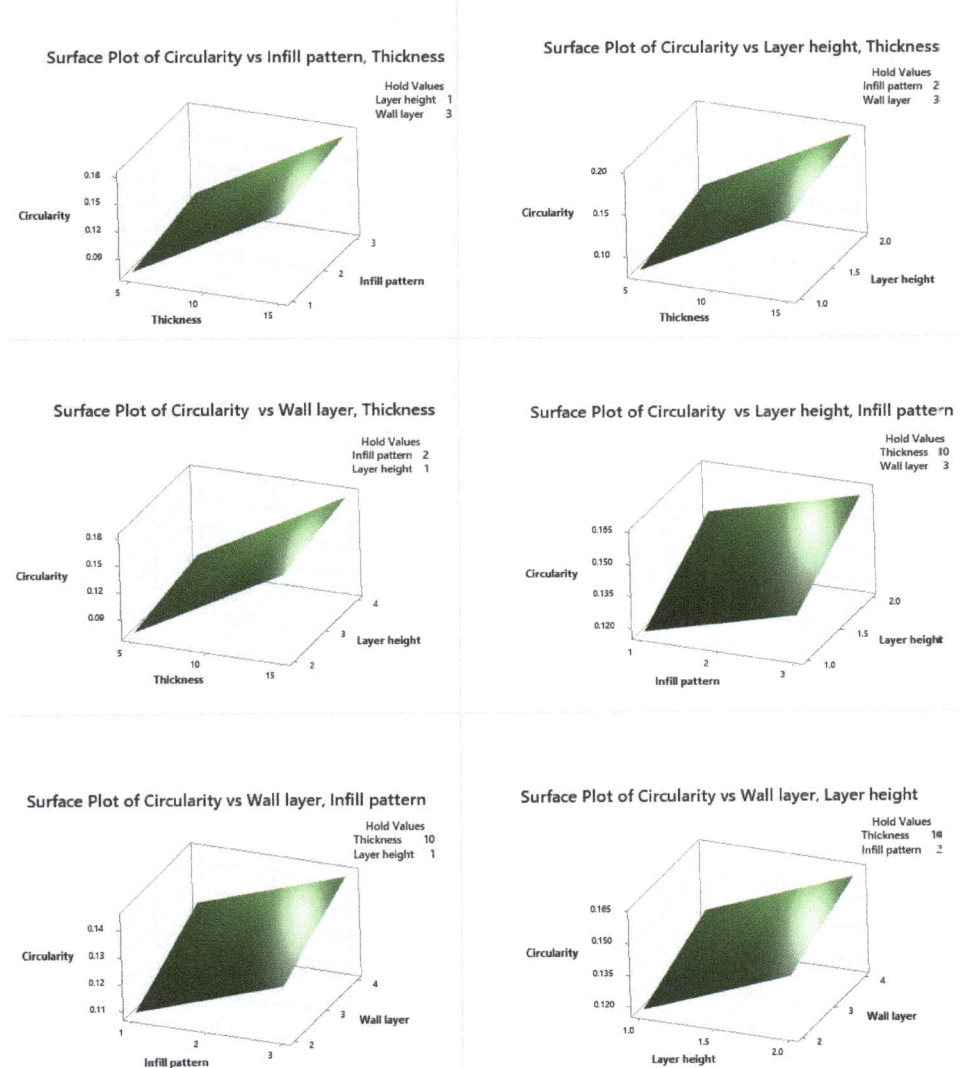

Figure 8. Circularity surface plots.

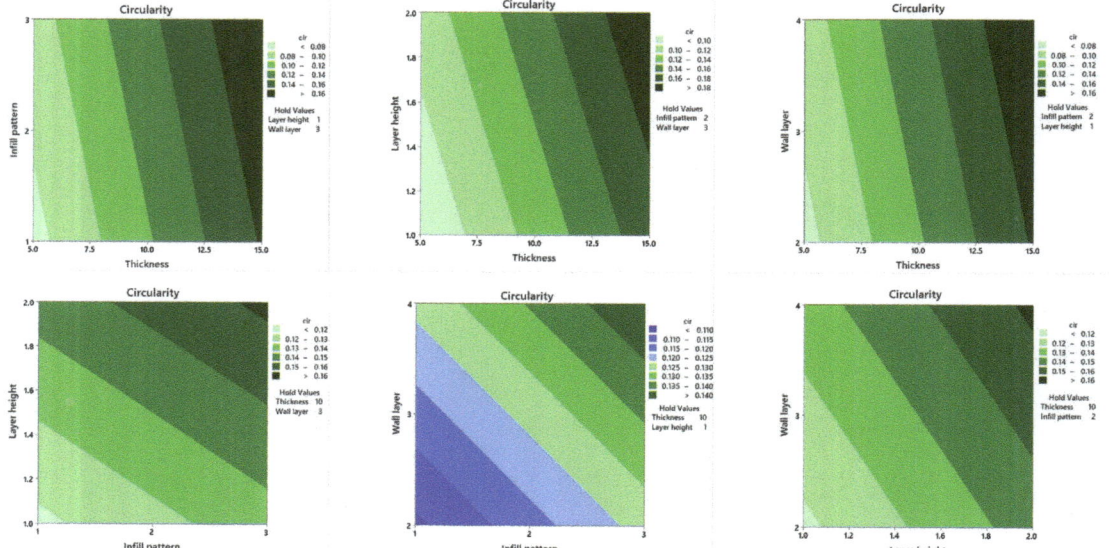

Figure 9. Circularity contour plots.

Taguchi design was used to create reproducible and valid results to investigate the effect of selected parameters on the geometrical error. This study revealed many unexpected findings. The impact of each is further analyzed by the impact of parameters on the responses and compared with related sources:

1. Parts with higher thickness have more material mass than the smaller diameter parts, and this causes the gravity force, which is one of the forces affecting the deformation, to be higher, and it will cause more geometric error so that the lowest thickness is shown better responses.
2. As it turns out, the amount of geometric error is minimal in the hexagonal infill pattern. This could be due to the polymer nodes. The more nodes, the tighter the piece and the less deformation. Since the number of polymer nodes increases in the hexagonal pattern, the deformation is less than other infill patterns. In the research mentioned in the introduction, the effect of the infill pattern was studied, and similarly, the hexagonal infill was introduced as the optimal infill pattern [28].
3. As the layer height is decreased, the number of deposited layers increases, and this will cause more interfaces and adhesion will be reduced. On the other hand, a high layer height causes higher thermal gradients between the layers, and more deformation will be accrued, and geometrical errors will be increased. But as mentioned in the literature of the article with different materials, it was observed that as the layer height decreased, the shape errors decreased. The reason may be the type of material that is used in this article. However, the layer height study results in other materials such as ABS and PLA showed that the lower the layer height, the lower the geometric and dimensional accuracy at lower layer heights. This difference in results between other research and the current research can be due to the limitation of the 3D printer used because values less than 1 and more than 2 cannot be selected as layer height [28,36,40].
4. According to the results, it was found that the amount of cylindricity and circularity is less for a lower number of wall layers. This may be because, in 3D printing of parts, the wall layers are first deposited on the platform, and then the internal infill will be deposited. During this short period, the wall layer will be fully solidified. Therefore, we will not have good adhesion in the interface of walls and infill sections compared

to the internal infill section. However, wall thickness (wall layer) studies on PLA revealed that the dimensional accuracy would be improved and optimized in average wall thickness values. The explanation for the variation in findings might be related to limitations in the design of the cylinders and 3D printers that were utilized. The type of material used may affect the results [33].

3.7. Confirmation Test

To validate the established empirical models, a confirmation experiment was carried out. The optimal process parameter condition was used to conduct a confirmation test. The optimal values for parameters were hexagonal infill pattern, two wall layers, 1.125 mm layer height, and 5 cm thickness. Therefore, a part with optimal parameters was 3 D printed, and the Solutionix D500 scanner measured the cylindricity and circularity values. Then the optimal values of the parameters were placed in the developed cylindricity and circularity formulas, and the obtained values were compared with the values obtained from the experiment. Table 8 shows these values. Because of the uncertainty, output response was predicted to fall within the confidence interval range. The confirmation test revealed that the suggested models for cylindricity and circularity were acceptable in 95% of the experimental domain's confidence interval (CI). According to the results presented in Table 8, it was found that the error of cylindricity and circularity is less than 20%, which is acceptable [41]. The results of the confirmation tests indicate that the optimization was successful.

Table 8. Comparison of cylindricity and circularity predicted by ANOVA and confirmation experiment.

Shape Error	Predictions (±95% CI) (mm)	Experimental (mm)	Error (%)
cylindricity	0.08705 (±0.1733)	0.10261	15.1
circularity	0.06851 (±0.1367)	0.07822	12.4

4. Conclusions

The present work leads to the following conclusions of experimental research on the effect of layer height, infill pattern, and the number of wall layers on shape deviation (cylindricity and circularity) in different thicknesses in the FFF process.

According to the L9 orthogonal array, the experiments were carried out by a MarkForged® Mark Two 3D printer.

The cylindricity and circularity indicator was measured using a Solutionix D500 scanner and Geomagic® Control X™ software. According to the results from ANOVA, the layer height and thickness influence is much more significant than the influence of the infill pattern and wall layers on cylindricity. Similarly, the effect of layer height and thickness on circularity is much more significant than the influence of the infill pattern and wall layers, according to DOE and optimization results.

Also, the results can help the designer to understand the phenomenological interactions between the parts' dimensions and the evolution of the geometric tolerances. From the (S/N) analysis for the cylindricity and circularity, it was found that layer height of 1.125 mm, hexagonal infill pattern, 5 mm thickness, and two wall layers were the optimal process parameters to minimize shape deviations. Also, a regression model was developed, and a confirmation test was applied to show that the predictions are in good agreement with experimental data.

Three-dimensional printing results can be influenced by unstable machine conditions, operator error, and other factors. Regression models were established to predict cylindricity and circularity.

Author Contributions: Conceptualization, Z.S., K.B., A.T. and M.S; methodology, Z.S.; software, Z.S.; validation, Z.S., K.B., A.T., M.A. and M.S.; formal analysis, Z.S.; investigation. Z.S.; resources, Z.S., K.B., A.T., M.A. and. M.S.; data curation, Z.S.; writing—original draft preparation, Z.S., K.B., A.T., M.A. and M.S writing—review and editing, Z.S., K.B., A.T., M.A. and M.S.; visualization, Z.S.;

supervision, K.B., A.T. and M.S.; project administration, K.B., A.T. and M.S. funding acquisition, K.B., A.T. and M.S. All authors have read and agreed to the published version of the manuscript.

Funding: This research received no external funding.

Institutional Review Board Statement: Not applicable.

Informed Consent Statement: Not applicable.

Data Availability Statement: The data presented in this study are available on request from the corresponding author.

Conflicts of Interest: The authors declare no conflict of interest.

References

1. Standard Terminology for Additive Manufacturing Technologies. In *91 on Terminology*; Astm International: West Conshohocken, PA, USA, 2012.
2. Zolfagharian, A.; Khosravani, M.R.; Kaynak, A. Fracture resistance analysis of 3D-printed polymers. *Polymers* **2020**, *12*, 302. [CrossRef] [PubMed]
3. Katarzyna, B.; Piesowicz, E.; Szymański, P.; Ślączka, W.; Pijanowski, M. Polymer composite manufacturing by FDM 3D printing technology. In *MATEC Web of Conferences*; EDP Sciences: Les Ulis, France, 2018; Volume 237, p. 02006.
4. Bárnik, F.; Vaško, M.; Sága, M.; Handrik, M.; Sapietová, A. Mechanical properties of structures produced by 3D printing from composite materials. In *MATEC Web of Conferences*; EDP Sciences: Les Ulis, France, 2019; Volume 254, p. 01018.
5. Bai, X.; Ding, G.; Zhang, K.; Wang, W.; Zhou, N.; Fang, D.; He, R. Stereolithography additive manufacturing and sintering approaches of SiC ceramics. *Open Ceram.* **2021**, *5*, 100046. [CrossRef]
6. Qadri, S. *A Critical Study and Analysis of Process Parameters of Selective Laser Sintering Rapid Prototyping*; Elsevier BV: Amsterdam, The Netherlands, 2021.
7. He, Q.; Xia, H.; Liu, J.; Ao, X.; Lin, S. Modeling and numerical studies of selective laser melting: Multiphase flow, solidification and heat transfer. *Mater. Des.* **2020**, *196*, 109115. [CrossRef]
8. Taylor, A.C.; Beirne, S.; Alici, G.; Wallace, G.G. System and process development for coaxial extrusion in fused deposition modelling. *Rapid Prototyp. J.* **2017**, *23*, 543–550. [CrossRef]
9. Polak, R.; Sedlacek, F.; Raz, K. Determination of FDM Printer Settings with Regard to Geometrical Accuracy. In Proceedings of the 28th DAAAM International Symposium, Zadar, Croatia, 5–12 November 2017; pp. 0561–0566.
10. Ford, S.L.N. Additive manufacturing technology: Potential implications for US manufacturing competitiveness. *J. Int. Com. Econ.* **2014**, *6*, 40.
11. Singh, B.; Kumar, R.; Chohan, J.S. Polymer matrix composites in 3D printing: A state of art review. *Mater. Today Proc.* **2020**, *33*, 1562–1567. [CrossRef]
12. Küpper, D.; Heising, W.; Corman, G.; Wolfgang, M.; Knizek, C.; Lukic, V. *Get Ready for Industrialized Additive Manufacturing*; DigitalBCG, Boston Consulting Group: Zürich, Switzerland, 2017.
13. Saleh Alghamdi, S.; John, S.; Roy Choudhury, N.; Dutta, N.K. Additive manufacturing of polymer materials: Progress, promise and challenges. *Polymers* **2021**, *13*, 753. [CrossRef] [PubMed]
14. Black, H.T.; Celina, M.C.; McElhanon, J.R. Additive Manufacturing of Polymers: Materials Opportunities and Emerging Applications. Available online: https://www.osti.gov/servlets/purl/1561754 (accessed on 21 October 2021).
15. Vidakis, N.; Petousis, M.; Tzounis, L.; Maniadi, A.; Velidakis, E.; Mountakis, N.; Kechagias, J.D. Sustainable additive manufacturing: Mechanical response of polyamide 12 over multiple recycling processes. *Materials* **2021**, *14*, 466. [CrossRef]
16. Krishna, S.; Sreedhar, I.; Patel, C.M. Molecular dynamics simulation of polyamide-based materials—A review. *Comput. Mater. Sci.* **2021**, *200*, 110853. [CrossRef]
17. Ahmadifar, M.; Benfriha, K.; Shirinbayan, M.; Tcharkhtchi, A. Additive Manufacturing of Polymer-Based Composites Using Fused Filament Fabrication (FFF): A Review. *Appl. Compos. Mater.* **2021**, 1–46. [CrossRef]
18. Anitha, R.; Arunachalam, S.; Radhakrishnan, P. Critical Parameters Influencing the Quality of Prototype in Fused Deposition Modeling. *J. Process. Technol.* **2001**, *118*, 385–388. [CrossRef]
19. Brajlih, T.; Valentan, B.; Balic, J.; Drstvensek, I. Speed and accuracy evaluation of additive manufacturing machines. *Rapid Prototyp. J.* **2011**, *17*, 64–75. [CrossRef]
20. Relvas, C.; Ramos, A.; Completo, A.; Simões, J.A. A systematic approach for an accuracy level using rapid prototyping technologies. *Proc. Inst. Mech. Eng. Part. B J. Eng. Manuf.* **2012**, *226*, 2023–2034. [CrossRef]
21. Fisher, R.A. Design of experiments. *Br. Med. J.* **1936**, *1*, 554. [CrossRef]
22. Bagchi, T.P. *Taguchi Methods Explained: Practical Steps to Robust Design*; Prentice-Hall: Hoboken, NJ, USA, 1993.
23. Kechagias, J.D.; Tsiolikas, A.; Petousis, M.; Ninikas, K.; Vidakis, N.; Tzounis, L. A robust methodology for optimizing the topology and the learning parameters of an ANN for accurate predictions of laser-cut edges surface roughness. *Simul. Model. Pract. Theory* **2022**, *114*, 102414. [CrossRef]

24. Maazinejad, B.; Mohammadnia, O.; Ali, G.A.; Makhlouf, A.S.; Nadagouda, M.N.; Sillanpää, M.; Sadegh, H. Taguchi L9 (34) orthogonal array study based on methylene blue removal by single-walled carbon nanotubes-amine: Adsorption optimization using the experimental design method, kinetics, equilibrium and thermodynamics. *J. Mol. Liq.* **2020**, *298*, 112001. [CrossRef]
25. Gunst, R.F.; Mason, R.L. Fractional factorial design. *Wiley Interdiscip. Rev. Comput. Stat.* **2009**, *1*, 234–244. [CrossRef]
26. Karna, S.K.; Sahai, R. An overview on Taguchi method. *Int. J. Eng. Math. Sci.* **2012**, *1*, 1–7.
27. Lee, B.; Abdullah, J.; Khan, Z. Optimization of rapid prototyping parameters for production of flexible ABS object. *J. Mater. Process. Technol.* **2005**, *169*, 54–61. [CrossRef]
28. Alafaghani, A.; Qattawi, A. Investigating the effect of fused deposition modeling processing parameters using Taguchi design of experiment method. *J. Manuf. Process.* **2018**, *36*, 164–174. [CrossRef]
29. Rizea, A.D.; Anghel, D.C.; Iordache, D.M. Study of the deviation of shape for the parts obtained by additive manufacturing. *IOP Conf. Ser. Mater. Sci. Eng.* **2021**, *1009*, 012050. [CrossRef]
30. Sood, A.K.; Ohdar, R.K.; Mahapatra, S.S. Parametric appraisal of mechanical property of fused deposition modelling processed parts. *Mater. Des.* **2010**, *31*, 287–295. [CrossRef]
31. Sheth, S.; George, P.M. Experimental investigation, prediction and optimization of cylindricity and perpendicularity during drilling of WCB material using grey relational analysis. *Precis. Eng.* **2016**, *45*, 33–43. [CrossRef]
32. Das, P.; Mhapsekar, K.; Chowdhury, S.; Samant, R.; Anand, S. Selection of build orientation for optimal support structures and minimum part errors in additive manufacturing. *Comput. Des. Appl.* **2017**, *14*, 1–13. [CrossRef]
33. Aslani, K.-E.; Chaidas, D.; Kechagias, J.; Kyratsis, P.; Salonitis, K. Quality Performance Evaluation of Thin Walled PLA 3D Printed Parts Using the Taguchi Method and Grey Relational Analysis. *J. Manuf. Mater. Process.* **2020**, *4*, 47. [CrossRef]
34. Chang, D.Y.; Huang, B.H. Studies on profile error and extruding aperture for the RP parts using the fused deposition modeling process. *Int. J. Adv. Manuf. Technol.* **2011**, *53*, 1027–1037. [CrossRef]
35. Eswaran, P.; Sivakumar, K.; Subramaniyan, M. Minimizing error on circularity of FDM manufactured part. *Mater. Today Proc.* **2018**, *5*, 6675–6683. [CrossRef]
36. Doli, B. Analysis of fused deposition modeling process for additive manufacturing of Abs parts. In Proceedings of the 10 International conference on precision, meso, micro and nano engineering (COPEN 10), Chennai, Tamilnadu, India, 7–9 September 2017; pp. 7–9.
37. Maurya, N.K.; Rastogi, V.; Singh, P. Fabrication of prototype connecting rod of PLA plastic material using FDM prototype technology. *Indian J. Eng. Mater. Sci. (IJEMS)* **2021**, *27*, 333–343.
38. Saqib, S.; Urbanic, J. An experimental study to determine geometric and dimensional accuracy impact factors for fused deposition modelled parts. In *Enabling Manufacturing Competitiveness and Economic—Sustainability*; Springer: Berlin/Heidelberg, Germany, 2012; pp. 293–298.
39. Myers, R.H.; Khuri, A.I.; Carter, W.H. Response surface methodology: 1966–1988. *Technometrics* **1989**, *31*, 137–157. [CrossRef]
40. Benfriha, K.; Ahmadifar, M.; Shirinbayan, M.; Tcharkhtchi, A. Effect of process parameters on thermal and mechanical properties of polymer-based composites using fused filament fabrication. *Polym. Compos.* **2021**. [CrossRef]
41. Cetin, M.H.; Ozcelik, B.; Kuram, E.; Demirbas, E. Evaluation of vegetable based cutting fluids with extreme pressure and cutting parameters in turning of AISI 304L by Taguchi method. *J. Clean. Prod.* **2011**, *19*, 2049–2056. [CrossRef]

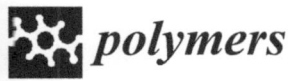

Article

Hybrid Bio-Inspired Structure Based on Nacre and Woodpecker Beak for Enhanced Mechanical Performance

Zhongqiu Ding, Ben Wang, Hong Xiao * and Yugang Duan *

State Key Laboratory for Manufacturing Systems Engineering, School of Mechanical Engineering, Xi'an Jiaotong University, Xi'an 710049, China; dingzhq@stu.xjtu.edu.cn (Z.D.); wangben@xjtu.edu.cn (B.W.)
* Correspondence: xiaohongjxr@xjtu.edu.cn (H.X.); ygduan@xjtu.edu.cn (Y.D.)

Abstract: Materials with high strength and toughness have always been pursued by academic and industrial communities. This work presented a novel hybrid brick-and-mortar-like structure by introducing the wavy structure of the woodpecker beak for enhanced mechanical performance. The effects of tablet waviness and tablet wave number on the mechanical performance of the bio-inspired composites were analyzed. Compared with nacre-like composites with a flat tablet, the strength, stiffness and toughness of the novel hybrid nacre-like composite with tablet wave surface increased by up to 191.3%, 46.6% and 811.0%, respectively. The novel failure mode combining soft phase failure and tablet fracture revealed the key to the high toughness of composites. Finite element simulations were conducted to further explore the deformation and stress distribution of the hybrid brick-and-mortar-like structure. It showed that the hybrid brick-and-mortar-like structure can achieve a much better load transfer, which leads to greater tensile deformation in tablet before fracture, thus improving strength and energy absorption. These investigations have implications in the design of composites with high mechanical performance for aerospace, automobile and other manufacturing industries.

Keywords: hybrid bio-inspired design; nacre; woodpecker beak; mechanical performance; failure mode; finite element simulation

1. Introduction

Many natural materials have an unusual combination of stiffness, low weight, strength and toughness beyond the reach of current engineering materials. The mechanical properties of natural materials far exceed the components that make them up. It is mainly due to the well-organized microstructure and abundant effective interface interactions on multiple length scales [1–4], which provides infinite inspiration for the manufacture of new biomimetic structural materials. Nacre is an excellent example of such materials, which is mostly made up of crystalline aragonite ($CaCO_3$) platelets (95% vol.) and bonded by a thin layer of biopolymer (5% vol.) [5–8]. Despite the high mineral content, nacre is almost 20–30 times tougher than aragonite alone [9,10]. The impressive mechanical property of nacre can be attributed to its brick-and-mortar structure [11,12], as shown in Figure 1a. In addition, the detailed sub-level structures, such as mineral bridges found in the organic matrix layers [13], nanoscale mineral islands found on the top and bottom surface of tablets [8] and tablet interlocks [7,14], also contribute to the toughness of the nacre. Many researchers have studied the effects of structure details on the mechanical performance of the composites inspired by nacre. Ghimire A et al. have designed nacre-like composites with interlocked tablets and analyzed the effects of waviness on mechanical properties of nacreous composites; results showed that increasing the tablet waviness can improve the stiffness, strength and toughness of the nacre-like composites [15–18]. Mirzaeifar et al. demonstrated that the existence of a hierarchical architecture in the designing of brick-and-mortar-like structures leads to superior defect-tolerant and structural properties [19,20]. Gu et al. systematically elucidate the effects of the density of mineral bridges on the mechanical response of nacre-inspired additive manufactured composites [21]. Jabir et al. [22] studied the influence of the material compliance gradient in mortar of nacre-like composites and proved the significant contribution of material compliance gradient to the mechanical performance. Although many researchers have studied the effects of brick-and-mortar structure and substructures on the mechanical performance which has provided guidance for the design of tough bio-inspired materials, they all revolve around the microstructure observed in the nacre itself.

However, only mimicking the brick-and-mortar structure and substructure of nacre makes a limited contribution to further enhancing the performance of such materials, because their properties are adapted to specific living environment. Actual service scenarios may have more complex and demanding performance requirements for engineering materials. Multibiological multiscale biomimetic design may be a promising design approach. Thus, we turn attention to other high-performance biomaterials, seeking inspiration for the design of high-performance composites with multiple strengthening and toughening mechanisms. Recently, the woodpecker beak has aroused the interest of scientists owing to its ability to withstand high impact [23]. In nature, a woodpecker beak repeatedly strikes into a tree trunk at a speed of 6–7 m/s, with an impact deceleration of 1000 g, without any recorded damage to the beak or brain [24,25]. Lee et al. suggested that the tightly packed keratin scales with wavy surface organized in an overlapping arrangement play an important role in resisting fracture during high-speed pecking [26], as shown in Figure 1b. Although the beak of other birds also shows this wavy structure, the waviness of those birds' beaks is smaller than that of the woodpecker beak [27]. This further highlights the role of the wavy surface structure in tuning the mechanical performance to suit biological functions. Ha et al. proposed a novel bio-inspired honeycomb sandwich panel based on the microstructure of the wavy structure in the woodpecker beak, which indeed exhibits superior energy absorption capability compared with the conventional honeycomb sandwich panel [28].

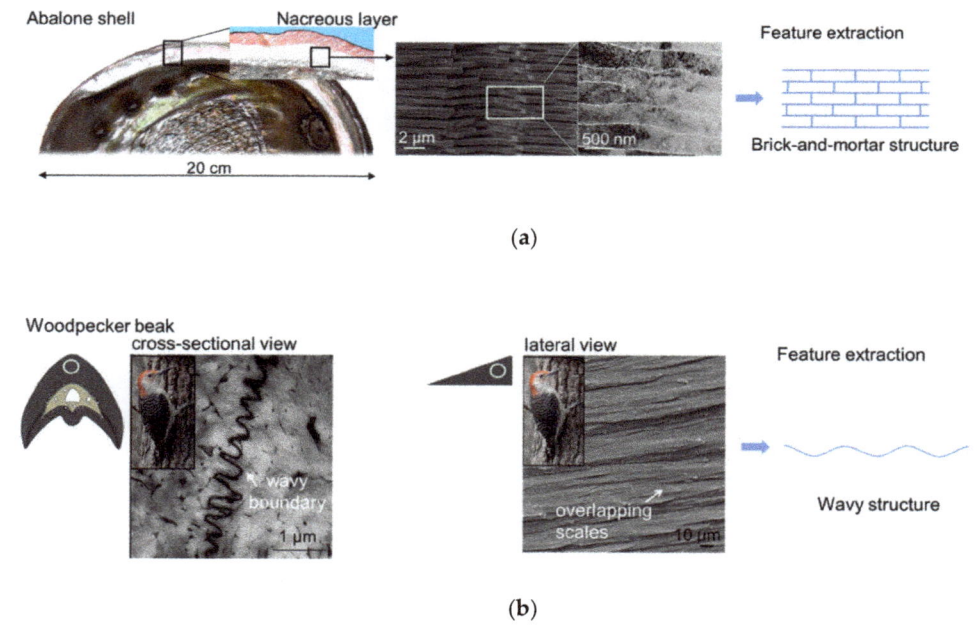

Figure 1. Biomimetic microstructures. (**a**) The typical microstructures of the nacre (adapted from Ref. [7]); (**b**) the typical microstructures of the woodpecker beak (adapted from Ref. [26]).

In this work, we aim to integrate different toughening strategies of the nacre and woodpecker beak to achieve higher mechanical performance amplification in the given material. A novel hybrid brick-and-mortar-like structure with wavy surface tablets was proposed. The multi-material 3D-printing technique allows us to exercise complete control over the tablet structure design. The influences of the tablet waviness and tablet wave number on the mechanical response of the brick-and-mortar-like structure and their behavior mechanism were studied experimentally and numerically. The results in this study can be used in the design of advanced tough composites.

2. Materials and Methods

2.1. Design of Hybrid Bio-Inspired Structures

In this study, the proposed brick-and-mortar-like structure combines the brick-and-mortar structure of the nacre with the wavy structure of the woodpecker beak, including discrete hard blocks bonded by soft interfaces, as shown in Figure 2. The height of the tablet h is 1.5 mm, the length of the tablet l is 7.5 mm, the overlapping length of the tablet is designed to be half the length of the tablet and the width of the tablet w is 3.14 mm. The horizontal surfaces of the tablet are sine wave-like interfaces with wavelength λ and amplitude A. To quantify the waviness of wavy surfaces in tablets, a non-dimensional geometric parameter wv is defined as $wv = A/\lambda$. Since the length of the tablet l is fixed, the wavelength λ is controlled by the wave number n. In order to investigate how the horizontal wavy surfaces of the tablet impact the mechanical performance of the composites, five waviness (0, 0.3, 1, 2 and 3) and five wave numbers (6, 7, 8, 9 and 10) were considered. The nomenclature for designs is NaWb, where N is the wave number with a being its value, and W means the waviness with b being its value. For example, N8W03 means that the wave number is 8 and the waviness is 0.3; N8W10 means wave number is 8 and the waviness is 1. For all designs, the in-plane thickness t of the soft interfacial layer fixed as 0.3 mm in consideration of the 3D printer limitations. However, the stiff phase volume f_v of the composites are not the same. More detailed dimensions of each design are listed in Table 1.

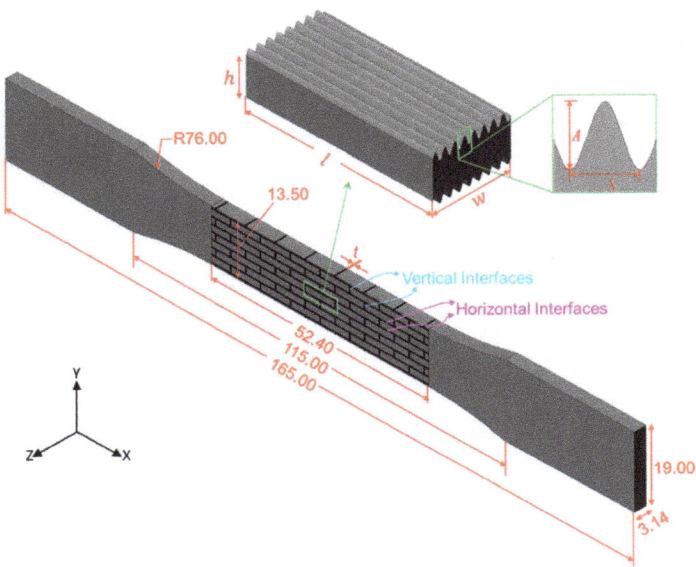

Figure 2. Geometric design of the hybrid bio-inspired structures in accordance with ASTMD368 standard and the unit-cell geometry of the samples with horizontal wavy interfaces.

Table 1. Dimensions of the designs.

Design	λ (mm)	A (mm)	h (mm)	l (mm)	t (mm)	w (mm)	f_v (%)
N8W00	0.3927	0	1.5	7.5	0.3	3.1415	78.994
N8W03	0.3927	0.1178	1.5	7.5	0.3	3.1415	78.993
N8W10	0.3927	0.3927	1.5	7.5	0.3	3.1415	78.992
N8W20	0.3927	0.7854	1.5	7.5	0.3	3.1415	79.001
N8W30	0.3927	1.1781	1.5	7.5	0.3	3.1415	79.876
N6W20	0.5236	1.0472	1.5	7.5	0.3	3.1415	79.635
N7W20	0.4488	0.8976	1.5	7.5	0.3	3.1415	79.373
N9W20	0.3491	0.6981	1.5	7.5	0.3	3.1415	79.071
N10W20	0.3141	0.6283	1.5	7.5	0.3	3.1415	78.990

The relevant geometries were generated using Solidworks (Dassault Systèmes SolidWorks Corporation, Waltham, MA, USA). The 3D models were created by extruding the 2D designs and rendering them into Stereolithography (.stl) files. In addition, the test models consisting of the above-mentioned hard blocks were designed with dog-bone-like ends in order to follow ASTM (American Society of Testing Materials) standards, giving them appropriate dimensions suitable for tensile testing, as shown in Figure 2.

2.2. Sample Fabrication

All specimens used in the study were fabricated using a Stratasys J750 multi-material 3D printer (Stratasys, Minneapolis, MN, USA), which makes complex geometry with around 100 µm printing resolution [29]. Two of Stratasys' commercial photopolymers, VeroWhite and TangoPlus, with strongly contrasting material properties, were used for the composites manufacturing a single print. VeroWhite is a white, stiff/rigid polymer representing hard tablets, and TangoPlus is a rubber-like transparent polymer in place of a biopolymer interface. Using Stratasys' technology, the two materials are sprayed simultaneously as liquid layers and then cured in situ by UV light. This instant curing ensures perfect interfacial adhesion between the two different materials [21]. At the same time, the intermixing of different liquid polymers before curing creates an interface between the two materials, resulting in the mechanical properties of the printed composite depending on the printing direction [21,30]. Thus, all specimens were printed along the same orientation to avoid the influence of the layer orientation on the mechanical properties of the specimens. Figure 3a shows images of a representative 3D-printed specimen. After printing, the water jet was used to remove the gel-like support material from the samples. For the saturation of the curing, the as fabricated specimens were kept at room temperature for 24 h before mechanical testing.

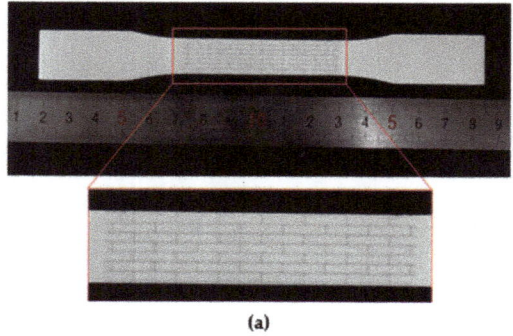

Figure 3. (**a**) A typical 3D-printed sample for tensile testing; (**b**) experimental setup for quasi-static tensile tests.

2.3. Mechanical Testing

To capture the mechanical response of the bio-inspired composites, quasi-static uniaxial tensile tests were performed using an universal testing machine (MTS Systems Corp, Minneapolis, MN, USA) endowed with a 25 kN load cell. The specimens were clamped in place using serrated steel grip faces attached to steel vice action grips. We employed a slow displacement rate 0.2 mm/min to overcome the viscoelastic effect on the mechanical properties, because the soft phase in our composite specimen is highly stretchable before fracture. The test proceeded until the crack propagates thoroughly through the specimen and the load dropped. Strain gages were used to measure the strains on the samples. Three specimens were fabricated for each type of design. Load–displacement curves from tensile tests were transferred to nominal stress–strain curves, while stiffness and strength were calculated based on these curves. The toughness here was defined as the area under the stress–strain curve.

2.4. Finite Element Analysis

Numerical simulations were conducted in ABAQUS (ABAQUS Explicit, version 2017, ABAQUS Inc., Providence, RI, USA) to study the mechanical response of the composites. The two constituent phases used in printing were considered as isotropic materials. For VeroWhite, a power law plasticity model was used to model the initial yield and hardening, and a linear damage evolution law defined by final fracture strain was used to capture the softening. A linear plastic hardening model with linear damage evolution was used to model the stress–strain behavior of TangoPlus. The detailed mechanical properties implemented in the ABAQUS software are presented in Table 2. Studies show no interfacial debonding occurs in the composites fabricated by 3D-printing, due to the perfect adhesion between two phases obtained from the in situ UV curing [31]. Thus, a 'tie' constraint was used for the connection between the tablets and the soft interfacial layer. General contact with a 'hard contact' relationship was used to prevent the penetration of the contact pairs into each other. All models were generated by 3D stress elements with reduced integration C3D8R and meshed after a convergence test. Displacement boundary conditions were applied in the loading direction to simulate the experimental conditions. The left side of the model was held fixed, and the right side was stretched.

Table 2. Mechanical properties of material used in numerical analysis.

Material	E (MPa)	σ_b (MPa)	v
VeroWhite	1927	35	0.3
TangoPlus	3.5	1.2	0.4

3. Results and Discussions

3.1. Experimental Tensile Test Response of Bio-Inspired Composites

Tensile tests were performed on all the presented bio-inspired composites. The results showed good repeatability in terms of stress–strain response and failure modes. For the sake of brevity, only one representative stress–strain curve and failure mode were reported for each sample series.

3.1.1. Influence of Tablet Waviness on Mechanical Behavior

Figure 4a shows the stress–strain curves from the tensile test results of bio-inspired composites with various tablet waviness, where the feature points were marked. Although all the specimens show uniform deformation at the initial tensile stage, the fracture behaviors of different specimens are obviously different and can be divided into three failure modes, as shown in Figure 4b. The first peak stands for the strength of vertical short interfaces, which is defined as σ_I; the last peak stands for the strength of horizontal interfaces, which is defined as σ_{II}. The mechanical behaviors of composites with the tablet waviness below 2 match Mode I: a two-stage fracture with low failure stress. Upon loading, the vertical short interface fractures first, leading to a softening stage in the stress–strain curve followed by a hardening stage; the shear deformation of soft material along the horizontal interface takes place, soon after which the composites rupture completely. Such two-stage fracture behavior is also reported for 3D-printed composites inspired by interlocks [16] and mineral bridges [19] of nacre and bone [31]. The mechanical behavior of composites with a tablet waviness of 2 meets Mode II: a three-stage fracture with high failure stress. Besides the softening and hardening stage, a tablet deformation stage appears due to the tensile deformation in the tablets. In addition, the mechanical behavior of composite with tablet waviness above 2 matches Mode III: a single-stage fracture with relatively high failure stress, which mainly corresponds to the tablet deformation. The corresponding deformation and stress distribution of these three different failure stages are shown in Figure 4c. These phenomena indicate that adjusting the tablet waviness can change the failure mechanisms of composites.

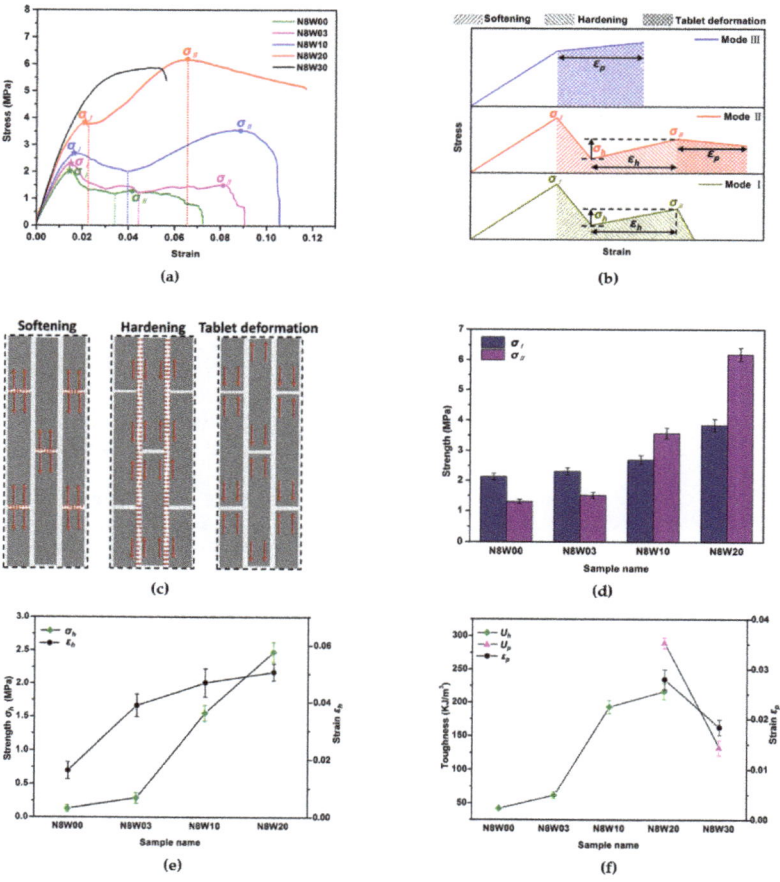

Figure 4. Tensile behaviors of composites with various tablet waviness. (**a**) Stress-strain curves of the composites with the tablet wave number of 8 and different waviness; (**b**) schematic of stress–strain curves for three different response modes; (**c**) schematic of failure patterns in the softening, hardening and tablet deformation stages; (**d**) comparison of the vertical short interface strength and horizontal interface strength of composites with different tablet waviness; (**e**) plot of increased stress and strain period of hardening stage; (**f**) plot of strain period of tablet deformation stage, energy absorption while hardening and tablet deformation of composites.

Figure 4d compares the σ_I and σ_{II} of composites with tablet waviness below 3. It can be seen that both the σ_I and σ_{II} show an increasing trend with the increase in tablet waviness. The σ_{II} increases significantly, from 1.3 MPa to 6.2 MPa, while the σ_I increases from 2.1 MPa to 3.8 MPa for the composites in the order N8W00, N8W03, N8W10 and N8W20. Notably, the value of σ_{II} for composite N8W10 firstly exceeds that of its σ_I. Moreover, the strain at the beginning of the softening stage also tends to increase, as shown in Figure 4a, which indicates that a larger tablet waviness is beneficial to the delay of crack generation and the energy absorption. Meanwhile, it can be seen from the strain–stress curves that as the tablet waviness increases, the hardening in composites with tablet waviness below 3 changes from a non-linear and unclear increase to a linear and obvious increase.

To further analyze the hardening behavior in relation to the tablet waviness precisely, we quantify three critical aspects of the stress–strain response, which are the strain period during the hardening stage (ε_h), the stress increased during the hardening stage (ε_h) and the energy absorption (U_h) during the hardening stage (the area under the stress–strain curve at hardening stage), as shown in Figure 4b. It can be found from Figure 4e that the values of σ_h, ε_h and U_h exhibit a strong dependence on the tablet waviness. Flat interfaces in N8W00 result in the lowest stress increase (σ_h = 0.13 MPa), whereas larger tablet waviness leads to a significant increase in the value of σ_h (σ_h = 0.29 MPa in N8W03, σ_h = 1.55 MPa in N8W10 and σ_h = 2.48 MPa in N8W20). This indicates that composites with larger tablet waviness can withstand higher stress and exhibit larger resistance to shear deformation resulting from tablet sliding. Compared with the ε_h value of composite N8W00 (ε_h = 0.016), the ε_h value of composite N8W03, N8W10 and N8W20 increases by 139.3% (ε_h = 0.039), 203.5% (ε_h = 0.047) and 262.5% (ε_h = 0.051), respectively. The composite N8W20 undergoes the largest hardening period and exhibits the highest hardening rate. In addition, the energy absorption during the hardening stage U_h displays an increasing trend upon the increase in the tablet waviness from 0 to 2. From the above results, it is clear that the rise in increased stress during the hardening stage of the composite with larger tablet waviness supports its increment in tablet deformation, leading to the increment of strain whilst hardening, thereby promoting the energy absorption. To further analyze the influence of tablet waviness on the deformation of composites, we quantify two critical aspects: the strain period during the tablet deformation (ε_p) (Figure 4b) and the energy absorption (U_p) during the tablet deformation (the area under the stress–strain curve at tablet deformation stage). It can be seen that both the ε_p and U_p decrease when the tablet waviness increases from 2 to 3, causing earlier rupture. This is because the hardening in the hardening stage of composite N8W20 is not complete and continues to the subsequent deformation stage. In other words, the deformation stage of composite N8W20 is the result of the combined effect of horizontal interface shear deformation and tablet deformation, while the deformation of composite N8W30 only comes from tablet deformation.

Figure 5a shows the normalized surface area (normalized by the surface area of composite N8W00) and the scale factor (defined as λ/w) of composites for various tablet waviness, Figure 5b–d show the strength, stiffness and toughness values of composites with different waviness. Obviously, the stiffness increases as the tablet waviness increases, which can be attributed to the increase in contact interface. Compared with the stiffness of composite N8W00 (159.09 MPa), the stiffness of composite N8W03, N8W10, N8W20 and N8W30 increases by 10.6% (175.98 MPa), 25.2% (199.14 MPa), 46.6% (233.17 MPa) and 69.4% (269.55 MPa), respectively. However, there is indeed an optimal structure design based on the geometry and mechanical properties of the constituents under the circumstance of strength and toughness. Among all composites, composite N8W20 has the highest strength (6.2 MPa) and toughness (562.8 KJ/m^3), which are 191.3% and 811.0% higher than that of composite N8W00. Interestingly, the increase in toughness of our nacre-like composite is obviously higher than that of other reported nacre-like composites [15–19,21]. Although the strength of composite N8W30 is slightly lower than that of composite N8W20, there is a sharp drop in the toughness of composite N8W30 compared with composite N8W20 due to smaller failure strain. This is essentially attributed the competition between interface hardening and stress concentration, which was also discussed by Horacio D et al. [16,18]. It should be noticed that the stiff phase volume in composites is not the same; composite N8W20 exhibits larger strength and toughness, although the stiff phase volume for composite N8W30 is higher than composite N8W20, which can be attributed to the wavy interface design.

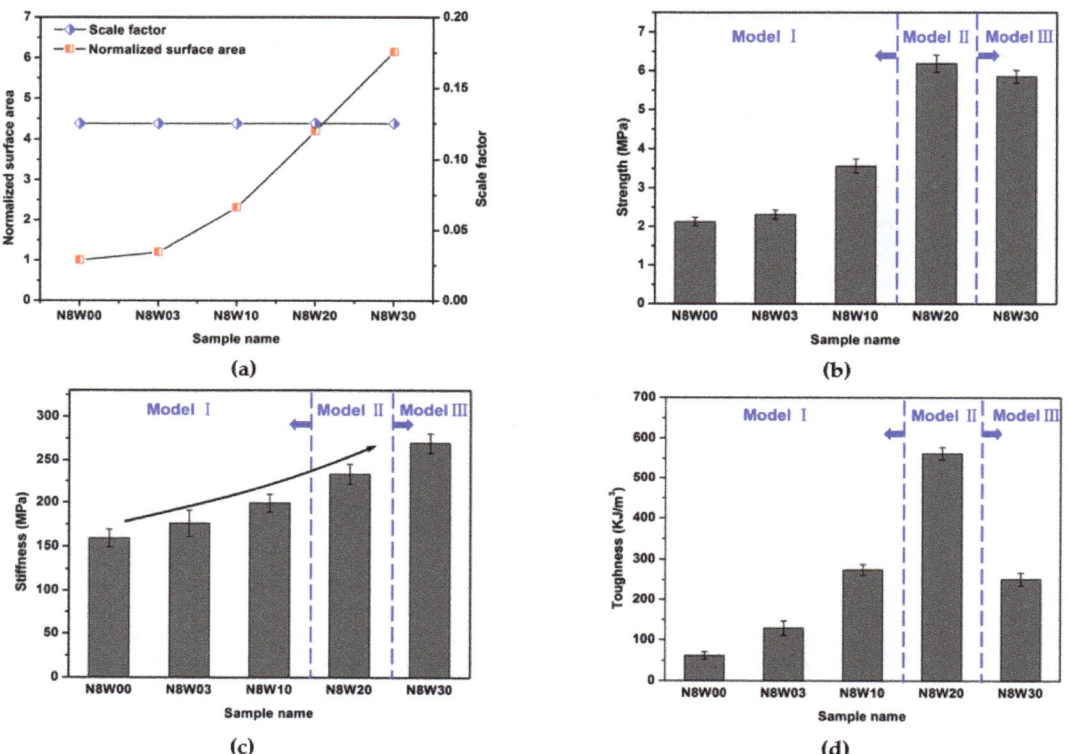

Figure 5. Effects of tablet waviness on the mechanical performance of composites. (**a**) The normalized surface area (normalized by the surface area of tablet in composite N8W00) and scale factor of composites; (**b**) strength; (**c**) stiffness; (**d**) toughness.

3.1.2. Influence of Tablet Wave Number on Mechanical Behavior

Figure 6a show the stress–strain curves from the tensile test results of composites with various wave number. It can be seen that these composites fail in a similar way, and their mechanical behaviors meet Mode II: a three-stage fracture with high failure stress. On the whole, although the tablet wave number does not have much effect on the mechanical behavior of composites compared with the tablet waviness, some regularities can still be observed. Figure 6c shows the σ_I and σ_{II} of composites with different tablet wave numbers. It is clear that the σ_I of composite N6W20, N7W20, N8W20, N9W20 and N10W20 are basically constant, while the σ_{II} of these composites have a slight decreasing trend as the tablet wave number increases. Figure 6d compares the increased stress during hardening σ_h and the strain period during hardening ε_h. When the tablet wave number increases to 10, there is an effective reduction in the value of σ_h, which indicates the composite N10W20 exhibits the lowest hardening rate. We also find that the ε_h increases first and then decreases as the tablet wave number increases, the composite N8W20 exhibits the largest ε_h (0.051). Additionally, the energy absorption during hardening U_h plotted in Figure 6e displays a similar changing trend with the increase in wave number. It is clear from Figure 6e that the tablet wave number has a greater influence on the deformation stage than the hardening stage. Both the strain period during deformation ε_p and energy absorption during deformation U_p display an increasing trend upon the increase in the tablet wave number. Compared with the composite N6W20, the ε_p and U_p of composites with other tablet wave numbers increase by up to 223.7% and 170.8%, respectively. To further analyze the influence of tablet wave number on the deformation of composites, we quantify another aspect of stress–strain response, the equivalent stiffness during the deformation stage (E_p) as shown in Figure 6b, to describe the deformation resistibility capacity of the structure. As discussed previously in failure mode II, the deformation stage of these composites is a combination of horizontal interface shear deformation and tablet deformation. Figure 6f evinces that the E_p of composites increases as the tablet wave number increases. This is likely due to the different degree of shear deformation of the horizontal interface and different stress in tablet at the deformation stage. As the deformation stage is the extension of hardening, the greater ε_h and σ_{II} means more adequate hardening, that is, larger horizontal interface shear deformation and greater stress in tablet. Thus, it inevitably leads to earlier failure of the horizontal interface and tablet at the deformation stage. The above discussion leads to a conclusion that although the increase in tablet wave number will reduce the load transfer, it will delay the failure of composites. In contrast, the decrease in tablet wave number will promote the load transfer but lead to premature failure of the composite.

Figure 7a shows the normalized surface area (normalized by the surface area of composite N6W20) and the scale factor (defined as λ/w) of composites for various wave number. Figure 7b–d show the strength, stiffness and toughness values of composites with different wave number. Although the contact interface area is the same, the mechanical properties are different due to the scale effect of the of microstructure design in nacreous composites. This result agrees with [32], where scale effect was studied for additively manufactured two-phase composites. It is clear that the strength displays a slight decreasing tend as the wave number increases. Compared with the strength of composite N6W20 (6.33 MPa), the strength of composite N7W20, N8W20, N9W20 and N10W20 reduces by 1.6% (6.23 MPa), 2.1% (6.20 MPa), 5.1% (6.01 MPa), 12.6% (5.54 MPa), respectively. In addition, the stiffness decreases as the tablet wave number increases, while there is indeed an optimum toughness. Among all the composites with same waviness, the composite N8W20 exhibits the greatest toughness, suggesting that composite N8W20 exhibits the optimal balance between interface hardening and stress concentration caused by geometrical scale.

The above results indicate that both the tablet waviness and tablet wave number can affect the mechanical response of the bio-inspired composite proposed in this work. By contrast, the tablet waviness has a greater impact on the composite than the tablet wave number. It is manifested in two aspects. On the one hand, the value of strength, stiffness and toughness of the composite changes more greatly due to the change of the tablet waviness. On the other hand, the transformation of failure mechanisms of the composite can be realized by tuning the tablet waviness, whereas tuning the tablet wave number cannot. The wavy interface design of the nacre-like composite offers additional resistance to shear effectively at macro scales, resembling the wavy surface of scales in woodpecker beaks, can boost the interface hardening and delay the fracture, leading to enhanced energy absorption.

3.2. Fracture Mechanisms and Morphologies

The typical failure patterns of the bio-inspired composites are shown in Figure 8a. The fracture patterns of the composites can obviously be divided into three types: the soft phase (interface) failure, the soft phase failure coupled with the tablet break and the tablet break. In composite N8W00, N8W03 and N8W10, soft phase failure in both the vertical and horizontal interface is the cause of the composite failure. Additionally, their zig-zag fracture path just proves that the shear deformation of soft phase during tablet sliding is the main mechanism of their energy absorption. Though similar fractures existed in composite N8W00, N8W03 and N8W10, we observed from Figure 8b that increasing the tablet waviness results in the increase in the fracture region area. Additionally, we quantified the fracture behavior of composites with tablet waviness below 2, as shown in Figure 8c. The increase in tablet waviness from 0 to 1 leads to an increase in the fracture region area up to two times. As discussed in Section 3.1.1, increasing the tablet waviness can improve the resistance of the composites to crack initiation and propagation. It can be seen from Figure 4a that the strain corresponding to the strength σ_I increases with the increase in tablet waviness, so the deformations spread to a larger area and cause tablet pullout in a larger region, as previous studies reported [15], leading to an increase in the fracture region of composite N8W00, N8W03 and N8W10.

Figure 8 W20 and other composites with the same tablet waviness, the overall failure mechanism changes: soft phase failure and tablet break act synergistically leading to the special fracture morphology, that is, the tablets are completely broken, while the fracture path still displays a small zig-zag tooth pattern. This partial failure of the soft interface is due to the limited tablet sliding during tensile deformation. In spite of this, the incomplete failure of soft phase promotes the delay of fracture and the increase in energy absorption. Meanwhile, a higher portion of load is transferred to the tablet due to the increased interface area, which is beneficial to improve the load bearing capacity of composites. However, the interfacial hardening strength increases as the strain increases and when it is greater than that of the tablets themselves, localized tablet break was observed. As the tablet waviness continues to increase, localized tablet break prevails and the function of soft phase failure declines in the failure of composites. This has been proven by the fracture pattern of composite N8W30. It can be seen that the fracture is almost neat, with obvious brittle fracture characteristics. The horizontal interface area in composite N8W30 is large enough that a higher portion of the load is transferred to the tablets, thus the shear deformation in soft phase is averted; at the same time, the narrower top of the wavier interface promotes the generation of stress concentration zones, which causes the tablet crack to expand rapidly leading to fracture. Although the tablets break in composite N8W20 and N8W30, the results in Figure 4f show that the strain period during tablet deformation stage ε_p of N8W20 is larger than that of N8W30. This can be correlated to the decrease in the strain during the hardening ε_h of N8W20 (Figure 4e); the horizontal interface of N8W20 does not completely fail in the hardening stage and continues in the tablet deformation stage, supporting the increase in strain during the tablet deformation stage.

In addition, Figure 8d shows the relationship between the failure mode of the composites and the structural feature. Through the above analysis, we can come to a conclusion that the key to high toughness of the composites lies in the balance of two failure mechanisms: soft phase failure and tablet break. The soft phase failure helps to increase the fracture strain and improve the energy absorption, making the failure of the composite show pseudoplasticity, while the tablet break can improve the bearing capacity (i.e., the strength and stiffness) of the composite before its failure.

3.3. Simulation Results

Since the stress distribution of tablet is the concrete manifestation of the difference in mechanical behavior of the different bio-inspired composites, we perform numerical analysis for composites with different tablet waviness and wave number and illustrate the stress distribution in the tablet before fracture.

The tensile stress distribution in a tablet selected at the same position of different composites at the same overall tensile strain 0.65% (elastic deformation region) is displayed in Figure 9a. Firstly, an efficient stress transfer between the hard tablets and soft interface can be observed, thus the tensile stress in bricks is much higher than that in interfaces. It is evident that the stress is seen to concentrate in the center region of the tablets; specifically, the stress concentrates on the top of the wave in the wavy tablet surface. Therefore, as the waviness in tablets increases, the stress concentration area on the tablet surface gradually changes from a single, continuous one to multiple, discontinuous ones, but the total area of the stress concentration area is decreases. Notably, the number of stress concentration areas is the same as the wave number of the tablet when the waviness is not zero. In addition, the number of stress concentration area increases as the tablet wave number increases, however, the total area of the stress concentration area increases first and then decreases. Among all composites with different wave number but the same waviness, the ratio of the stress in the middle area of the tablet to that on the top of the wave in composite N8W20 is the smallest, which means that the tablet of composite N8W20 can withstand greater tensile deformation.

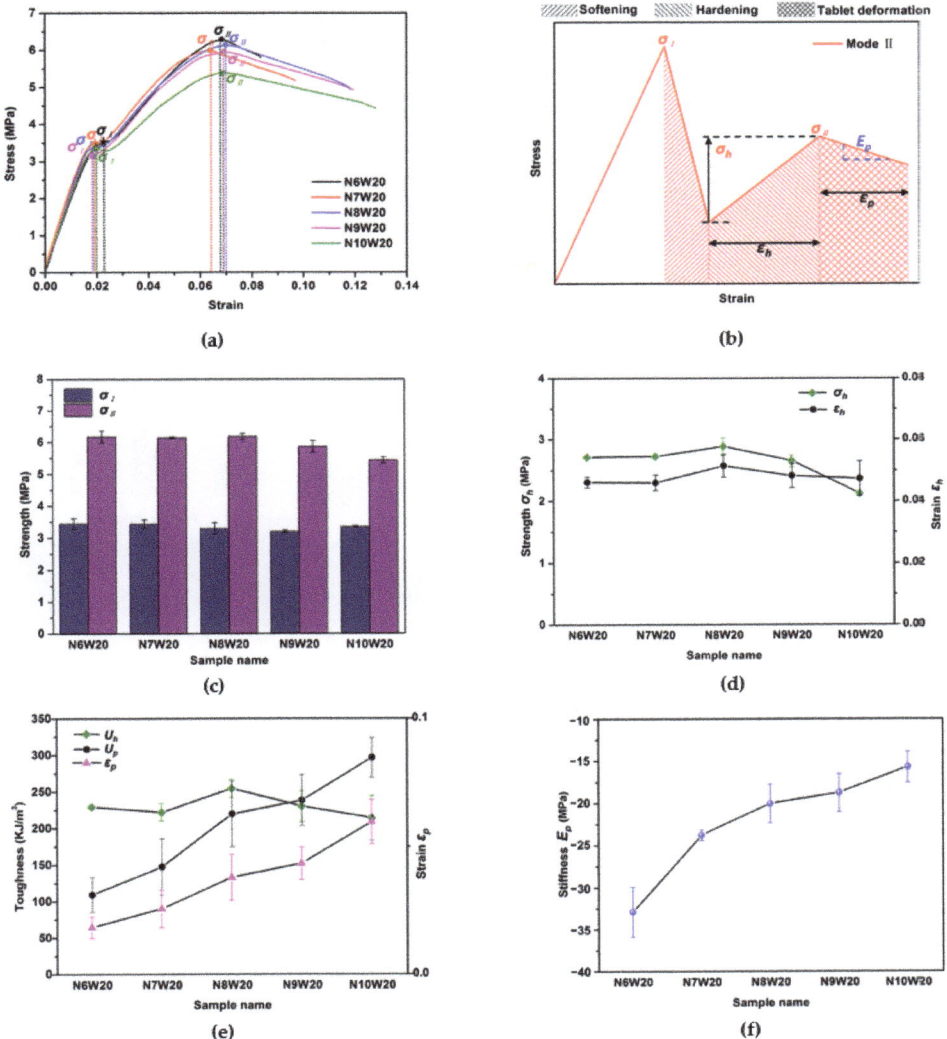

Figure 6. Tensile behaviors of composites with various tablet wave number. (**a**) Stress–strain curves of the composites with Table 2. and different wave number; (**b**) comparison of the vertical short interface strength and horizontal interface strength of composites with different tablet wave number; (**c**) plot of increased stress and strain period of hardening stage; (**d**) plot of strain period of tablet deformation stage, energy absorption while hardening and tablet deformation of composites; (**e**) plot of strain period of tablet deformation stage, energy absorption while hardening and tablet deformation of composites; (**f**) plot of the equivalent stiffness during the deformation stage.

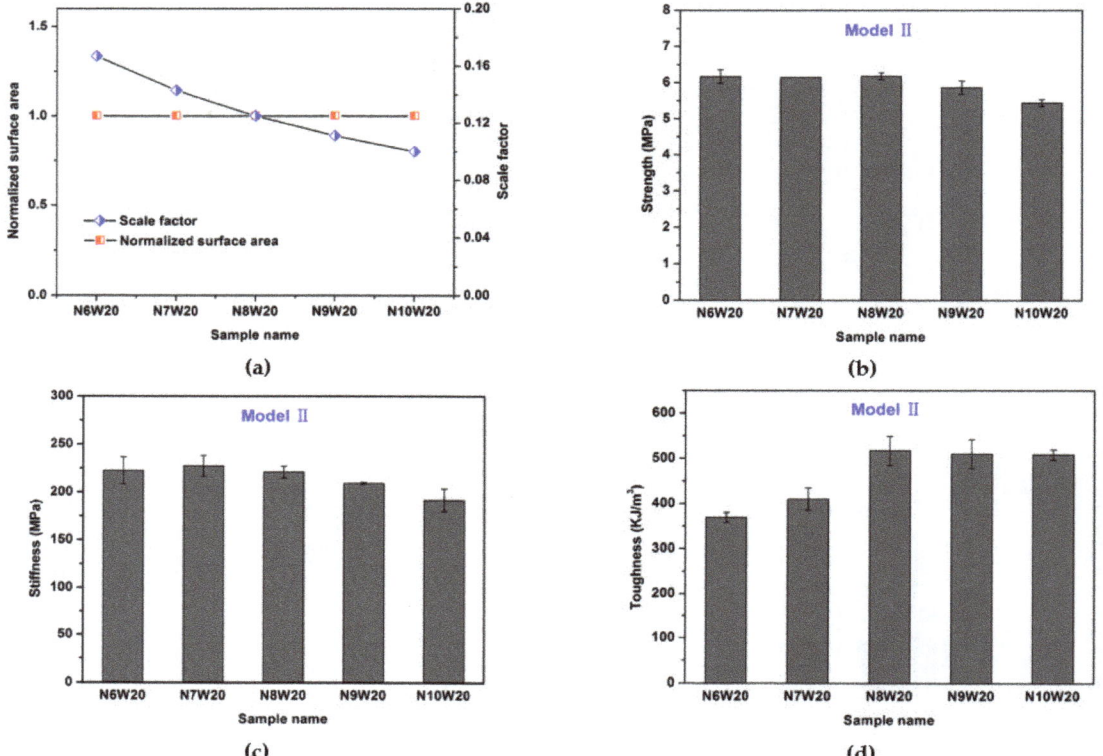

Figure 7. Effects of tablet wave number on the mechanical performance of composites. (**a**) The normalized surface area (normalized by the surface area of tablet in composite N6W20) and scale factor of composites; (**b**) strength; (**c**) stiffness; (**d**) toughness. Schemes follow another format.

To further analyze the influences of tablet waviness and wave number on the stress distribution in composite while tensile, we compare the stress on the wave surface in the tablet, as shown in Figure 9b,c. It is clear that the stress increases as the tablet waviness increases, suggesting that more load is transferred to the tablets. Compared with the maximum stress at the top of the wave of tablet in composite N8W00, that of composite N8W30 increases by up to about nine times. The maximum stress of composite N8W30 quickly reaches the strength limit as the strain increases, leading to the earliest fracture of the tablet before the strong interface fails, which has been observed in experimental analysis of the composites. In contrast, the maximum stress at the stress concentration of composite N8W20 is second to that of composite N8W30, indicating that the stress increases at a slower rate, which is consistent with the delayed fracture of tablet of composite N8W20. The load transfer of composites with tablet waviness below 2 are relatively small, thereby the damage mainly occurs in the soft interface. Moreover, the maximum stress at the top of the wave of tablet decreases as the wave number increases. Compared with the maximum stress at the top of the wave of tablet in composite N6W20, that of composite N7W20, N8W20, N9W20 and N10W20 reduces by 3.9%, 3.1%, 9.0% and 14.1%, respectively. This indicates that the smaller wave number, the higher level of stress concentration, leads to easier crack growth as we have observed in the experimental analysis.

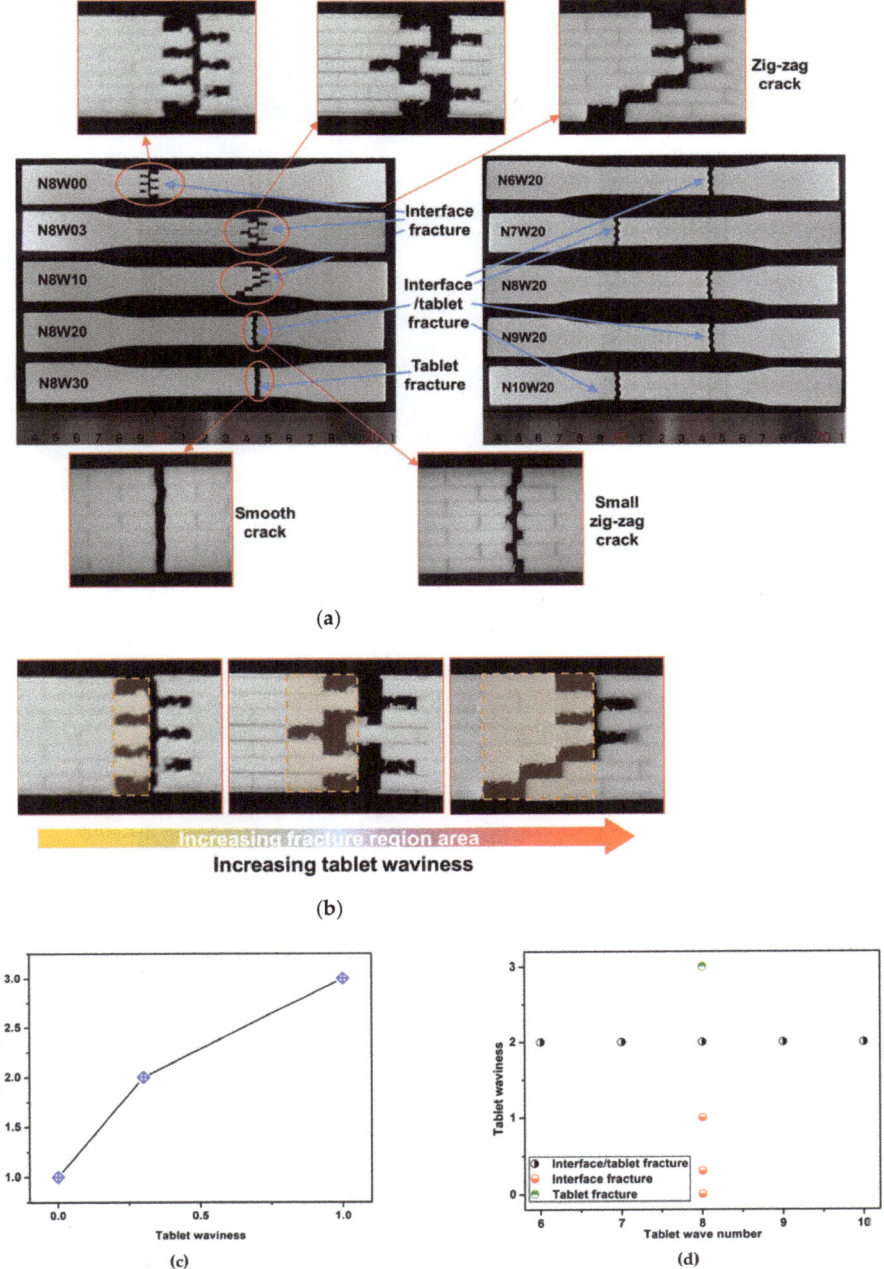

Figure 8. (a) Failure morphologies of the bio-inspired composites after tensile test; (b) schematics displaying that the increase in tablet waviness leads to the increase in of fracture region area; (c) quantification of fracture morphologies of composite with tablet waviness wv = 0, 0.3, 1, wave number of 8; (d) map of identified failure modes as they relate to the tablet waviness and wave number.

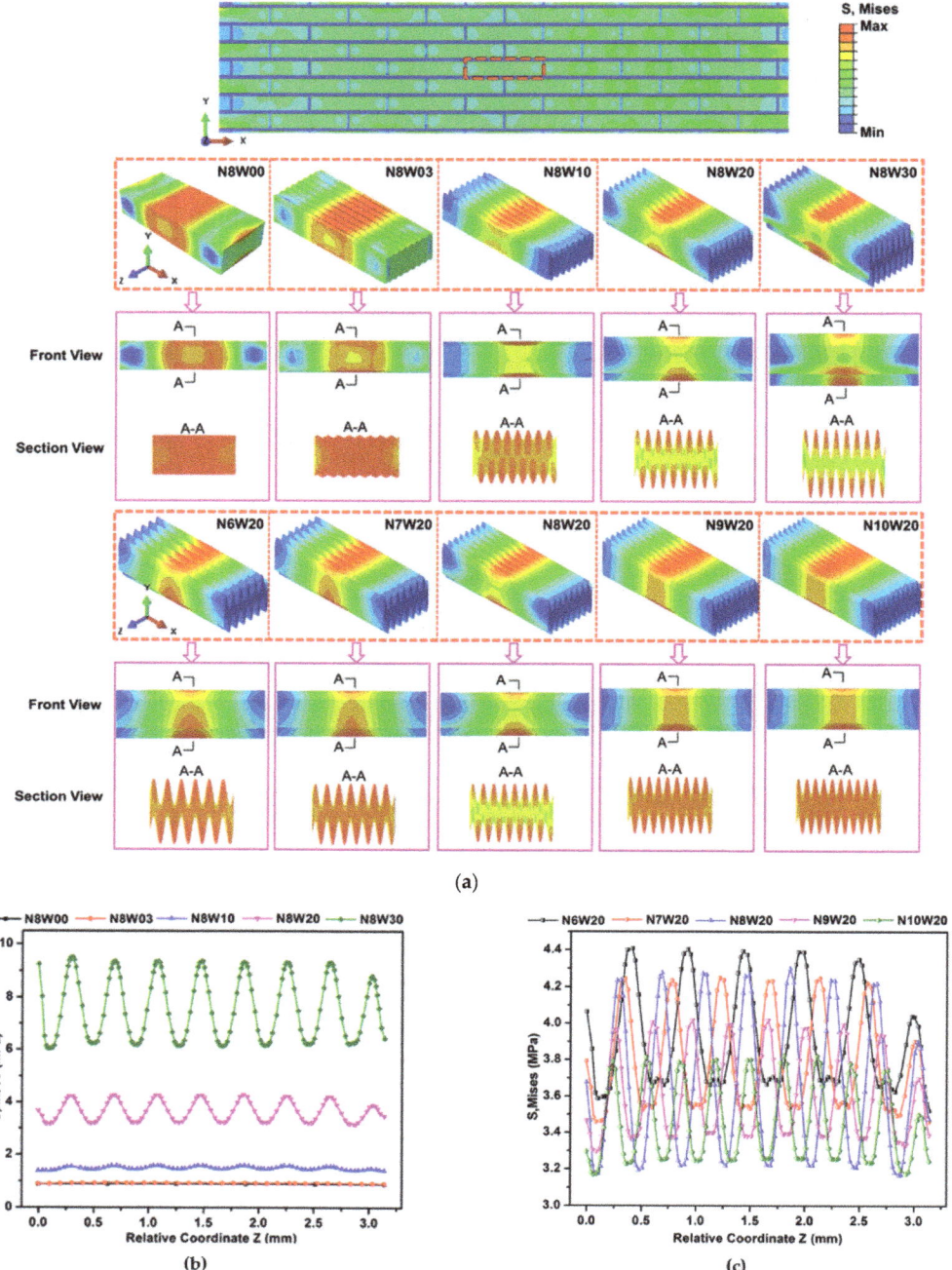

Figure 9. The simulation results of composites. (**a**) The stress distribution in tablets of composites with vari-ous tablet waviness at the same overall tensile strain based on specific table; (**b**) the comparison of the maximum stress at the stress concentration located in the center of tablet of composites with various tablet waviness; (**c**) The comparison of the maximum stress at the stress concentration lo-cated in the center of tablet of composites with various tablet wave number.

4. Conclusions

In this study, we proposed a hybrid brick-and-mortar-like structure by introducing the wavy suture structure of the woodpecker beak into the brick-and-mortar structure of the nacre. Compared with nacre-like composites with flat tablet, the strength, stiffness and toughness of the nacre-like composite with tablet of wave surface (N8W20) increase by up to 191.3%, 46.6% and 811.0%, respectively. This unusual combination of mechanical properties is an exciting result, especially the improved toughness achieved by the wave microstructure design, which is higher than that of the interlocking tablet design of the mineral bridge design [15–18,21]. Through an approach that integrates finite element simulations and experiments, we systematically investigated the influences of the tablet waviness and wave number on mechanical performance of the composites. Results show that the tablet waviness significantly affects the mechanical properties and failure patterns of the composite, while the tablet wave number only has a certain effect on the mechanical properties of the composite. By increasing the tablet waviness, the contact interface area of the tablet increases, thus providing a larger area for shear deformation during the hardening and promoting the transfer of load to the tablet. However, exaggerated tablet waviness may cause severe stress concentration, leading to localized brittle fracture of the tablet in composite. Three failure modes are observed in the tensile tests of bio-inspired nacreous composites: soft phase failure, soft phase failure coupled with tablet break and tablet break. Composites with the highest strength and toughness are contributed by the combination of soft phase failure and tablet break: providing adequate tablet sliding to delay fracture and enhancing the hardening to improve the energy absorption. The analysis of fracture path reveals that in the case of soft phase failure, the higher tablet waviness is conducive to the spread of cracks throughout the composite, thus promoting the overall deformation of the composite. In contrast, the change in geometric scale of wave-shaped design caused by the change of tablet wave number may also somewhat affect the load transfer to the tablet, and there is an optimal geometric scale of wave microstructure for the toughness of composite. The simulation verifies the stress distribution of the composites and proves that the high stress concentration area is confined to the peak of the wave in the center of the tablet. In this study, we conclude that the key to high strength and high toughness is to achieve the optimal balance between load transfer and stress concentration, transferring as much load to the tablet as possible while delaying the failure of the composite as much as possible. Thus, given the critical role of design revealed in this study, tuning fracture mode through design optimization in bio-inspired composites can improve mechanical properties of synthetic composites and can bolster the search for new functional advanced materials.

Author Contributions: Conceptualization, Z.D.; Formal analysis, Z.D.; Funding acquisition, H.X.; Investigation, Z.D.; Project administration, Y.D.; Resources, Y.D.; Supervision, Y.D.; Validation, Z.D.; Writing—Original draft, Z.D.; Writing—Review and editing, B.W. and H.X. All authors have read and agreed to the published version of the manuscript.

Funding: This research was funded by the National Natural Science Foundation of China, grant number 51875440.

Data Availability Statement: Data available on request due to privacy. The data presented in this study are available on request from the corresponding author. The data are not publicly available due to these data are also part of ongoing research.

Conflicts of Interest: The authors declare no conflict of interest.

References

1. Dunlop, J.W.C.; Fratzl, P. Biological composites. *Annu. Rev. Mater. Res.* **2010**, *40*, 1–24. [CrossRef]
2. Studart, A.R. Biological and bioinspired composites with spatially tunable heterogeneous architectures. *Adv. Funct. Mater.* **2013**, *23*, 4423–4436. [CrossRef]
3. Meyers, M.A.; Chen, P.-Y.; Lin, A.Y.-M.; Seki, Y. Biological materials: Structure and mechanical properties. *Prog. Mater. Sci.* **2008**, *53*, 1–206. [CrossRef]
4. Connections, M. Structural biological materials: Critical mechanics-materials connections. *Science* **2013**, *339*, 773–780.

5. Lin, A.Y.M.; Meyers, M.; Vecchio, K.S. Mechanical properties and structure of *Strombus gigas*, *Tridacna gigas*, and *Haliotis rufescens* sea shells: A comparative study. *Mater. Sci. Eng. C* **2006**, *26*, 1380–1389. [CrossRef]
6. Kotha, S.P.; Li, Y.; Guzelsu, N. Micromechanical model of nacre tested in tension. *J. Mater. Sci.* **2001**, *36*, 2001–2007. [CrossRef]
7. Barthelat, F.; Tang, H.; Zavattieri, P.; Li, C.-M.; Espinosa, H. On the mechanics of mother-of-pearl: A key feature in the material hierarchical structure. *J. Mech. Phys. Solids* **2007**, *55*, 306–337. [CrossRef]
8. Wang, R.Z.; Suo, Z.; Evans, A.G.; Yao, N.; Aksay, I.A. Deformation mechanisms in nacre. *J. Mater. Res.* **2001**, *16*, 2485–2493. [CrossRef]
9. Barthelat, F.; Espinosa, H.D. An experimental investigation of deformation and fracture of nacre–mother of pearl. *Exp. Mech.* **2007**, *47*, 311–324. [CrossRef]
10. Wegst, U.G.; Ashby, M.F. The mechanical efficiency of natural materials. *Philos. Mag.* **2004**, *84*, 2167–2186. [CrossRef]
11. Wegst, U.G.; Bai, H.; Saiz, E.; Tomsia, A.P.; Ritchie, R.O. Bioinspired structural materials. *Nat. Mater.* **2014**, *14*, 23–36. [CrossRef]
12. Wang, J.; Cheng, Q.; Tang, Z. Layered nanocomposites inspired by the structure and mechanical properties of nacre. *Chem. Soc. Rev.* **2011**, *41*, 1111–1129. [CrossRef]
13. Song, F.; Zhang, X.H.; Bai, Y.L. Microstructure and characteristics in the organic matrix layers of nacre. *J. Mater. Res.* **2002**, *17*, 1567–1570. [CrossRef]
14. Katti, K.S.; Katti, D.; Pradhan, S.M.; Bhosle, A. Platelet interlocks are the key to toughness and strength in nacre. *J. Mater. Res.* **2005**, *20*, 1097–1100. [CrossRef]
15. Ghimire, A.; Tsai, Y.-Y.; Chen, P.-Y.; Chang, S.-W. Tunable interface hardening: Designing tough bio-inspired composites through 3D printing, testing, and computational validation. *Compos. Part B Eng.* **2021**, *215*, 108754. [CrossRef]
16. Askarinejad, S.; Choshali, H.A.; Flavin, C.; Rahbar, N. Effects of tablet waviness on the mechanical response of architected multilayered materials: Modeling and experiment. *Compos. Struct.* **2018**, *195*, 118–125. [CrossRef]
17. Liu, F.; Li, T.; Jia, Z.; Wang, L. Combination of stiffness, strength, and toughness in 3D printed interlocking nacre-like composites. *Extreme Mech. Lett.* **2019**, *35*, 100621. [CrossRef]
18. Espinosa, H.D.; Juster, A.L.; Latourte, F.J.; Loh, O.Y.; Grégoire, D.; Zavattieri, P. Tablet-level origin of toughening in abalone shells and translation to synthetic composite materials. *Nat. Commun.* **2011**, *2*, 173. [CrossRef]
19. Mirzaeifar, R.; Dimas, L.S.; Qin, Z.; Buehler, M.J. Defect-tolerant bioinspired hierarchical composites: Simulation and Experiment. *ACS Biomater. Sci. Eng.* **2015**, *1*, 295–304. [CrossRef] [PubMed]
20. Henry, J.; Pimenta, S. Bio-inspired non-self-similar hierarchical microstructures for damage tolerance. *Compos. Sci. Technol.* **2020**, *201*, 108374. [CrossRef]
21. Gu, G.X.; Libonati, F.; Wettermark, S.D.; Buehler, M.J. Printing nature: Unraveling the role of nacre's mineral bridges. *J. Mech. Behav. Biomed. Mater.* **2017**, *76*, 135–144. [CrossRef] [PubMed]
22. Ubaid, J.; Wardle, B.L.; Kumar, S. Bioinspired compliance grading motif of mortar in nacreous materials. *ACS Appl. Mater. Interfaces* **2020**, *12*, 33256–33266. [CrossRef] [PubMed]
23. McKittrick, J.; Chen, P.-Y.; Bodde, S.G.; Yang, W.; Novitskaya, E.; Meyers, M. The structure, functions, and mechanical properties of keratin. *JOM* **2012**, *64*, 449–468. [CrossRef]
24. Liu, Y.; Qiu, X.; Zhang, X.; Yu, T.X. Response of woodpecker's head during pecking process simulated by material point method. *PLoS ONE* **2015**, *10*, e0122677. [CrossRef] [PubMed]
25. Wang, L.; Cheung, J.T.-M.; Pu, F.; Li, D.; Zhang, M.; Fan, Y. Why do woodpeckers resist head impact injury: A biomechanical investigation. *PLoS ONE* **2011**, *6*, e26490. [CrossRef] [PubMed]
26. Lee, N.; Horstemeyer, M.; Rhee, H.; Nabors, B.; Liao, J.; Williams, L.N. Hierarchical multiscale structure–property relationships of the red-bellied woodpecker (*Melanerpes carolinus*) beak. *J. R. Soc. Interface* **2014**, *11*, 20140274. [CrossRef] [PubMed]
27. Seki, Y.; Kad, B.; Benson, D.; Meyers, M.A. The toucan beak: Structure and mechanical response. *Mater. Sci. Eng. C* **2006**, *26*, 1412–1420. [CrossRef]
28. Ha, N.S.; Lu, G.; Xiang, X. Energy absorption of a bio-inspired honeycomb sandwich panel. *J. Mater. Sci.* **2019**, *54*, 6286–6300. [CrossRef]
29. Lee, J.-Y.; An, J.; Chua, C.K. Fundamentals and applications of 3D printing for novel materials. *Appl. Mater. Today* **2017**, *7*, 120–133. [CrossRef]
30. Zhang, P.; To, A.C. Transversely isotropic hyperelastic-viscoplastic model for glassy polymers with application to additive manufactured photopolymers. *Int. J. Plast.* **2016**, *80*, 56–74. [CrossRef]
31. Libonati, F.; Gu, G.X.; Qin, Z.; Vergani, L.; Buehler, M.J. Bone-inspired materials by design: Toughness amplification observed using 3D printing and testing. *Adv. Eng. Mater.* **2016**, *18*, 1354–1363. [CrossRef]
32. Su, F.Y.; Sabet, F.A.; Tang, K.; Garner, S.; Pang, S.; Tolley, M.T.; Jasiuk, I.; McKittrick, J. Scale and size effects on the mechanical properties of bioinspired 3D printed two-phase composites. *J. Mater. Res. Technol.* **2020**, *9*, 14944–14960. [CrossRef]

MDPI
St. Alban-Anlage 66
4052 Basel
Switzerland
Tel. +41 61 683 77 34
Fax +41 61 302 89 18
www.mdpi.com

Polymers Editorial Office
E-mail: polymers@mdpi.com
www.mdpi.com/journal/polymers

www.ingramcontent.com/pod-product-compliance
Lightning Source LLC
LaVergne TN
LVHW070746100526
838202LV00013B/1318